Watershed Management
Issues and Policies for 21st Century

Editor
K. Palanisami
D. Suresh Kumar
B. Chandrasekaran
Water Technology Centre
Tamil Nadu Agricultural University
Coimbatore-641 003

2016
Associated Publishing Company
A Division of
Astral International Pvt. Ltd.
New Delhi - 110 002

First Published, 2002
Reprinted, 2016

ISBN: 978-93-5130-989-5 (International Edition)

Published by : **Associated Publishing Company**
A Division of
Astral International Pvt. Ltd.
– ISO 9001:2008 Certified Company –
4760-61/23, Ansari Road, Darya Ganj
New Delhi-110 002
Ph. 011-43549197, 23278134
E-mail: info@astralint.com
Website: www.astralint.com

Lasertypeset : **CES, Delhi-110 093**

Digitally Printed at : **Replika Press Pvt. Ltd.**

Foreword

Indian economy is substantially based on agriculture, as around 65 per cent of the population depend on agriculture. Agriculture contributes 30 per cent of India's GDP and 60 per cent of employment potential. India has made great strides in agricultural production during the last three decades, culminating in self-sufficiency and even surplus production in case of several crops. However, objectives such as poverty alleviation, higher growth in food grain consumption (more than two per cent) will continue, thus resulting in definite stress on food and allied agricultural production systems in the years to come.

Driven by growth in income (four to five per cent), the demand for fruits, vegetables, edible oils, livestock and dairy products, semi-processed and processed foods, will grow rapidly in the domestic market. However, the availability of agricultural land will remain a constraint and may even decline, thus warranting up-scaling of research efforts for realizing a jump in agricultural productivity particularly in rainfed areas. Hence, the need for enhancing the productivity of rainfed areas which constitute nearly 68 per cent of the net arable land in the country.

Land, water, climate, flora and fauna are the basic natural resources for agricultural development, which are subjected to various processes leading to their deterioration particularly in rainfed areas. Hence, it is imperative to develop strategies for conservation and improvement of these resources. It is in this context, watershed has become the pivotal point and plays a prominent role for soil and land development works.

Watershed development, in recent times, emerged as one of the important sub-sectors of agricultural development. In the past few decades, the watershed research and development was given the required attention by the state and central governments, developmental agencies and policy makers and planners. The Government of India has initiated different programmes to promote watershed development activities in different phases. A number of international donors have come forward to invest in watershed development activities.

Despite the concerted efforts and wide ranging developmental activities relating to watershed development, a host of issues remain to be addressed. The issues such as rainfed agriculture and watershed development research, watershed planning through decision support system, participatory watershed management, socio-economic issues and new challenges in watershed management in the 21st century, are crucial that require in depth analysis. It is in this context, a national level workshop on "**Watershed Management Strategies : A New Outlook for 21st Century**" was organized by the Water Technology Centre, Tamil Nadu Agricultural University, Coimbatore, to critically examine these issues and formulate policies for the future watershed development programmes.

This edited book contains papers on different themes presented and discussed in the workshop. I hope that the book with elaborate policy recommendations will be useful to watershed development planners, researchers and extension agents from both government and non-governmental organizations and teachers and students involved in watershed development programmes.

Prof. Dr. S. Kannaiyan
Vice-Chancellor
Tamil Nadu Agricultural University
Coimbatore 641 003

Preface

Restoring the ecological balance and increasing productivity in rainfed areas necessitates development to be taken on watershed basis. Watershed management is a holistic approach to optimize the use of land, water and vegetation in an area and thereby alleviate drought, floods, prevent soil erosion, improve water availability and increase agricultural production.

In this context, the Government of India initiated three special programmes known as Drought Prone Area Programme (DPAP), Desert Development Programme (DDP) and Intensive Watershed Development Programme (IWDP) for overcoming the problems of drought, desertification and degradation of land, respectively. Since 1995, programmes have adopted a common watershed approach with stakeholder participation at the planning, implementation, decision making, financial control, management and maintenance stages of the project.

Though watershed development has been accepted as the prime approach to area development, conservation and development of natural resources and restoration of ecological balance on a sustained basis, the benefits of this approach are restricted largely to land development and water harvesting and its impact on optimal use and conservation of natural resources and production systems has not been to the desired extent. There are also reports that participatory approaches in watershed development have led to institutional conflicts at the village level and other socio-economic issues have not been adequately addressed.

This book is edited from excerpts of different experts in watershed development and management from a national level workshop on ***Watershed Management Strategies: A New Outlook for 21st Century*** organized by Water Technology Centre, Tamil Nadu Agricultural University, Coimbatore. This book encompasses valuable papers on various themes such as watershed development research, decision support system for watershed development and planning, socio-economic and institutional issues, participatory watershed development and management, monitoring and evaluation, new challenges and strategies and experiences and case studies. This book has tried to provide a forum for discussing various problems affecting watershed management. It is hoped that this effort of ours will enhance understanding of the issues involved and aid in proper policy framework. Also this book would be very much useful to the researchers, scholars, NGOs, PIAs, government officials, donor agencies and policy makers and planners.

We are highly thankful to the sponsors of the national workshop. The sponsors include State Land Use Board, State Planning Commission, Government of Tamil Nadu, Winrock International India, New Delhi, National Bank for Agriculture and Rural Development, Danish International Development Assistance (DANIDA), Tirunelveli, National Academy of Agricultural Sciences, New Delhi, State Bank of India, TNAU Branch, Coimbatore, Department of Environment, Chennai.

We also thank all the staff members of Water Technology Centre, Tamil Nadu Agricultural University, Coimbatore for extending full support in successful conduct of the workshop. The assistance rendered by R. V. Selvi, SRF, R. Santhi, JRF and K. Vinupriya, JRF in typesetting the entire draft is greatfully acknowledged.

K. Palanisami
D. Suresh Kumar
B. Chandrasekaran

List of Contributors

Abraham, Philip
Danida Advisor
Comprehensive Watershed Development Project, DANIDA
Tirunelveli, Tamil Nadu

Alemu, Bisrat
Planning Department
Government of Ethiopia
Bahirdar

Ashwin, M. A.
Centre for Environmental Education
Bangalore
Karnataka

Bhat, Mahadev G.
Environmental Studies
Florida International University
Miami, FL 33199

Chandrakanth, M. G.
Principal Investigator
Team of Excellence for Natural Resource Economics (NATP)
Department of Agricultural Economics
University of Agricultural Sciences
GKVK, Bangalore
Karnataka

Chandrasekaran, B.
Professor (Agronomy)
Water Technology Centre
Tamil Nadu Agricultural University
Coimbatore
Tamil Nadu

Daniel, Anand Raj
Area Team Coordinator
Agriculture Man Ecology
Tiruchirapalli, Tamil Nadu

Das, M.
Water Technology Centre for Eastern Region
Bhubaneswar
Orissa

Erappa, S.
Assistant Professor
Institute for Social and Economic Change
Bangalore University, Bangalore
Karnataka

Ganapathysankaran, S.
Assistant Director of Agriculture
TANWA, Directorate of Agriculture
Chennai, Tamil Nadu

Haffis, Shaik
Central Research Institute for Dryland Agriculture
Hyderabad
Andhra Pradesh

Jayabalakrishnan, M.
District Rural Development Agency
Coimbatore
Tamil Nadu

Jayakumar, S. A.
Asst. Geologist (Selection Grade)
Public Works Department
Groundwater Circle, WRO
Taramani
Chennai, Tamil Nadu

Kannan, K.
Scientist (Agronomy)
Water Technology Centre for North-eastern Region
Bhubaneswar
Orissa

Karuppasamy, N.
Project Executive
DHAN Foundation
Madurai, Tamil Nadu

Kerr, John
International Food Policy Research Institute
Washington DC
USA

Kolavalli, Shashi
International Food Policy Research Institute
Washington DC
USA

Krishnan, J.
Agricultural Specialist
Action For Food Production (AFPRO)
Bangalore
Karnataka

Kulandaivelu, R.
Professor of Agronomy (Retd.)
Tamil Nadu Agricultural University
Coimbatore

Mahendran, K.
Assistant Professor (Agrl. Economics)
Department of Agricultural Economics
Tamil Nadu Agricultural University
Coimbatore

Mallari, Shoba S.
Institute for Social and Economic Change
Bangalore University
Bangalore
Karnataka

Manoharan, M.
Professor and Head
Training Division
Tamil Nadu Agricultural University
Coimbatore

Mohamed, Raja
Association for Sarva Seva Farms (ASSEFA)
Sivagangai, Tamil Nadu

Mohanty, R. K.
Water Technology Centre for Eastern Region
Bhubaneswar, Orissa

Mohanty, S.
Water Technology Centre for Eastern Region
Bhubaneswar
Orissa

Nanda, P.
Scientist (Agrl. Economics)
Water Technology Centre for Eastern Region
Bhubaneswar
Orissa

Oppilamani, A.
Superintending Engineer(AE)
Project Manager
Comprehensive Watershed Development Project (CWDP)
Tirunelveli, Tamil Nadu

Palanichamy, N. Venkatesa
Assistant Professor (Agrl. Economics)
Department of Agricultural Economics
Tamil Nadu Agricultural University
Coimbatore

Palanisami, K.
Director
Water Technology Centre
Tamil Nadu Agricultural University
Coimbatore, Tamil Nadu

Pangare, Ganesh
International Food Policy Research Institute
Washington DC
USA

Pangare, Vasudha Lokur
International Food Policy Research Institute
Washington DC
USA

Paramasivam, P.
Assistant Professor (Agrl. Economics)
Water Technology Centre
Tamil Nadu Agricultural University
Coimbatore

Pattnaik, S.
Water Technology Centre for Eastern Region
Bhubaneswar
Orissa

Puhazhendhi, V.
National Bank for Agriculture and Rural Development (NABARD)
Mumbai
Maharashtra

Rajakesavan, K.
Project Assistant
CEE South
Bangalore
Karnataka

Ramana, D. V.
Deputy General Manager
Agricultural Finance Corporation Ltd.
Hyderabad
Andhra Pradesh

Ramasamy, C.
Director
Centre for Agricultural and Rural Development Studies
Tamil Nadu Agricultural University
Coimbatore, Tamil Nadu

Ramasubramanian, M.
Ph.D. Scholar
Department of Agricultural Extension
Tamil Nadu Agricultural University
Coimbatore

Ramaswamy, K.
Associate Professor
Horticultural College and Research Institute
Periyakulam-625 604
Tamil Nadu

Ravikumar
Area Team Member
Agriculture Man Ecology
Tiruchirapalli
Tamil Nadu

Reddy, G. P.
Principal Scientist (Agrl. Economics)
Water Technology Centre for Eastern Region
Bhubaneswar
Orissa

Reddy, Y. V. R.
Principal Scientist and Head, SDA
Central Research Institute for Dryland Agriculture
Hyderabad
Andhra Pradesh

Rout, Shishir
Water Technology Centre for Eastern Region
Bhubaneswar
Orissa

Sahoo, Deepika
Water Technology Centre for Eastern Region
Bhubaneswar
Orissa

Sahoo, N.
Principal Scientist (Soil & Water Engg.)
Water Technology Centre for Eastern Region
Bhubaneswar
Orissa

Sastry, G.
Principal Scientist
Central Research Institute for Dryland Agriculture
Hyderabad
Andhra Pradesh

Selvi, R. V.
Senior Research Fellow
Water Technology Centre
Tamil Nadu Agricultural University
Coimbatore-641 003
Tamil Nadu

Shailaja, R.
CEE South
Bangalore, Karnataka

Sikka, Alok K.
Head
Central Soil & Water Conservation Research and Training Institute, Research Centre
Udhagamandalam-643 004, Tamil Nadu

Singandhupe, R. B.
Senior Scientist
Water Technology Centre for Eastern Region
Bhubaneswar, Orissa

Singh, H. P.
Central Research Institute for Dryland Agriculture
Hyderabad, Andhra Pradesh

Sivanappan, R. K.
International Agricultural Consultant
Coimbatore, Tamil Nadu

Sivaraman, K. R.
Assistant Director (Geophysics)
PWD, Groundwater Circle
Madurai, Tamil Nadu

Sivasubramaniyan, K.
Research Associate
Madras Institute of Development Studies
Chennai, Tamil Nadu

Sreedharan, C. K.
Chief Conservator of Forests
Tamil Nadu Afforestation Project
Chennai, Tamil Nadu

Srivastava, R. C.
Principal Scientist
Water Technology Centre for Eastern Region
Bhubaneswar, Orissa

Suresh Kumar, D.
Assistant Professor (Agrl. Economics)
Water Technology Centre
Tamil Nadu Agricultural University
Coimbatore-641003
Tamil Nadu

Tagat, V.
National Bank for Agriculture and Rural Development (NABARD)
Mumbai
Maharashtra

Thangaiah, P.
Project Director
Association for Sarva Seva Farms (ASSEFA)
Sivagangai, Tamil Nadu

Thiruvenkataswamy, K. R.
Former Superintending Engineer
Department of Agricultural Engineering
Government of Tamil Nadu

Vaidyanathan, V.
Professor Emeritus
Madras Institute of Developmental Studies
Chennai, Tamil Nadu

Venkatraman, A.
I.A.S. (Retd.)
Expert Committee on Rain Water Harvesting
Government of Tamil Nadu
Chennai, Tamil Nadu

Verma, H. N.
Water Technology Centre for Eastern Region
Bhubaneswar
Orissa

Contents

Watershed Management: Issues and Policies for 21st Century pp. 1-15

Watershed Development: Concepts and Issues

K. Palanisami, D. Suresh Kumar and B. Chandrasekaran

INTRODUCTION

During the past five decades of planned development, India has achieved spectacular increase in food grain production. Green revolution enabled us to enter into a new era in input use. The use of high yielding variety seeds, fertilizers, plant protection chemicals have stepped up manifold. Development of major and minor irrigation projects had gone up. But still, the overall gains had not been equitably shared by the farming community. This is mainly because two thirds of agriculture in our country is rainfed. According to one recent estimate, the availability of water for irrigation would emerge as a critical bottleneck. The requirements of ground water exceed its availability for irrigation by nearly 30 per cent. Since, the requirement will exceed the replenishable supplies for irrigation at the existing rate of recharge of ground water, it will lead to over exploitation of ground water in several states and by 2020 as much as 71 per cent of irrigated area is likely to be from ground water sources (Chopra and Golder, 2000). Closely related to this is the problem of degradation of land in the rainfed areas. As it is, the rate of degradation of land in such areas in the 1990s is likely to have proceeded at more than twice the rate observed in 1980s, basically on account of soil erosion from run-off (Reddy, 2000). This is happening despite all the official and NGO intervention for soil and moisture conservation through watershed development which may not have covered more than 10 per cent of rainfed areas needing treatment. The acute famine for drinking water experienced in several parts of the country, despite a series of favourable monsoons preceding it, shows that droughts are becoming increasingly severe in the recent period. Clearly, the existing interventions for drought proofing have failed to neutralize the overall processes of degradation of natural resources underway in the rural areas. This calls for a thorough reappraisal of our agricultural policies with a view to evolving an overall policy framework that provides adequate incentives and opportunities for soil and moisture conservation through watershed development as well as for judicious use of scarce water resources in rainfed areas.

Rainfed agriculture in India is characterized by low productivity, degraded natural resources and widespread poverty. Most of the millions of people living in our country depend on agriculture and natural resource management for their livelihoods, so development planners are eager to implement productive, environmentally sustainable land and water management. It is in this context, the concept of watershed development has been introduced in our country. A watershed is a geographical area that drains to a common point, which makes it an attractive unit for technical efforts to conserve soil and maximize the utilization of surface and subsurface water for crop production. It is an ecosystem or bio-geographical unit in which the interdependence is internalized. In the words of Barrow, "Within a watershed physical and biological resources are linked by a complex of processes..." (Barrow, 1987). These resource regions can thus appear in different forms or sizes and are not easily amenable to standardization in terms of mini, micro or macro typologies. Watershed development projects are designed to harmonise the use of water, soil, forest and pasture resources in a way that conserves these resources while raising agricultural productivity, both through in situ moisture conservation and increased irrigation through water harvesting. Watershed development has been conceived basically as a strategy for protecting the livelihoods of the people inhabiting the fragile eco-systems experiencing soil erosion and moisture stress. The aim has been to ensure the availability of drinking water, fuel wood and fodder and raise income and employment for farmers and landless labourers through improvement in agricultural production and productivity.

Conceptually the present strategy of watershed development is prompted by the need to protect the inhabitants of the fragile ecosystem from acute distress caused by recurring droughts. In this sense, it is a strategy for survival, even though in quite a few cases the successful implementation of the watershed development programme has led to a substantial increase in income and employment of the people. Even the many cases of outstanding success pioneered by the eminent NGOs reveal that such efforts were originally driven by acute distress caused by recurring droughts. It is necessary to distinguish this strategy for protection or survival from a holistic strategy for agriculture in dry land or rainfed areas. Such a wider strategy would involve issues of agricultural technology suited to rainfed areas, price and credit policies, infrastructure development for processing, marketing and transportation of agricultural produce and trade policies to afford reasonable protection to dry land farmers and to provide opportunities for export of their products. Some of their supporting measures are no doubt present in certain dry land areas especially in the watershed areas run by some of the distinguished NGOs, where there is a sustained increase in agricultural productivity and income. But this is nowhere comparable to the policy framework and infrastructure available to the areas in the country which have experienced green revolution to the rainfed area is essential not only for deriving full benefit from soil and moisture conservation already accomplished in some area through the watershed development programme which Anil Shah has aptly described as "Watershed plus" (Shah 2000). But more fundamentally to provide the necessary economic incentives for turning watershed development into truly spontaneous and demand driven movement of the people in dry land areas not covered so far, and to sustain watershed development already undertaken when the current official programmes comes to an end.

WATERSHED DEVELOPMENT: CONCEPT AND CLASSIFICATION

Watershed development refers to the conservation, regeneration and the judicious use of all the resources like natural (land, water, plants, animals) and human within a particular watershed. Watershed management tries to bring about the best possible balance in the environment between natural resources on the one side and human and other living beings on the other. People and their environment are interdependent. Any change in the surrounding environment directly affects the people living there. A degraded environment results in a degraded quality of life of the people. Thus efforts to reduce poverty and improve the standard of living of the people must aim at improving the environment they live in. The environment does not recognize man made administrative boundaries. A watershed provides a natural environmental unit for planning a developmental initiative.

The entire geographical area of the country is divided into six water resources regions. They are (*i*) Indus River Systems in India, (*ii*) Ganga River System (*iii*) Brahmaputra River System, (*iv*) All the rivers flow to Bay of Bengal (except Ganga and Brahmaputra), (*v*) All the rivers flow to Arabian Sea (except Indus River) and (*vi*) Desert River System (Rajasthan).

Water resource region into river basin: Prominent rivers such as Narmada, Tapti, Kaveri, Godawari, Yamuna etc as river basins and in between rivers are clubbed and named. There are 66 river basins in India.

Division of river basins into sub river basin: The division is based on area, the division starts from down water right and left of the river course.

Division of subriver basins to watershed: The classification based on the streams and areas of catchment.

Division of Watershed into subwatershed: Bifurcation of main stream with area.

Division of subwatershed to micro watershed: Early manageable hydro-geological unit.

Area restriction in the above classification based on Mansinghal Associates (1958).

Water Resources Region	:	500000 sq.km
River Basins	:	50000 sq.km
River sub-basin	:	5000 sq.km
Watershed	:	500 sq.km
Sub watershed	:	50 sq.km
Micro watershed	:	5 sq.km

The concept of watershed development has universal relevance as the natural resources particularly land and water, whether they are the very high rainfall areas of north-east, the Himalayas and the sub-continents region or the Peninsular part of the country. The watershed development programme, by virtue of their involving the entire community and natural resources, particularly land and water favourably influences the following:

- Socio-economic trends in income generation, employment pattern, household assets, health care, education, energy use and impact on women and the landless.
- Perceptions, attitudes and behaviour of the people towards project activities and their participation in planning and execution of the project and management of resources proposed to be developed under the project together with constraints.
- Productivity/production of crops and forage, changes in cropping pattern and land use, adoption of improved cropping practices, survival and growth of perennial crops as non-arable lands, increase in wood and non-wood products, improvements in livelihood, milk production and policy.
- Improvement of environment by build up of vegetative cover, reduction of run-off and soil loss, improvement in soil fertility, micro nutrients, organic matter and optimal use of land.
- Use of land, water, vegetation, man, animal and environment.

- Soil erosion, moderation of floods, alleviation of drought, water availability in soil, in surface and underground storage, availability of fuel, fodder and agricultural production.
- Development of institutions for planning, implementation, management and training.
- Ensuring sustainability of improvements brought about on farmers land. It is thus clear that water and soil conservation through an integrated approach is a key to sustainable production of food, fibre, fodder and fuel and to meaningfully addressing the social, economic, cultural and other concerns of the rural areas and the people. As such watershed development has to be an integral part of the strategy of development.

Watershed projects are a favourable alternative as most of the unskilled rural women could be drawn into agriculture. Unfortunately, agriculture is not helping people to cross the poverty line because of poor agricultural production levels. Therefore, agriculture has to be made a more paying enterprise. This will also check rural migration which is causing environmental hazards. It is in this context, watershed development is being pursued. The potential for achieving food security and alleviating poverty through improved livelihood and generation of employment opportunities is indeed significant. In addition, there are compulsions such as demographic compulsion and ecological compulsions that force the need for watershed development. Unless there is a check on population, the gains of development will be regularly neutralized. Population control efforts call for involvement of the village community. Moreover, ecology is fragile which can be preserved only through planned efforts of the people and the development agencies. Today a multitude of organizations are involved in watershed development works. These organizations have different approaches to handle the ecology. Care has to be taken that the ecology is not disturbed.

WATERSHED DEVELOPMENT IN INDIA

Watershed development based activities were initiated in 1956 at the Central Soil and Water Conservation Research and Training Institute, Dehradun. Ministry of Rural Development initiated the programmes of Drought Prone Area Programme, Desert Development Programme and Wasteland Development Programme. The National Afforestation and Ecological Development Board started a programme of restoring degraded forest lands. Agencies such as World Bank, DANIDA, EEC, Swiss Development Corporation also came forth for a number of watershed development projects. Based on the experiences gained during the Sixth Plan, through a Pilot Project spread over 19 watersheds in the country, NWDPRA was launched during VII Plan in 99 selected watersheds of the country. The emphasis was on increasing crop production of arable lands. During the Seventh Plan, the scheme was modified to provide finance for both arable and non-arable lands, with 100 per cent finance. During Eight Plan, the allocation was Rs.11000 crores spread over 2479 watersheds covering 350 districts located in 25 States and 2 Union territories. Community development blocks having less than 30 per cent of arable land under assured irrigation were qualified for inclusion in the project.

The experiences of all those watershed development projects show that there has been a marked improvement in the access to drinking water in the project areas even during drought years. Crop yields have risen and there has been a substantial increase in area under cultivation in the rabi season, leading to a rise in employment and reduction in migration of labour. Availability of fodder for animals has also improved leading to a rise in yield of milk. All this might came as a pleasant surprise to many who, for understandable reasons, continued to remain skeptical about the prospects for ensuring people's participation on a wider scale under the government sponsored programmes, even as they were convinced from the experience of success stories that when people's participation is ensured, watershed development leads to a substantial improvement in their livelihoods (Hanumantha Rao, 2000). In spite of these success stories, the overall impact has however, not always been encouraging. Though there are a few glowing successes, the macro objectives have not been achieved to a satisfactory level. The poor often got less benefits and rich farmers, with lands in the lower toposequence, received more. For example, the River Valley Project was launched in 1962-63 to control siltation of reservoirs. Even after 38 years, reservoirs are still silting at an alarming rate. The DPAP was launched in 1972-73 to mitigate effect of drought and drought proof vulnerable areas. But not even a single district/block is drought proof. Droughts continue to cause immense suffering and hardship to human beings and livestock. Similarly, the NWDPRA was launched in 1985-86 to increase and stabilize production of rainfed crops like pulses, oilseeds and millets. However, production still depends upon the monsoons. This calls for a thorough review and analysis of various issues in watershed development to make this a successful concept.

ISSUES IN WATERSHED DEVELOPMENT

Despite the noticeable improvement in performance, the experience outlined above raises a number of important issues which have a bearing on the sustainability of watershed development when the official programme comes to an end. The important issues that are emerging in relation to watershed development and programme are: (*i*) Rainfed agriculture and watershed development research, (*ii*) Decision support system for watershed development, (*iii*) Socio-economic and institutional issues, (*iv*) Participatory watershed development, (*v*) sustainability and replicability, (*vi*) Equity issues, (*vii*) Supply driven to demand driven models, (*viii*) Capacity building, (*ix*) Institutional issues, (*x*) Administrative issues, (*xi*) Cost sharing (*xii*) Conflict resolution and (*xiii*) Monitoring and evaluation.

Rainfed agriculture and watershed development research

In the irrigated areas, given the technology, the decision and efforts of farmers as individuals were centred to the spread of green revolution, whereas in the dryland areas such individual farm level decisions have to be preceded by community effort for soil and moisture conservation which is a painstaking and time consuming process. Besides, dryland areas are characterized by greater agro-climatic diversity than the traditional irrigated areas, necessitating location specific research for evolving viable technologies and practices. However, dryland areas have not received so far the priority they deserve in respect of agricultural research. Green revolution was characterized by land saving but water using technologies whereas dryland areas need water saving enterprises and practices which optimize output per unit of scarce water. In the absence of such technologies, the farmers in the dryland areas go in for water intensive crops like rice when water resources are conserved giving rise to conflicts on water and scarcity of drinking water. However, given the necessary shift in the policy framework in respect of agricultural research, enterprises and supply of inputs like seeds, infrastructure development and price and credit policies, dryland areas hold considerable promise for the development of overall agriculture.

Developmental efforts in the technology of rainfed agriculture were directed more towards soil conservation experiments. The impact of these efforts was quite encouraging but the magnitude of the task involved as well as the thin spread of these experiments indicated need for more concentrated efforts. The results of the trials conducted at experimental stations were quite encouraging but failed to replicate the demonstrated yield gap on farmers' field. As a result, the adoption rates of these practices as well as crop varieties were quite low. One of the prominent features in deciding the low response was the location specificity of the technology and the nexus of risk aversion with adoption of technology. All this led to a unique solution of an integrated planning approach to the development of rainfed agriculture (Jodha, 1979).

Any holistic planning approach for agriculture will involve the area in the form of a resource region where the central resources will be soil and water and in which the planning exercise will have to be undertaken. These works were taken independently under various institutional arrangements and across a large number of region in the country. Such multiplicity of the experiments was not very conducive for evolving a common framework, however, the mosaic of experiments across the country offers a great deal of diverse situations essential for management.

If the watershed management has to become successful, use to be simple so that farmers and villagers can understand and internalize it, low cost so that resource poor farmers in rainfed areas can adopt and replicate with minimal external support and should be based on vegetative measures which are self regenerative. However, present technology of constructing engineering structures and planting vegetative structures to stabilize would have to be reversed. The better approach would be to construct low cost structures to help the establishment and growth of the vegetation. The approach should be to shift from soil conservation to water conservation, indigenous and introduce technology, low cost, simple and vegetative conservation measures based technologies.

Watershed planning and decision support system

Any watershed development would have an essential component of an action plan. This is formulated based upon the facts, requirements and standards of watershed programme by the PIAs, put up before the sanctioning authorities and thereafter becomes the guidelines of the watershed development process. Thus, it is the only component which would be responsible for success or failure of all efforts that go into the exercise. The planning exercise should focus on producing several alternative solutions to be identified so as to help eventually pick up those items which are technically feasible, approved by all technical sciences and at the same time solutions offered be acceptable to people from the concerned rural area. Mapping up of data is essential for proper watershed planning. The assistance of Geographical Information System is very much needed right from the level of data illustration to manoeuvering mapped data, processing to generating final maps as per the requirements etc.Computerised GIS comes very handy for performing several maps merging exercises which are extremely difficult manually and at the same time the automation has sufficient capacity for human control and intervention.

In recent years, Satellite Remote Sensing Techniques have proved to be of immense use in watershed development programme in many parts of the world. Satellite images by virtue of its repetitive (temporal), multi-spectral scanning and synoptic coverage can provide sequential images which will enable us to evaluate and for optimum management of natural resources for sustainable development of the watershed. Satellite sensors record many features of the land surface including natural and artificial cases. Land features can be interpreted on the basis of image elements. Remote Sensing Techniques can be used in watershed development projects in selection of watersheds through prioritization, generation of natural resources information and integrating these through GIS for site specific recommendations for sustainable development and also for monitoring and evaluation of the watershed projects.

Socio-economic and institutional issues

Though the watershed development has considerable merit in economic, agricultural and environmental impacts and socio-economic conditions of the people who belong to it, watershed development has not produced desired results in many parts of the country. The watershed interventions need hitherto in many situations have failed to make any discernible impact on adoption of technologies by the farmers even in the adjoining villages. There are several socio-economic constraints responsible for the poor performance of watershed development programme. The factors which are responsible for poor performance of watershed development programmes in the country are: (*i*) Poor socio economic status of people, (*ii*) Low literacy and conservatism, (*iii*) Remote locations, (*iv*) Socio-political conflicts, (*v*) Inadequate credit facilities, (*vi*) Subsistence orientation, (*vii*) Inadequate marketing facilities, (*viii*) Absentee Landlordism, (*ix*) Subdivision and fragmentation of land holdings, (*x*) Inadequate storage facilities, (*xi*) Lack of proper infrastructural facilities, and (*xii*) Lack of legal mechanism. In addition to these factors, there are several issues centered around the watershed development which include, financial, technological, people's participation, criteria for selection of watersheds, capacity building, institutional support, awareness creation, monitoring and evaluation, energy and co-ordination.

The other village level issues concerned are: Some people live outside the watershed area but their lands are inside the watershed area, some people live inside the watershed area and use several of its resources, but their lands are outside the watershed area, some people are landless but use the resources in the watershed area, some people have clear titles to their lands, while others do not and may be unwilling to make investments on the lands they are cultivating, some lands are publicly owned by parties with whom the people may or may not be able to negotiate, lack of transparency in work and financial issues and some people are solely dependent on their lands and are willing to invest on it, others may not be as concerned, as they may have other major livelihood sources. These issues should be looked into by the nodal agency and the project implementing agencies.

Participatory watershed management

Like all other development programmes, watershed development is banking heavily on participatory approach. In fact watershed development programme envisages an integrated and comprehensive plan of action for the rural areas. Therefore, people's participation at all levels of its implementation is very important. This is so because the watershed management approach requires that every field/piece of land located in watershed be treated with appropriate soil and water conservation measures and used according to its physical capability. For this to happen, it is necessary that every farmer having land in the watershed accepts and implements the recommended watershed development plan. If these factors are looked into in an efficient manner, the watershed development could be hastened in the country. Hence, people's participation is the key element in the success of any development project, because they are going to be the beneficiaries and promoters of any development activity. As the issue of sustainable natural resource management becomes more and more crucial, it has also become clear that sustainability closely linked to the participation of the communities who are living in close association with these natural resources. The slowing down, arresting and reversing the degradation of the important natural resources like land and water will be possible only if it becomes a people's movement involving the rural community and all those who work with them and have any stake in their development. This requires sustained effort to inform and educate the rural community, demonstrate to them the benefits of watershed development and that the project can be planned and implemented by the rural community with expert help from government and non-government sources.

Over a period of time, people have become dependent on the state for all investment for development in the form of subsidy. The excessive reliance on the state will have to be replaced with a spirit of self-reliance and the people have to find resources for the development of their own land, leaving work on common lands to be done by the state with some contributions from the people. The rural society is plural and stratified. There are divisions based on gender, castes, land relations, religions etc. Homogenous groups need to be identified on the basis of commonality of interests, gender, etc such as Self-Help Groups, User Groups and Women's Groups.

However, participatory watershed management is constrained by several factors which affect the effective planning and implementation of the programme. The peoples participation depends on attitude of the people, rural environment, attitude of the government functionaries, government approach, lack of capacities, lack of women's participation, formulation of local institutions, ignoring local technologies and cost sharing. Thus, a watershed development project to become successful, the various issues concerned should be addressed.

Sustainability and replicability

The basic issue of watershed management projects is that of enabling the beneficiaries to sustain and maintain the assets created with or with little or no support after the implementation of the project. It is also important that the activities are replicable with minimal government support. First of all, communities must be willing to participate and share costs in the implementation of project activities. The best way to achieve sustainability is to hand over the ownership of the programme to the community. However, since most watershed programmes are implemented with the poorer sections, the cost sharing should primarily be in the form of labour and not necessarily cash. To effectively achieve the sustainable management of watershed development projects, the issues such as cost sharing, reform process, replicability, linking with other institutions, ownership rights are flooding into the minds of economists.

The will to share the cost would depend on the benefits expected from any activity. Usually, the interest of the community in developing a common land is low and their share could be as nominal as responsibility of watch and ward. On the other hand, in activities where the benefits are more direct to individual families, contribution could be as high as possible. The quantification of cost sharing should, however, be kept flexible. Experiences show that if the activities reflect the priority of the communities, cost sharing is not a problem. Reforms in terms of reduced subsidy or increased cost could gradually be raised. Constituting a high power committee at the state level to review the performance of watershed development projects will ensure efficient use of available financial resources and a positive coordination. This would also help the replicability of the project activities.

Donors should join hands with government programmes and implement the projects on agreed common guidelines. Rather than running projects independently, implementing projects in collaboration with government will help improve replicability. There are different institutional arrangements followed by different agencies for different programmes. The groups are formed on the basis of beneficiaries and communities. With 73rd and 74th constitutional amendments, it is now essential to link these groups with the Panchayat Raj Institutions (PRIs) to ensure long term sustainability.

Equity issues

Equity is perhaps the most difficult objective to achieve within the context of a watershed development programme, and even more difficult to sustain. Watershed development is aimed to improving the productivity of land and water. The benefits therefore primarily go to those who own or control the land and the water resources in the watershed. Those who do not see these sections are helped to maximize such benefits. It is not possible to redistribute rights over land in the short term, so this avenue towards equity is not really open. The same applies to rights over water except where it is common property (e.g., for drinking and domestic use and for livestock). In the planning and execution of watershed treatment works it should be possible to try and ensure that benefits go to the assetless people and women. Once a decision is taken to take up a watershed for development, the processes of building awareness of the project in the community should reach the disadvantaged and the women. Similarly for the processes of promoting community organizations, one should try and ensure that these sections are also represented in "user groups" and the watershed committees.

The idea is to work with one village in the watershed with perhaps two or three small adjoining hamlets/settlements. For these processes to be meaningful, one should not take a village panchayat area unless there is only one village in the panchayat. We need to work with a reasonably cohesive community, that is where different sections feel they are part of the community, and make sure that the needs of all sections are understood and taken into account in planning the watershed treatment and development. Equity is important in the matter of cost sharing contributions. For the common lands which generally benefit the assetless and the marginal sections, the community can be persuaded to decide upon relatively smaller contributions, related to the benefits that they may expect to get in the short term. In the execution only local labour should be employed, unless the latter agrees to get labour from outside or the use of machines in exceptional cases. Technology choices for the treatment measures should be such that the productivity of poorer lands can be very significantly improved because of their implications to the poor. This approach would call for examining conventional technology options specifically from this point of view.

In the process of community organization one may try to see that even in the post-project maintenance and management phase the disadvantaged sections have a clear role. As the population increases, and expectations rise, the watershed resources will not provide an adequate livelihood base for all. Therefore, the consideration for equity might imply the need for those with little or no productive assets to move to non-farm occupations.

Capacity building

Capacity building is an important component in watershed management. Different people have different roles and responsibilities in project implementation and there is a need to train people involved in the watershed. Training enhances knowledge, skill, attitude and relationships. The objectives of the training programme are: (*i*) to strengthen those processes and skills that help in the delivery of programmes/activities, (*ii*) to convey technical subject matter, (*iii*) to develop communication skills and (*iv*) to increase community participation. The broad areas can be technical information and skills, managerial skills, conceptual skills, human resource development, decision power and communication.

Most of the training programmes in watershed development target men than women. Some Project Implementing Agencies (PIAs) hold meeting with women to explain the objectives of the watershed development programme to be implemented in the village. Women as a rule are given no technical training in watershed development. Even when training is made available for women, a definite gender bias exist in the kind of training and awareness programmes planned for men and women. Women are typically isolated from the scientific and technical aspects of watershed development. Due to lack of access to technical training, women generally never hold supervisory positions on land development sites. They only contribute to the labour components. Women are also prevented from participation in decision making related to the selection of sites and identifying appropriate conservation structures due to lack of training. In most cases, they are simply not consulted, on the assumption that women do not have the necessary skills for participating in these decisions.

In addition, capacity building focused only on watershed development team members, training primarily addressed to theoretical and technical aspects, poor performance in administration and monitoring of the programmes, functionaries at administrative level and village levels neglected, and issues of sustainability, equity, gender and community organizations have received limited attention. In view of these observed drawbacks of the watershed development programmes, the training needs must be redesigned according to the roles and responsibilities of the various actors in the programme. These actors are also the target groups for the capacity building process.

Institutional issues

Major changes taking place in the overall context of watershed development include: (*i*) Increased government commitment, evident in the increased resources available for central government programmes, (*ii*) Increasing demand for improved transparency and accountability in the implementation process, and (*iii*) Increasing pressure towards clarity and transparency in decentralization of decision-making. The major issues in the changing scenario are: Inter-relationship between Panchayat Raj Institutions and community based organisations, involvement of state governments and line departments and scope for creating competition among service providers.

The relationships are very vital for smooth functioning of a watershed programme. New roles of PRIs have developed new relationship equations within the watershed. In spite of the weaknesses, the PRIs are here to stay. Fundamental requirements are: to enhance the technical capabilities and linkages of PRIs and to increase downward accountability by helping some members of Water User Association and Watershed Committees or of other community based organisations to get elected to Gram Panchayat.

State Governments are generally by-passed by funding direct from the Central Government to DRDAs. Hence, the States may not show any interest in ensuring that their line departments respond to watershed development needs, planning future locations of watershed development, and monitoring performance. Therefore, clear efforts to work with State Governments and line departments are needed. It is not difficult to place funds in the hands of the resource users or Project Implementing Agencies and allow them to "buy in" services.

All the government developmental programmes had been top down i.e. planned at the top level not based on felt needs and thrust upon the communities. Communities were not consulted in the design and implementation. Their needs and priorities were not given adequate importance. At the most, people were involved only as labourers. But in the recent watershed approach the project action plan has to emerge from people to who are actually facing the production related problems. The action plan with estimated budget will only be sent up for formal approval and release of the funds through a registered local institution. Even though there are proper guidelines to implement the watershed development projects, there are all possibilities of failure of the programme due to the following issues:

- A top down approach still prevails. For example, the DRDA controls the entire programme in the district. It makes its own rules and regulations overlooking the guidelines.
- Most of the district watershed advisory committees are defunct because the heads of line departments have no time for attending these meetings regularly.
- The PIA with a team of limited members is unable to implement 10-12 watersheds with participatory approach as envisaged in the guidelines. The PIA is unable to cope up with the work load of the community mobilization and participatory implementation.
- Immediately after allotment of watersheds, DRDA insists on action plans. So without organizing the community, watershed association and watershed committee are formed hurriedly.
- No user groups are formed. Action plans are prepared by PIA/WDT with the help of office bearers of watershed committee. Most of the times farmers are not involved in deciding the activities.
- SHGs are formed much late to distribute the revolving fund. They do not play any role in watershed programme except thrift and credit.

- The ch airpersons and secretary execute the activities. No transparency in financial transactions. All the records are maintained by WDT. Watershed Committee meetings are not held.
- Local decisions have no place. Decisions are mostly taken by WDT/PIA.
- The watershed association meeting never take place. Nobody is aware of its powers and functions.
- The structures are constructed by outside skilled labourers.

Administrative issues

Implementation of watershed projects has remained poor because of the low capacity of the field machinery and the related administrative issues. They are (*i*) there is little impact assessment or evaluation of physical progress after the scheme has run for a couple of years, (*ii*) horizontal linkages between various line agencies at the district level are very weak, (*iii*) there is no arrangement for handing over of structures and maintenance of plantation after a project is completed, (*iv*) government functionaries have neither time nor training to nurture democratic leadership, (*v*) the hierarchial, institutional, culture and pre-established, inflexible project guidelines limit the drawing of lessons or learning from experience, (*vi*) government staff have subcontracted all work related to participatory process to NGOs without developing any internal capacity and (*vii*) staff and community have not been exposed to successful projects, nor been provided effective training.

Cost sharing

Cost sharing is the contributions made by the local community which is in addition to the assistance provided by an outside agency in implementing any programme. The investment made by farmers/user groups is important. It ensures an increase in their stake in the project and as eventual ownership of the assets created in the process. It also means that farmers would only be interested in appropriate and essential structures to be constructed. This in turn ensures cost effectiveness of the project. Contribution made by poor and landless families enables them to lay a claim to the sharing of benefits from the work that have been carried out on village commons. When the members of the community contribute on mutually agreed principles and norms, there is a feeling of equity and trust among the villagers. This helps in generating enthusiasm for the project and thus, their willing participation.

It is not that poor farmers can not contribute or do not have the willingness to do so. It is just that they may not have the means to contribute upfront or in cash. In order to facilitate their contributions, they should be given the flexibility to contribute in terms of voluntary labour. The new paradigm envisages the watershed management should become a mass movement with outsiders providing a helping hand. The new paradigm firmly advocates that no one would be more interested in natural resource upgradation than local people. So the concept of cost sharing is being advocated. The following guiding principles are emerging:

- People are interested in development and in sustaining natural resources. Governmental Organisations, NGOs, donors, etc. should provide only enabling assistance. This is considered as an outsiders' contribution to the local initiatives.
- Nothing should be given free and nothing should be done free for the people by the outsiders. People would own and sustain assets and institutions only when they pay for it.

- If people make real and substantial contribution, they would ensure that what is actually needed is planning and implementation. Substantial contributions by people would reduce the cost of construction of less useful structures.
- Cost sharing should not be uniform at the state level. We are dealing with a diverse environment. Therefore, the direct benefit, paying capacity of the local people and ecological needs of the areas should be the basis for cost sharing. The people's share should not be given back in any form; otherwise it becomes a pseudo contribution. Villages and communities showing good results, both in social and biophysical areas, should be rewarded with a special "revolving fund". The maintenance cost should come from water harvesting structures, collecting fodder from silvipastoral systems, etc.

Conflicts resolution

The neglect of heterogeneity (rich and poor, varying interest on resource use, castes etc) in rural areas has led to the failure of many watershed development projects. The outcome of this neglect is evident in the form of projects ridden with conflicts or the differential flow of project benefits among the poor and rich, with the poor, resource dependent farmers becoming worse off. This lack of understanding results in further degradation of the natural resources, poor representation of the interests of the marginal sections in the decision making and the reinforcement of the power of elite leaders through their appropriation as leaders in village institutions (Manta Vardhan, 2000). The conflicts arise due to (*i*) Village Panchayats rights over natural resources are still inconsistent with rights granted to other bodies under other legislations, (*ii*) Village Panchayats cover several villages and several natural resource user groups - conciliate or play off one against the other, (*iii*) Likely domination by the elite group of farmers and (*iv*) Sometimes, village Panchayats do not correspond with watershed/ natural resources boundary, therefore, potential lack of coherent management of natural resources. Conflict arises due to cost sharing is another major concern. The willingness to share the cost would depend on the benefits expected from any activity. Usually the interest of the community in developing a common land is low and their share could be as nominal as responsibility of watch and ward. On the other hand, in activities where the benefits are more direct, the individual families (eg.) developing a water source for drinking or introduction of new crop varieties, contribution could be as high as 50 to 90 per cent.

There is a lot of confusion and conflicts on the rights of communities on forest/common lands. Community rights on these lands need to be reviewed and publicized to ensure the community's involvement in judicious use and effective maintenance of such resources. Socio-economic relationships among people in a watershed create conflicts in introducing seemingly straight forward technical improvements. This is because a watershed contains multiple decision makers whom watershed development affects unequally. When a watershed project is introduced, often the bulk of the work is done in the upper reaches while the benefits accrue primarily in the lower reaches. For example, revegetating the upper reaches involves banning grazing and felling trees so that plants can establish. As a result, the people who utilise the upper watersheds, typically relatively poor people with little or no lands bear the brunt of the costs of watershed development, which mainly benefit wealthier farmers in the lower watershed. Those who are made worse-off by a watershed project can undermine its efforts if they refuse to go along with it. Herders, for example, might refuse to abide by grazing bans and trespass on the common lands if they are able to. In general watershed technologies are likely to fail if they divide benefits unevenly but require near universal co-operation to make them work. In this case, equity becomes a prerequisite to efficiency (Kerr and Sanghi, 1992).

Experience from watershed development project shows that holdings in the upper reaches belong to marginal groups. Therefore, the project must be able and willing to invest in improving the depth and soil quality in the upper reaches. This requires a higher level of investment than what is adequate for lower areas. It would also require new measures like land shaping which is not accepted in the official guidelines for watershed treatment. Training of farmers in management practices which may be new to them and may require much more time and efforts on their part than they are presently investing, is required. Unless, these decisions are taken and implemented, no significant increase in productivity in these areas can be achieved. Consequently, the gap in sustainable income between farming groups in the upper and lower reaches will increase and equity will suffer. In most of the watershed development activities (soil and water conservation measures) women's role is still considered unskilled while men are the key players. Though women's and men's contributions in terms of resource involvement is equal, the economic compensation women receive is only 40 per cent of the total where as men receive 60 per cent. But in the case of forestry there is significant change. Women have been identified as the key player in this sector. The resource involvement of men is 30 per cent and that of women is 70 per cent. The economic compensation of women is 70 per cent and that of men is 30 per cent (Fernandes, 2000).

Monitoring and evaluation

Watershed management is a unique development approach in which watershed is used as a basic unit of planning and management of land, water and other resources. This approach is holistic and multidisciplinary. It enables the planners and managers to consider together various physical, biological, socio-cultural, economic and institutional factors operating within a watershed, and its surrounding environment, thereby they can formulate a comprehensive and integrated watershed development plan to achieve social objectives. The watershed management programme aims at improving the conditions of natural resources within a watershed. Local level institutions have to take charge of the programme so that it addresses the site-specific needs of the resources and of the people and sustain the development initiative. Monitoring the watershed development programme entails checking if these aims are being met successfully. It may not always be possible to measure the results that have been achieved because they may be intangible or it may be too costly to measure them effectively. In such cases indications that success is being achieved will make good proxies. Such indicators, however, must be chosen carefully so that they are reliable substitutes to direct measurement and are easy to measure in terms of time and effort. The choice of indicators is determined by who the end-user is.

Despite wide coverage of the programme in the country launched over the years, there is virtually no institutional mechanisam put in place at the national level for evaluation or for undertaking studies of a basic and long term nature. A programme like watershed development, which involves a hierarchy of administration and communities at the grass roots level in highly varying agro-climatic and socio-economic conditions, invariably requires mid-course corrections of varying degrees in the light of the ongoing experience. Typically, an implementing agency would see a greater value in spending an extra few crores of rupees for undertaking works in the field rather than spending this money for monitoring and evaluation. However, according to some observers, mid-course corrections can improve the programme benefits substantially, in some cases upto 100 per cent. But even if we consider the improvement to be very modest, say, 10 per cent, then a 1 per cent of programme outlay on meaningful monitoring and evaluation would have a very high pay-off in terms of achieving the programme objectives. It is of utmost importance, therefore, to put in place an institutional mechanism for research and monitoring and evaluation in the field of watershed development by involving reputed institutions in the country for upgrading the quality of evaluation. For example, evolving methodologies for capturing the participatory processes and convergence of complementary interventions, by guaranteeing to them adequate funding on a long-term basis and by ensuring autonomy to them in the pursuit of monitoring and evaluation.

CONCLUSION

Today watershed development has become the main intervention for natural resource management. Watershed development programmes not only protect and conserve the environment, but also contribute to livelihood security. The importance of watershed development as a conservation programme is being recognized, not only for rainfed areas, but also for high rainfall areas, coastal regions, and the catchment areas of dams. With the large investment of financial resources in the watershed programme, it is important that the programme becomes successful. For achieving the best results, it is necessary to keep in mind those issues to which attention have been drawn in this book.

References

Fernandes, Al (2000). Equity in Watershed Management : The Myrada Experience, Social and Institutional Issues in Watershed Management in India, 2000.pp:203-207.

Anil Shah (2000). Eloquent Silent Revolution, Development Support Centre, Ahmedabad.

Barrow, Chris (1987). Water Resources and Agricultural Development in Tropics, Longman Scientific and Technical Publications, Essex, U.K.

Chopra, Kanchan and B.Goldar (2000). Sustainable Development Framework for India: The Case of Water Resource, (Mimeo), Institute of Economic Growth, New Delhi.

Hanumantha Rao, C.H., (2000). Watershed Development in India: Recent Experiences and Emerging Issues" Economic and Political Weekly **35** (45): 3943-3947.

Jodha, N.S. (1979). Dry Farming Technology: Achievements and Obstacles, In Ch.Shah and C.N.Vakil (ed.) Agricultural Development in India, Orient Longmans, Bombay.

Kerr, John and Sanghi (1992). Indigenous Soil and Water Conservation in India's Semi Arid Tropics. Substantial Agricultural Programme Gatekeeper Series paper No.34, (London: International Institute for Environment and Development).

Manta Vardhan (2000). Watershed Development: Levelling Uneven Grounds, Social and Institutional Issues in Watershed Management in India, pp.198-202

Reddy, R. V. (2000). Land Degradation in India: Extent, Costs, Determinants and Trends, (Mimeo), Centre for Economic and Social Studies, Hyderabad.

Watershed Management: Issues and Policies for 21st Century pp. 16-25

Reviving Decaying Systems: A Case for Rainwater Harvesting

A. Vaidyanathan

INTRODUCTION

Traditional water harvesting systems are by no means dead. In various stages of decay across the country, these can still be revived. However, their full potential will only be realised if local community institutions are revitalized. Though they are not replacements of other forms of harnessing water, these systems nevertheless have a major role to play in conserving land and water resource and increase land productivity on a sustainable basis.

Nearly ten years ago the Centre for Science and Environment had organised a national conference on traditional water harvesting techniques. The publication *Dying Wisdom* documented extensive, diverse and centuries old traditions of water harvesting in different parts of the country; the important role played by local community effort in constructing and maintaining them; the deterioration of these systems and institutions; and, the neglect on part of the state to sustain and build upon these traditional ways of harnessing water. The conference made an impassioned plea on the need to preserve and protect these traditional techniques.

I would like, specifically, to comment on (*a*) current state of traditional water harvesting systems and the factors which have brought them to their present predicament and measures need to resuscitate them; and, (*b*) the scope for, and means of, extending local water harvesting on a much wider scale.

DECLINE OF TRADITIONAL SYSTEMS

The *khadins* of Rajasthan, the community wells of Gujarat and the 250,000 irrigation tanks spread all over the country are outstanding examples of traditional water harvesting. There is a widespread impression that these systems have become derelict—physically and institutionally. Borewells, tubewells and piped public water systems drawing water from rivers and surface reservoirs have indeed distracted attention from traditional systems, overshadowed them and, in many cases, even put them out of use. The progressive reduction in the area irrigated by tanks and other minor surface sources reported by official statistics would seem to corroborate the impression of widespread and progressive decay.

But this picture may be overdrawn. Studies of a number of specific tanks in Tamil Nadu give a rather mixed picture. Physical degradation such as siltation and encroachment of feeder channels, encroachment of foreshore area, siltation of tank beds, and extensive damages to sluices, regulators, surplus weirs and distribution networks were found to be widespread. Some of the tanks have indeed gone out of use or have suffered a significant reduction in irrigated area. But this is not the case for all tanks—a majority of those investigated are very much in use; in many, there is no indication of a significant reduction in the area irrigated by them, and a large number continue to function despite significant changes in local institutions and power structures. Community-managed institutions have been found to be active in a sizeable proportion of the tanks which were studied.

The deterioration is no doubt partly due to government negligence of regular repairs and maintenance. There are innumerable instances of damage to systems cause by floods and other natural calamities. Investment in rehabilitation of existing works and construction of new tanks, local division works and the like is meagre. Practically, all public resources allocated for expansion and improvement of irrigation facilities have gone into developing large canal systems and groundwater. At the same time, the expansion of roads, railways, and residential settlements has also increased. Such development projects have interrupted natural drainage courses and adversely affected flow of water into the traditional water harvesting systems. Planning of such projects has been unmindful of this consequence.

Having said this one must also recognize that the linking of existing tanks to larger reservoirs (which are called system tanks in Tamil Nadu) has, at least in some areas, helped increase the volume and assurance of supplies to local community storage. It must also be supplied to local community storage. It must also be said responsibility for their deterioration cannot be laid wholly at the doors of government. We need to examine the reasons for the weakening of interest among beneficiary communities in keeping the traditional systems in good,repair. Several factors have contributed.

The massive increase in demographic pressure on land has contributed significantly to extension of cultivation to marginal lands, encroachment of common lands, government and forest lands, and deforestation. Deforestation of upper catchments aggravates soil erosion and siltation of natural drainage channels and reservoirs. Encroachments of tank foreshore areas and feeder channels and conversion of tank beds to residential and industrial sites are again due to growing demographic pressure.

Second, the capacity of the traditional systems being constant, perhaps even declining, have proved to be inadequate to meet the needs of a growing population. This has forced communities and governments to look for alternatives. New technologies have become available which, on the face, make it possible to augment water supply beyond the limits of existing traditional devices. Over most parts of the country, rainfall is concentrated in 3-4 months and is very unevenly distributed across space. Productivity of rainfed agriculture over vast areas remains low because of inadequate and unreliable rainfall. Under these conditions, larger reservoirs are needed to impound surplus monsoon precipitation in high rainfall areas for providing larger, more assured supplies for utilisation during the main crop season and also to extend the crop season. Since irrigation increase the productivity of land manifold, interest in and pressure for construction of reservoirs is understandable. Nor can it be faulted so long as the resulting increase in productivity is much larger than can be obtained even after locally available water resources are fully harnessed and the benefits accrue to areas with low productivity and relatively low incomes.

The introduction of drilling methods and energised pumping permit tapping deeper strata of groundwater, which is impossible with traditional devices. This is one reason why public water supply systems have gone in for tubewells and piped water supply systems. In the case of irrigation, tubewells and electrical diesel pumps enable larger volumes of water to be lifted per unit of time. Moreover, they give individual farmers much greater control over when and how much to irrigate, and flexibility to adjust irrigation according to the needs of crops. Farmers are naturally attracted to go in for wells and tubewells because they enable them to irrigate larger areas and realise higher productivity.

All this does not, of course, warrant unqualified acceptance of the way in which these 'modern' techniques have been exploited. On the contrary, the scale and manner in which they have been applied and managed is seriously flawed in several ways. The development of canal and pump irrigation does not justify the scant attention given to the restoration and improvement of existing local systems, and to exploring possibilities of similar additional work for fuller, more effective harnessing of local water resources. At the same time, modern techniques have been promoted indiscriminately, without proper assessment of costs and benefits. Costs have invariably been underestimated and environmental costs altogether ignored even as benefits have been consistently overstated. There are far too many instances of bad management and indiscriminate irrigation leading to land degradation. Policies (like cheap electricity and diesel, low interest credit and lax loan recovery) have encouraged inefficient use of water and over-exploitation of groundwater in most parts of the country. Correction of these deficiencies is essential.

The deterioration of traditional local systems is also due, to an important degree, to the weakening of local institutions for maintaining the existing facilities and managing them. It would be erroneous to suppose that in the past local institutions were everywhere and always strong or effective. But there is ample evidence that they did function depending crucially on the cohesion of the community and the existence of a village leadership capable of enforcing compliance with customary rules with a numerically small, dominant upper caste group controlling much of the land and the absence of factions in the caste. That this group stood to benefit significantly from irrigation gave them a strong incentive in ensuring proper management of tanks and other land community works.

The conditions, however, did not obtain everywhere or in equal measure at all time. The effectiveness of the local institutions, therefore, always varied a great deal. Nor were they 'equitable': one should not forget that scheduled castes were not allowed access to village common wells/ponds for drinking, washing and bathing. The question of access to irrigation water from tanks made no difference to them because they have little or no land to cultivate.

The situation has changed materially in several respects. In some parts of the country ownership of land which used to be concentrated in numerically small upper caste groups has passed (for a variety of reasons, including migration of upper castes to urban area, land reform legislation and rising political consciousness of the other castes and communities) to the numerically large, middle and so-called 'backward' castes. The earlier congruence between social and economic hierarchy in the village and the resulting concentration of local power in the upper caste has, therefore, weakened.

The strength of local power centres was also dependent on the village being a relatively self-contained socio-economic entity. Increasing commercialization of agriculture and growing importance of government bureaucracy, legislators and political parties have progressively weakened the traditional centres of power within the community. With growing political consciousness, lower castes with little or no land, artisans and service castes are no longer willing to accept local authority based on caste status and land ownership. The sense of collective interest has been further weakened by the emergence of individually-owned wells and tubewells especially among stronger farmers who are community leaders.

The cumulative effect of all this, taken together with the fact that available facilities had to serve an increasing number of claimants, also played an important role in weakening the ability of local institutions to make and enforce rules concerning community matters. Correcting these institutional weaknesses must receive as much attention as additional investments in any programme for local, community-controlled water harvesting systems to become a far more important part of India's water control systems.

REVIVING WATER HARVESTING

In terms of the future role of water harvesting, the first obvious priority is the rehabilitation of 'traditional' water harvesting systems to their full potential. These systems are numerous. According to the Minor Irrigation Census of 1986-87, nearly 250,000 tanks and close to 450,000 small-scale diversion and lift irrigation works are in active use, spread all over the country. Some 50,000 works are reported out of use. The number of communities served is very large and, for a good proportion of them, these works are the main source of water other than rain. The area irrigated by them, though small in relation to other sources, is substantial. More importantly, the declining trend in the area irrigated, come down from seven million hectares (mha) in the 1960s to an estimated 5.5 mha currently, is indicative of their deterioration. Getting these existing works back to their potential should, therefore, be the first order of business.

Though programmes for restoration and modernisation of tanks and other traditional local systems exist, these are minuscule in relation to needs. Schemes decided and implemented entirely by government's engineering establishments tend to be preoccupied with physical improvements; the works themselves are not related to the problems specific to each tank; they are very expensive in terms of outlays per unit area; and there is little knowledge (or even concern to find out) about the performance of the works after they are completed and their impact on productivity. This approach must change.

More resources, of course, need to be allocated for rehabilitation of existing water harvesting systems. It is possible to find substantial resources for this purpose from existing rural development and poverty alleviation programmes if only they were better organised. For more important is the effective use of the resources. This calls for several measures. Standardised packages will just not do. Rehabilitation schemes must be tailored to the local needs and circumstances of specific systems. This is not possible with the information currently available. A phased programme for assessing the condition of existing facilities individually is essential. An inventory of these facilities is a necessary first step. This can be done expeditiously with the help of records maintained by government departments verified by satellite imagery. A more precise assessment of the condition of specific works, and needed remedial measures to restore them to their full potential, will require a phased programme for detailed ground-level investigation.

Such investigations will, of course, need technically qualified personnel. But these personnel need not all be from government. Non-governmental organizations (NGOs) can play an important complementary role. More importantly, the communities served by each tank need to be involved in the process. This will begin a process of getting communities interested, making them aware of the rationale of improvement and eventually involving them in the process of rehabilitation. It will help bring local knowledge of problems and ideas regarding possible solutions based on long, first hand experience—knowledge which no amount of sophisticated surveys by outsiders can match—to bear on preparation of schemes suited to the specific needs of each location.

While government personnel and NGOs will have an important role in the planning process, it does not follow that the implementation of schemes and their management should rest wholly with government. Local water harvesting works will not realise their potential unless the beneficiary communities have a sense of stake in them and manage them on their own. Involving the communities in the process of planning and construction helps in this, the more so if they are required to contribute part of the costs. It is likely to be strengthened by divesting the public works department of the power to decide on resource allocation priorities and entrusting this function to elected Panchayati Raj institutions, leaving them to mobilise additional resources of their own.

On the other hand, the government has an important role in creating and sustaining an environment conducive to community-base institutions. Specifically, it needs to enunciate a framework by which local harvesting systems are to be managed individually and sharing of water among communities in a common watershed is to be related and the mechanisms for sorting out differences among them. Strong and impartial enforcement of these principles, especially in clearing encroachments and other impediments to proper use, are critical. Legislation should, however, confine itself to enunciating general principles of community management leaving sufficient flexibility for details to be settled by the concerned communities through mutual negotiation.

The NGOs have a key role in (*a*) helping individual communities to evolve workable arrangements of common management based on their specific traditions, custom and circumstances; and, (*b*) improving water literacy of the user communities so that they will better understand the role of water, the relation between the way its use is managed, and the outcome in terms of quality, productivity and equity.

EXPANDING WATER HARVESTING

Besides rehabilitating existing traditional water harvesting systems, there is scope for enlarging their role. One way is to explore and take advantage of the possibilities of linking more of the existing rainfed tanks to largest reservoirs. The policy, followed by some states, of prohibiting such linkages must be reversed. Using large reservoirs in combination with tanks dispersed in their command should be accepted as a guiding principle, both in existing canal irrigation systems and in designing new systems.

Such a strategy has the double merit of reducing the scale of large reservoirs needed to harness available water even as it increases the flexibility of users and user communities to ensure more effective use of water. This is evident from the experience of system tanks in South India. It has also been shown that by linking the tanks in Saurashtra to the Narmada reservoir, and operating them in an integrated manner, the storage capacity needed at Sardar Sarovar can be substantially reduced.

WATERSHED DEVELOPMENT

By far the most important means for expanding the role of local water harvesting is integrated watershed development. Nearly two-thirds of the country's cultivated land still depends exclusively on rainfall. Even if the estimated potential for large storage and groundwater were fully exploited, a major part of the area will remain dependent on rainfall. Inadequate water availability for domestic use, animals and for agriculture is, and will remain, an endemic problem in a large part of the country, especially in areas with low rainfall.

A drawback of most soil and moisture conservation programmes is that these focus mainly on the nature of the works, without assessing the difference they will make to the quantum of effective availability of water rainfall retained in the soils of the catchments and has increased surface runoff. Both have adverse effects on productivity. The available evidence suggests that in most states there has been little, if any, increase in productivity of rainfed land over the last two decades. In sharp contrast, productivity of irrigated lands has everywhere shown a sustained rise. These trends imply widening disparities between farmers/ communities with and without irrigation which is obviously a cause for concern. The interests of achieving and sustaining an adequate growth in agricultural productivity of rainfed areas as well as considerations of equity, demand a bigger, more effective programme to raise productivity of drylands. Improved soil and moisture conservation is of critical importance for this purpose.

The central purpose and rationale of watershed development is to ensure that as much as possible of the local precipitation is harnessed and retained for use in the watershed. This can be done by restoring adequate vegetative cover in the upper slopes to reduce the rate of runoff; soil conservation measures to increase the quantum of precipitation absorbed and retained in the soil; and, creation of small ponds to store part of the precipitation and to facilitate larger recharge into the local aquifers.

In the process, availability of water for domestic and agricultural uses will increase. This would enable farmers to cope better with fluctuations in rainfall, get more assured and higher yields during the main crop season, and perhaps also sow more of their land in the dry season which follows.

How much more water can be retained in a watershed depends on numerous factors. The scope for increasing the quantum of moisture retained depends on the existing soil cover status (in terms of depth and structure) and the extent to which soil conservation measures can improve these parameters. This potential is likely to be more in watersheds with heavily eroded soils and degraded land and less when rainfall is in excess of the capacity of the soil to store moisture or when it comes in infrequent and intense spells. The possibilities of retention are greater if surplus monsoon rains can be stored on the surface of underground. The potential for increasing the effective retention of local rainfall for use and the means appropriate for the purpose will naturally vary between regions.

A drawback of most soil and moisture conservation programmes (including integrated watershed development schemes as presently conceived) is that these focus mainly on the nature of the works (restoration of vegetative cover in the upper catchment, contour bunds, check dams and farm ponds), without assessing the difference they will make to the quantum of effective availability of water by way of carryover storage in the soil, in tanks/ponds and underground. Even less attention is given to planning the mix of interventions in relation to costs involved.

Notwithstanding these caveats there can be no doubt that there is substantial scope for capturing, through local water harvesting, a larger proportion of rainfall. With accumulating knowledge from experience and further research, the limits can be further extended. Exploiting this potential and managing the additional supplies judiciously is imperative if there is to be a significant and sustained improvement in the productivity of rainfed agriculture.

LACUNAE WITHIN

India has had a long history of research and other programmes for soil conservation and improved methods of dry farming but, unfortunately, they have not been particularly successful. Mere interventions on agricultural lands are not enough.

The idea of integrated watershed development was accepted as official policy nearly 25 years ago. But fragmented programmes based on concepts of older vintage—such as soil conservation on agricultural land, soil conservation of river catchments, wasteland development schemes continue. The concept of integrated watershed development has been tried out in some areas by the Indian Council of Agricultural Research in addition to several projects implemented with funding from various multilateral donors. The government itself has launched a programme called National Watershed Development Programme for Rainfed Area. But, in reality, planning, management and integration are confined only to the title.

The way watershed programmes are currently planned and implemented, there is little hope that these programmes will make any significant or sustained impact-in terms of raising productivity levels. Arresting the degradation of land and ensuring that they are brought to their full production potential is even less likely. The main deficiencies are:

- Proliferation of programmes and agencies for land improvement and soil and moisture conservation;
- Persistence of piecemeal and fragmented intervention aimed at particular classes of land (such as rainfed farmlands, command areas irrigation, saline and alkaline lands, wastelands and forests);
- Absence of arrangements to ensure coordinated action (in design and implementation) among concerned government agencies;
- Planning or interventions by government agencies without meaningful consultations, not to speak of involvement, of affected communities;
- Lack of follow-up on the part of agencies to monitor the condition of the works after completion, and evaluate their actual impact in relation to expected impact over an appropriately long period of time;
- Preoccupation with physical works (check dams, contour bunds, farm ponds, and afforestation) without any attempt at spelling out the expected impact in terms of the increase in quantum of moisture retained in the watershed, the sources of the increase, the nature and extent of the vegetative cover, and magnitude and time profile of impact on landuse, cropping and output;
- Near total indifference to defining the arrangements (rules, responsibilities and mechanisms) for continuing maintenance of works, on the basis of which access to and use of additional water and biomass on public and common lands will be regulated, and the distribution of costs and benefits between the state and different segments of the community;
- The task of setting up appropriate institutions to ensure proper management of watershed according to the rules remains unaddressed;

- The beneficiary communities are unaware of the rationale of the schemes or the benefits to be expected; nor have the community institutions (including the panchayat) been assigned a role, much less a well defined role in the continuing management of common facilities created in the watershed; and,
- Rights of ownership and powers to regulate use of common facilities and resources vested with the community are inadequate for them to manage developed watersheds. On the contrary, most of the non-private land is legally owned by different government departments (revenue, forests, public works department) which are loathe to part with their ownership and control. Community participation and control, so crucial for effective watershed planning and management, are not possible under the existing legal framework.

THE WAY AHEAD

The effective resuscitation and more widespread adoption of water harvesting techniques calls for drastic changes in the way these programmes are conceived, planned and organised. To begin with, we need to broaden our concept of water harvesting. Water harvesting, in its broadest sense, is essentially a way of harnessing, through various forms of collection and storage, precipitation in seasons and regions which are in excess of their requirements for use in seasons/regions where rainfall is short of requirements. In this perspective, even large reservoirs are a form of water harvesting.

At the same time, experience has highlighted the dangers—by way of degradation, loss of valuable land, forests and other resources and disruption.

Many of the traditional forms and practices need change in the light of changing demographic, socio-political and economic contexts.

Dispossession of communities displaced of excessive preoccupation with larger reservoirs, their poor planning and management. While it is incorrect to suggest large and small water storages of surface and groundwater as mutually exclusive alternatives, the balance between them is a very important issue. Much greater attention to local water harvesting is essential for raising the productivity of rainfed lands, for reducing dependence on larger reservoir, for exploiting the overall water potential, and for more efficient management of water in large systems.

There is much to be learnt from both 'engineering' techniques and institutional arrangements of traditional systems. That there is a need for a greater effort in this respect is also beyond doubt. We must, however, pay more attention to their underlying principles rather that get fixated on the particular forms and practices associated with them. Many of the traditional forms and practices need change in the light of changing demographic, socio-political and economic contexts. Many new techniques of engineering, construction and management are available. We should not, in the name of preserving traditions, deny ourselves the benefit of using these new techniques. We must at the same time be conscious of the dangers of blind and indiscriminate adoption of the modern techniques. The latter must be used selectively, modified contexts, which are highly variable.

All the existing schemes affecting land, water and forests should be merged into a single, unified programme for watershed development. Starting with integrated planning and management of land and water in micro-watershed at the village level, these plans should be aggregated through basins of small rivers and tributaries of large rivers to sub-basins of major rivers and eventually covering entire basins.

At each level, responsibility of preparation of programmes, implementing and monitoring them should be unified under agriculture, forests or irrigation departments (depending on the nature of the watershed) with officials from each department being part of the team and reporting through the land and water development hierarchy. The restructured arrangement should be such that the watershed committee comprising community representatives should be an integral and effective part of the decision-making mechanisms at every level. The diverse interests within each watershed community need to be involved to facilitate formulation of generally acceptable rules regarding contributions to creation and maintenance of facilities, use of the facilities and sharing of the benefits. These should evolve by consensus. Failing that, a mediated process of negotiation and compromise should be followed. The latter requires a well-defined framework of general principles regarding property rights, cost and benefit-sharing to be legislated and enforced by government.

Participatory planning is still no more than a slogan in government programmes. The tradition of consultation, leave alone participation, is weak; and, the problems are numerous and difficult. NGOs, which appreciate the importance of this, have had to put in patient and sustained efforts to bring it about. These efforts are limited, their strategies vary and so do their accomplishments. But attempt to promote community-managed watershed development on a wide scale must learn from these experiments.

Reorganization of administration along lines suggested involves upsetting existing arrangements in terms of division of function and associated powers, both at the political level between ministries, between political functionaries and official and between different segments of the bureaucracy. It is understandable that the bureaucracy is not particularly enthusiastic about it. Less easy to understand, and correct, is the deep-seated scepticism bordering on distrust among politicians and bureaucrats about community participation. Indeed, their attitudes even about consulting affected communities are lukewarm.

Calculated strategy is needed to break this logjam. It has to recognise and accept that reforms of such a fundamental nature cannot be accomplished in one stroke but will have to be engineered in a phased manner.

A practical strategy, I dare suggest, might be prepared along the following lines:

- Start by getting rid of the needless and counterproductive proliferation of land and water development programmes and replacing them with a unified administration for integrated planning and implementations of watershed development;
- Affirm government's commitment to participatory, integrated watershed development and spell out the nature of the organisational structure and mechanisms which are ultimately sought to be put in place;
- Encourage official watershed development agencies to get into the habit of consulting community representatives, making them better informed about the rationale of the programmes, and involving them in collecting and interpreting information and exploring alternative interventions and their implications;
- Demonstrate, in selected watersheds, what a properly planned and managed watershed programme can do by way of increasing overall production and employment;

- Experiment in active collaboration with NGOs, innovative technologies to design more effective interventions, reduce costs and increase the returns by securing high biomass output and adding value to them;
- Identify and remedy legal and socio-political impediments to community participation; and,
- Devise measures to create stake and incentives for community participation in watershed management and for equitable distribution of benefits to resource-poor segments.

All this is bound to be a long haul, but without it the resources and effort spent with much fanfare on land and water development will be wasted. What is more, the prospects of arresting and reversing land degradation, of restoring land to its potential, and progressively increasing it by the application of technology will be in serious jeopardy.

Watershed Management: Issues and Policies for 21st Century pp. 26-36

Watershed Development Programmes in India

D. V. Ramana

INTRODUCTION

The post Green Revolution period in the country has witnessed a major thrust in rainfed agriculture for sustained production and productivity and food security in the rainfed areas. The Government of India has taken major initiatives for implementation of a national level programme for development of rainfed areas through adoption of watershed approach. The watershed approach aimed at augmentation and stabilisation of production and productivity, minimising ecological degradation, reduction in regional disparity and generation of employment for the rural poor in these rainfed areas. The watershed approach has also been adopted for developing the drought prone desert areas and wastelands for sustained crop production and productivity and to provide sustainable means of livelihood in these areas. The Drought Prone Area Programme (DPAP), Desert Development Programme (DDP), the Integrated Watershed Development Programme (IWDP) and National Watershed Development Programme (NWDP) have all focussed at integrated development for land and water resources for sustained livelihood. The Technical Committee constituted by Ministry of Rural Development in 1994 under the chairmanship of Prof. C. Hanumantha Rao have recommended common set of operational guidelines, objectives, strategies and expenditure norms for the watershed approach based area development programmes in the country. While the objectives under DPAP, DDP, IWDP more or less similar primarily focussed at sustainable levels of agricultural production in these areas, the study has taken five States covering one State in East, one in West, one in North, one in South, and one State in central part for studying the impact of the Watershed Development Programme under different programmes in the country.

OBJECTIVE

The objective of the study is primarily focussed on the impact of the watershed approach adopted under different area development programmes in the country under different agro-climatic conditions. The sub-objectives of the study are:

- Study the physical and financial achievements of different components in the selected States;
- Assess the impact of the watershed projects on irrigation/water resources development;
- Quantify the impact of the watershed approach on crop productivity;
- Evaluate the impact of the developmental approach on increasing the household income and generation of employment in rural areas.

METHODOLOGY

The Agricultural Finance Corporation Ltd at the instance of Ministry of Rural Development, Government of India has conducted a quick evaluation of Watershed Development Projects in the entire country covering 20 States. Out of the 20 States covered by AFC for quick evaluation of watershed projects, five States representing one each in East, West, Central, North and South namely Orissa, Gujarat, Madhya Pradesh, Uttar Pradesh and Tamil Nadu have been considered to generalise the impact for the entire country. The sample design showing number of districts, number of watersheds under different programmes and number of sample households covered in each State are shown in Table 1. The study covered a total of 269 watersheds under different programmes namely Integrated Watershed Development Programme, Drought Prone Area Programme and Desert Development Programme spread over 28 selected districts in five selected States.

Table 1. Number of sample households covered in selected districts

Sl.No	State	No. of Districts Selected	No. of Watersheds Selected				Households covered
			IWDP	DPAP	DDP	Total	
1.	Orissa (East)	4	6	42	–	48	288
2.	Gujarat (West)	4	–	17	10	27	162
3.	Madhya Pradesh (Central)	8	41	51	–	92	552
4.	Uttar Pradesh (North)	8	4	58	–	62	372
5.	Tamil Nadu (South)	4	13	27	–	40	240
	Total	28	64	195	10	269	1614

THE STUDY AREA

The study area comprise five States each representing Orissa in East, Gujarat in West, Madhya Pradesh in Central, Uttar Pradesh in North and Tamil Nadu in South. The general features of the study area are presented in Table 2.

PHYSIOGRAPHY AND SOILS

The eastern State namely Orissa is divided into four physiographic regions namely, Northern Plateau, Central Table Land, Eastern Ghat Region and Coastal Regions. The soils are lateritic, red yellow, coastal alluvial, delta alluvial, mixed red and black soils and brown forest soils. The presence of iron oxides moderate acidic generally deficient in nitrogen, organic matter are the characteristics of red soils. Fixation of phosphate is a problem in these soils. The soils are light textured, single grained, highly porous. The western part represented by Gujarat is divided into three regions namely the main land plains, the hilly peninsular region and the hill track on the north east and east. It has a coastal line of 1500 kms on the west. The soils are black soils, mixed red and black soils, residual sandy soils, alluvial soils, saline and alkali soils, desert soils, lateritic soils, hilly soils and forest soils. The central India represented by Madhya Pradesh falls under the catchment of Yamuna, Ganga, Narmada and Godavari rivers. It has northen low lying plains. Malwa and Vindhya plateau, Narmada valley, Satpura stretch, Chattishgarh plains and Bastar plateau and it is 50-1200 mts above MSL. The soils are alluvial deep black, medium black, shallow black, mixed red and black, mixed red and yellow and skeletal soils. On the northern part, the State of Uttar Pradesh has three regions namely, Himalayan, Gangetic plain and Southern plateau. A wide range of soils both of residual and alluvial origin are predominant. The broad soils groups are hill soils, bhabhar soils, tharoid soils, alluvial soils, vindhan soils, bundelkhand soils and aravalli soils. Alluvial soils occupy 62 per cent of total area. The Tamil Nadu State representing the southern part of India comprise coastal plains, eastern ghats, central plateau and western ghats. The soils are black, red, laterite and alluvial type. Red soils occupy 61 per cent, alluvial 24 per cent and the balance are other soils. About 12 per cent of the cultivated area is effected by soil salinity and sodicity.

Table 2. General features of selected States

Sl.No	Particulars	East Orissa	West Gujarat	Central Madhya Pradesh	South Tamil Nadu	North Uttar Pradesh
A.	GENERAL					
1.	No. of Districts	30	25	61	28	83
2.	No. of Blocks	314	230	459	385	904
3.	Total Area (lakh ha)	155.71	19602	443.46	130.04	297.94
4.	Cultivated Area (lakh ha)	59.68 (38.22 per cent)	112.45 (57.37 per cent)	199.40 (44.96 per cent)	70.26 (54.03 per cent)	174.75 (58.65 per cent)
5.	Uncultivated Area (lakh ha)	18.09	8.52	14.82	30.39	11.88
6.	Irrigated Area (lakh ha)	17.32 (29.02 per cent)	26.55 (23.61 per cent)	65.3 (32.76 per cent)	29.03 (41.32 per cent)	104.85 (60.0 per cent)
7.	Forest Area (lakh ha)	55.34	18.87	145.97	21.44	51.50
8.	Total Population (lakhs)	316.60	413.00	661.36	558.59	1391.12
	Rural (lakhs)	274.25	271.56	507.88	367.81	1115.06
	Urban (lakhs)	42.35	141.44	153.48	190.78	276.06
9.	Live stock population (lakhs)	110.74	68.03	482.00	263.66	701.83
B.	CLIMATE					
	Rainfall (Av. Annual in mm)	1497 mm	300-1500 mm	800-1600 mm	947 mm	594-2173 mm

CLIMATE AND RAINFALL

The Orissa state representing east lies in sub-tropical belt with hot and dry summer and the temperature ranges between 22°-42° centigrade from March to May. The average annual rainfall is 1497 mm. In case of Gujarat representing western part has a tropical climate with high variations in rainfall distribution starting from average annual rainfall of 300 mm in Kutch to 1500 mm in southern part of Valsad and Dang districts. In case of Madhya Pradesh which represents central part of India, has the annual rainfall varying between 800 mm in north and western regions and 1600 mm in the eastern districts and the average annual rainfall is 1040 mm. In case of Uttar Pradesh in the north has sub-tropical climate with average annual rainfall ranging between 594 to 2170 mm. The mean minimum temperature ranges between 2.5° to 8° centigrade in winter and highest temperature ranging between 32.5° to 42.5° centigrade in summer. Tamil Nadu at the southern part of the country experiences tropical climate, the temperature varying between 8° to 24° centigrade in winter and 29° to 38° centigrade in summer. The average annual rainfall is 947 mm.

WATERSHED DEVELOPMENT PROGRAMME

Orissa implemented Drought Prone Area Programme in 8 districts covering 212 watersheds with an area of 42,000 ha and Integrated Watershed Programme in 7 districts covering an area of 83,000 ha during 1995-98 by incurring an expenditure of Rs. 134.3 millions. Gujarat has implemented Drought Prone Area Programme in 10 districts covering 858 watersheds, Desert Development Programme in 6 districts covering 845 watersheds and Integrated Watershed Development Programme in 9 districts covering 312 watersheds. Madhya Pradesh has implemented Drought Prone Area Programme in 25 districts covering 1253 watersheds with a total area of 8.24 lakh ha and Integrated Watershed Programme in 22 districts covering 247 watersheds with an area of 1.17 lakh ha during 1995-98 and incurred a total expenditure of Rs. 786.4 millions during 1995-98. Uttar Pradesh in 1995-98 has implemented Drought Prone Area Programme mostly covering 18 districts with a total area of 1.5 lakh ha at a cost of Rs. 609.5 millions. Integrated Watershed Development Programme was also implemented in Agra district. Tamil Nadu implemented Drought Prone Area Programme in 16 districts covering a total area of 1.49 lakh ha and Integrated Watershed Development Programme in 3 districts covering a total area of 0.12 lakh ha and has incurred a total expenditure of Rs. 617.37 millions during 1995-98.

PROJECT INTERVENTIONS

The major works carried out in different States comprise land development, treatment of drainage lines, water resources development, biological measures, afforestation, pasture development and community assets, livestock development including fisheries and poultry. The physical and financial target and achievements of various components/project interventions are presented in Table 3.

Highest achievements are observed in southern State namely Tamil Nadu with 95.3 per cent expenditure while the lowest being in Orissa with 54.4 per cent expenditure over the planned budget upto March, 1999. Gujarat has made a financial achievement of 80.8 per cent, Uttar Pradesh 78.4 per cent and Madhya Pradesh 56.4 per cent over the planned budget. The componentwise target and achievements are discussed below.

Land development

The land development measures undertaken are contour/ graded bunding with vegetative hedges, bench terracing, gully stabilisation with vegetative checks. In physical terms Uttar Pradesh has achieved maximum and Tamil Nadu has reached over 99 per cent. In terms of financial achievements except Madhya Pradesh almost all States have achieved more than 65 per cent. The overall achievement of all the five States under land development is 55.5 per cent of physical targets and 66.8 per cent of financial targets.

Table 3. Activity wise target and achievement in five States (For selected Watersheds upto March,1999)

No.	State	Physical (in Hectares)			Financial (upto March1999) (in Rs. lakhs)		
		Target	Achievement	Percent	Target	Achievement	Percent
I	LAND DEVELOPMENT						
1	Orissa	29491.1	6621.1	22.5	45.8	30.0	65.5
2	Gujarat	27426.8	11481.3	41.9	17.3	15.2	87.4
3	Madhya Pradesh	599644.0	274328.0	45.7	221.0	89.1	40.3
4	Uttar Pradesh	13320.6	79204.0	594.6	357.3	272.2	76.2
5	Tamil Nadu	851.0	846.0	99.4	21.3	36.4	171.5
	Total B I	670733.5	372480.4	55.5	662.7	442.9	66.8
II	TREATMENT OF DRAINAGE LINES						
1	Orissa	13817.0	1714.0	12.4	31.9	10.4	32.6
2	Gujarat	193.20	87.2	45.1	6.5	36.5	558.1
3	Madhya Pradesh	33412.0	16138.0	48.3	208.4	92.9	44.6
4	Uttar Pradesh	2058.0	1667.0	81.0	136.3	81.6	59.8
5	Tamil Nadu	528.0	395.0	74.8	10.9	9.4	86.0
	Total B II	50008.2	20001.2	40.0	394	230.8	58.6
III	WATER RESOURCES DEVELOPMENT						
1	Orissa	5167.0	1114.0	21.6	83.1	46.1	55.4
2	Gujarat	435.10	320.1	73.6	142.5	98.8	69.3
3	Madhya Pradesh	7765.0	3537.0	45.6	389.6	295.7	75.9
4	Uttar Pradesh	604.0	426.0	70.5	162.7	141.0	86.7
5	Tamil Nadu	305.0	292.0	95.7	74.2	69.6	93.8
	Total B III	14276.1	5689.1	39.9	852.1	651.2	76.4
IV	AFFORESTATION, PASTURE DEVELOPMENT, COMMUNITY ASSETS						
1	Orissa	1259.3	1380.6	109.6	31.0	23.7	76.5
2	Gujarat	1206.1	5704.3	473.0	62.0	34.1	55.1
3	Madhya Pradesh	139731.3	120117.2	86.0	272.2	136.8	50.3
4	Uttar Pradesh	1360.1	1162.2	85.4	87.2	77.1	88.4
5	Tamil Nadu	588.5	866.3	147.2	34.0	18.1	53.3
	Total B IV	144145.3	129230.6	89.7	486.4	289.8	59.6
V	BIOLOGICAL MEASURES						
1	Orissa	815.1	25.6	3.1	19.5	5.8	29.8
2	Gujarat	1179.7	1377.2	116.7	12.9	12.0	92.9
3	Madhya Pradesh	105.2	57.1	54.3	18.2	13.5	73.8
4	Uttar Pradesh	174.5	289.2	165.7	15.0	22.0	146.6
5	Tamil Nadu	18.0	20.0	111.1	1.2	1.3	113.0
	Total B V	2292.5	1769.1	77.2	66.8	54.6	81.7
VI	LIVESTOCK, FISHERIES, POULTRY DEVELOPMENT						
1	Orissa	586.1	141.2	24.1	5.1	1.0	19.0
2	Gujarat	0.0	0.0	0.0	0.0	0.0	0.0
3	Madhya Pradesh	3146.0	40316.0	1281.5	4.9	0.4	8.3
4	Uttar Pradesh	173.0	183.0	105.8	1.4	2.3	163.6
5	Tamil Nadu	0.0	0.0	0.0	0.0	0.0	0.0
	Total B VI	3905.1	40640.2	55.5	11.4	3.7	55.5

Table 3 (*Contd. on* page 31)

Table 3 (*Contd. from* page 30)

VII	SHELTER BELT AND SAND DUNE STABILISATION						
1	Orissa	105.0	118.5	112.9	0.2	0.8	410.0
2	Gujarat	5775.0	4476.0	77.5	8.4	5.3	62.4
3	Madhya Pradesh	303.0	222.0	73.3	8.7	5.1	57.9
4	Uttar Pradesh	0.0	0.0	0.0	0.0	0.0	0.0
5	Tamil Nadu	4.0	5.0	125.0	0.4	0.5	127.5
	Total B VII	6187.0	4821.5	77.9	17.7	11.7	5.6
VIII	ALL COMPONENTS TOTAL						
1	Orissa	51240.6	11114.9	21.7	216.5	117.7	54.4
2	Gujarat	36215.9	23446.1	64.7	249.7	201.8	80.8
3	Madhya Pradesh	784106.5	454715.3	58.4	1123.0	633.3	56.4
4	Uttar Pradesh	17690.2	82931.5	468.8	760.0	596.2	78.4
5	Tamil Nadu	2294.5	2424.3	105.6	141.9	135.3	95.3
	Total B VIII	891547.7	574632.1	64.5	2491.1	1684.3	67.6

Treatment of drainage lines

The treatment of drainage lines include the measures related to nala bank stabilisation, live check dams, loose boulder check dams with vegetative hedges, earthern check dams with vegetative hedges. In physical terms overall achievement is 40 per cent and in financial terms 58.6 per cent in all the selected States. Except Uttar Pradesh and Tamil Nadu the achievement is less than 50 per cent in other three States and Orissa has achieved only 12.4 per cent in physical terms. Gujarat has made highest percentage of financial expenditure followed by Tamil Nadu and Uttar Pradesh which have spent maximum expenditure while the other two States namely Madhya Pradesh and Orissa have achieved less than 50 per cent. However, in reality actual expenditure incurred in Madhya Pradesh is Rs. 92.9 lakhs followed by Uttar Pradesh with Rs. 81.6 lakhs and the other States have spent less on this component.

Water resources development

Water Resources Development measures include farm ponds, nala bunds, check dams and percolation tanks. In physical terms maximum achievement is observed in Tamil Nadu with 95.7 per cent followed by Gujarat 93.6 per cent, Uttar Pradesh 70.5 per cent, Madhya Pradesh 45.6 per cent and Orissa 25.6 per cent and the overall physical achievement is about 40 per cent. The combined financial achievement is 76.4 per cent. Except Orissa almost all States have expended about 70-94 per cent of the planned budget.

Afforestation, pasture development and community assets

The measures under this component include fodder and timber development, fuel wood, agro forestry, block plantations, pasture development, repairs to community assets and crop demonstrations. The overall physical achievement of all the States is about 90 per cent and in terms of physical achievement it is about 60 per cent. In physical terms the achievements are satisfactory in almost all States. Madhya Pradesh alone incurred an expenditure of Rs. 136.8 lakh out of Rs. 272.2 lakh budget and Uttar Pradesh spent Rs. 77.1 lakh upto March, 1999.

Biological measures

The biological measures namely horticulture plantations and cultivation of fodder crops and the overall physical achievement is 77 per cent and the financial achievement is about 82 per cent. The achievement of biological measures both in physical and financial terms is satisfactory in all the States except Orissa.

Livestock, fisheries and poultry development

The overall achievement both in physical and financial terms is about 56 per cent and there are no proposals under this component in Gujarat and Tamil Nadu. Maximum achievements are observed in Madhya Pradesh and Uttar Pradesh in physical terms and Uttar Pradesh in financial terms.

Shelter belt and sand dune stabilisation

The overall physical achievement under this component is 78 per cent and financial achievement is 65.6 per cent. The component is not implemented in Uttar Pradesh. Gujarat and Madhya Pradesh have spent maximum though the physical achievement is ranging between 58-62 per cent.

Overall achievement

The overall physical achievement of all components in the five States in the country is 64.5 per cent over the planned targets and 67.6 per cent in financial achievement over the budgeted amount. All the States together expended Rs. 168.45 millions while the maximum being Madhya Pradesh and Uttar Pradesh with Rs. 63.33 and 59.62 millions respectively. The programme interventions under various components both in physical and financial terms is presented in Table 4 and the Statewise details are in Appendices I to V.

PROGRAMME IMPACT

The programme impact is assessed in terms of increase in area under irrigation, crop productivity and average annual income to the beneficiary households.

Water resources development

The impact of water resources development interventions is substantial with significant increase in irrigated area over the pre-project conditions. The overall area increase is from 22,930 ha in pre-project condition to 32,092 ha by post-project period showing an increase of about 40 per cent of area under irrigation. The measures are dug wells, nalas and streams and ponds, tube wells and the groundwater levels have significantly increased invariably in all the States wherever the programme is implemented. The increase in area under irrigation is maximum in Orissa in terms of percentage with 67 per cent and the lowest being Tamil Nadu with 17 per cent. In terms of area Uttar Pradesh has achieved maximum from 7,000 ha to 11,600 ha with an increase of 65 per cent followed by Tamil Nadu, Madhya Pradesh and Gujarat.

Table 4. Status of water resources during pre and post project periods

States	Source	Number			Irrigated Area (in Hectares)		
		Pre-Project	Post-Project	Per cent	Pre-Project	Post-Project	Per cent
Orissa							
	Dug wells	658	660	100.30	77.97	76.00	97.47
	Nalas/Stream	30	72	240.00	54.13	111.00	205.06
	Others	35	120	342.86	48.00	80.00	166.67
	Ponds	126	126	100.00	157.10	295.00	187.78
	Tube Well	172	102	59.30	3.50	6.00	171.43
	Sub-Total	1021	1080	105.78	340.70	568.00	166.72
Gujarat							
	Dug wells	683	2133	312.30	2,226.00	3,107.00	139.58
	Nalas/Stream	10	10	100.00	558.00	570.00	102.15
	Ponds	28	37	132.14	743.20	628.00	84.50
	Tube Well	171	213	124.56	229.57	509.62	221.99
	Sub-Total	892	2393	268.27	3,756.77	4,814.62	128.16
Madhya Pradesh							
	Dug wells	1727	1798	104.11	1,776.12	335.01	131.47
	Nalas/Stream	36	40	111.11	283.41	656.00	231.47
	Others	3	8	266.67	12.00	42.00	350.00
	Ponds	140	322	230.00	591.89	1,328.93	224.52
	Tube Well	75	89	118.67	315.00	430.07	136.53
	Sub-Total	1981	2257	113.93	2,978.42	4,792.01	160.89
Uttar Pradesh							
	Dug wells	38	62	163.16	161.90	343.70	212.29
	Nalas/Stream	23	41	178.26	724.40	1,124.30	155.20
	Others	880	1783	202.61	5,523.00	8,063.50	146.00
	Ponds	35	276	788.57	222.83	823.83	369.71
	Tube Well	29	65	224.14	440.00	1,245.00	282.95
	Sub-Total	1005	2227	221.59	7,072.13	11,600.33	164.03
Tamil Nadu							
	Dug wells	773	833	107.76	4,268.00	4,870.00	114.10
	Nalas/Stream	156	182	116.67	1,935.00	2,132.00	110.18
	Others	0	15	0.00	0.00	174.00	0.00
	Ponds	165	170	103.03	2,389.00	2,602.00	108.92
	Tube Well	112	207	184.82	190.00	539.00	283.68
	Sub-Total	1206	1407	116.67	8,782.00	10,317.00	117.48
All Total							
	Dug wells	3879	5486	788	8510	10732	695
	Nalas/Stream	255	345	746	3555	4593	804
	Others	946	1963	944	6326	8988	747
	Ponds	637	1107	1346	3590	5559	1113
	Tube Well	388	463	587	949	2220	875
	Total	6105	9364	826	22930	32092	737

Table 5. Productivity of dominant crops during pre & post project periods*

State	Crop	Pre-Project	Post-Project	Per Cent
Orissa	Barley	12.00	14.01	16.75
	Jowar	7.15	10.00	39.86
	Maize	14.44	15.00	3.88
	Paddy	13.64	10.74	-21.26
	Sesamum	8.50	9.50	11.76
	Spices	9.00	7.18	-20.22
	Wheat	5.71	4.52	-20.84
Gujarat	Arhar	7.71	9.91	28.53
	Bajra	13.93	15.57	11.77
	Castor	5.61	7.50	33.69
	Cotton	7.90	13.37	69.24
	Groundnut	8.61	15.01	74.33
	Jowar	7.91	11.15	40.96
	Wheat	25.00	28.50	14.00
Madhya Pradesh	Arhar	7.93	8.59	8.32
	Bajra	6.41	11.32	76.60
	Gram	7.31	7.91	8.21
	Jowar	11.94	11.25	-5.78
	Lentil	6.41	6.76	5.46
	Maize	6.65	8.76	31.73
	Millets	6.45	7.40	14.73
	Moong	6.31	6.93	9.83
	Paddy	8.60	10.70	24.42
	Soyabean	9.11	9.57	5.05
	Wheat	18.00	28.00	55.56
Uttar Pradesh	Arhar	6.90	11.99	73.77
	Bajra	8.21	8.28	0.85
	Barley	10.25	11.72	14.34
	Gram	12.31	13.85	12.51
	Lentil	4.67	8.46	81.16
	Maize	10.00	11.20	12.00
	Mustard	17.08	18.51	8.37
	Pea	5.00	6.08	21.60
	Sugarcane	385.22	543.35	41.05
	Urad	7.50	7.51	0.13
	Wheat	30.73	33.10	7.71
Tamil Nadu	Bajra	7.00	7.00	0.00
	Cotton	6.96	7.08	1.72
	Groundnut	14.46	17.54	21.30
	Gram	11.00	11.75	6.82
	Jowar	13.33	19.00	42.54
	Maize	12.00	16.00	33.33
	Paddy	29.19	33.32	14.15
	Ragi	6.16	8.34	35.39

* In quintal/per hectare

Crop productivity

Major crops grown in the study area are paddy, wheat, castor, bajra, groundnut, jowar and cotton. The yield increase in almost all the cases is observed at different intensities. The paddy yield increase is varying between 5-33 per cent in different states. Jowar 11.8-42.5 per cent, maize 3.8-33 per cent, bajra 14-76.6 per cent, wheat 7.3-55.5 per cent, arhar 4.7-31 per cent, groundnut 21-73 per cent and cotton 12-69 per cent. In other words almost all the crops both irrigated, irrigated wet and rainfed crops have shown significant yield increase due to provision of life saving irrigation through watershed development approaches (Table 5).

Table 6. Average annual income per beneficiary household during pre and post project periods

State	District	Programme/ Project	Income Pre-Project (Rs.)	Income Post Project (Rs.)	Per Cent
Orissa					
	Bolangir	DPAP	20165	27725	37.49
	Kalahandi	DPAP	19755	18621	-5.74
	Phulbani	DPAP	19979	26453	32.40
	Bolangir	IWDP	17630	23800	35.00
	Sub-Total		19382	24150	24.60
Gujarat					
	Banaskantha	DDP	66083	75429	14.14
	Kutch	DDP	78768	112851	43.27
	Panchmahal	DPAP	52945	58505	10.50
	Bhavnagar	DPAP	46222	49378	6.83
	Sub-Total		61005	74041	21.37
Madhya Pradesh					
	Bastar	DPAP	27411	32370	18.09
	Betul	DPAP	28115	30829	9.65
	Dhar	DPAP	13085	21280	62.63
	Rewa	DPAP	23336	31128	33.39
	Shahdol	DPAP	15059	27853	84.96
	Neemuch	IWDP	25150	31550	25.45
	Raigarh	IWDP	13946	17983	28.95
	Raipur	IWDP	29347	45980	56.68
	Sub-Total		21931	29872	36.21
Uttar Pradesh					
	Almora	DPAP	18846	18870	0.13
	Pithoragarh	DPAP	15931	18982	19.15
	Tehri Garhwal	DPAP	12226	13409	9.68
	Jalaun	DPAP	29632	40031	35.09
	Lakhimpur Kheri	DPAP	95500	100665	5.41
	Lalitpur	DPAP	61621	66091	7.25
	Sitapur	DPAP	58438	84079	43.88
	Agra	IWDP	83902	88360	5.31
	Sub-Total		47012	53811	14.46
Tamil Nadu					
	Periyar	IWDP	22256	24719	11.07
	T. Samyurayur	IWDP	35574	40372	13.49
	Poddukotal	DPAP	35733	41260	15.47
	Tuticorin	DPAP	34050	38602	13.37
	Sub-Total		31903	36238	13.59
	Grand Total		35739	43113	20.63

Income

The average annual household income among the overall households has shown an increase of 20.63 per cent over the pre-project conditions. The average household income of Rs. 35,739 during pre-project period has increased to Rs. 43,113 by post project period. The increase of income in different states is presented in Table 6.

Employment

The programmes have generated a total of 20.53 lakh mandays of additional employment under various project components. The statewise employment generation through watershed development programmes during 1995-98 is as in Table 7.

Table 7. Employment generation

(in lakh mandays)

S.No	State	Employment Generation
1.	Orissa	1.411
2.	Gujarat	1.731
3.	Madhya Pradesh	7.680
4.	Uttar Pradesh	7.170
5.	Tamil Nadu	2.536
	Total	20.58

CONCLUSION

The watershed approach under different ongoing area development programmes have no doubt created substantial impact on water resources development, crop productivity, rural household income and employment. Since the objectives of different programmes/projects related to desert development programme, drought prone area programme and integrated watershed development programme are different, the impact of different components are not similar because of various priorities accorded under each. However, the common approach adopted in these programmes minimises the duplication of similar efforts by different agencies. What is required in these programmes is providing necessary technical support by adoption of available local technologies to a great extent. Besides the normal components, to make them more integrated, the relevant components in livestock development for fisheries, sheep rearing, piggery and poultry development may also be included as part of the watershed approach. Despite different priorities accorded under each developmental programme the watershed approach has proved to provide dependable methodologies for sustainable crop production and sustained levels of household incomes.

Watershed Management: Issues and Policies for 21st Century pp. 37-42

New Challenges for Watershed Development in 21st Century

A. Venkatraman

INTRODUCTION

Reflection and experience pave way to change and progress. We have been implementing watershed development programmes all over the country for more than three decades. What is our track record? Are we progressing in the right direction? Are we addressing adequately key issues of concern? The rationale of using watershed as a unit for development is sound and is rarely disputed. There is growing interest and increasing willingness to commit public resources for watershed development, though prompted by different perceptions and considerations. What are the challenges facing us? Will business as usual do? How should we design watershed development strategies in the coming decades, in the backdrop of lessons learned from past performance? In my view, aims of watershed programmes have been multiple, strategies diffused, approaches varied, and gains rather patchy. Considerable course correction is needed to make watershed a purposeful and productive strategy for development. In this paper, I propose to discuss three key issues - objective, strategy and institution development that need considerable refinement.

THE CHALLENGES

Our developmental efforts are primarily technology driven. Focus, in the past, has invariably all along been on promoting new technologies and handy products for overcoming production constraints, and using market opportunities for securing immediate gains. In this approach conservation of natural resources—land and water, two basic factors of production, have received limited, rather inadequate, attention. These natural resources have been used, and in many cases over used, to such an extent that the natural capital today stands badly depleted and environmental problems threatening our very socio-economic fabric are surfacing at an appalling pace.

Demographic changes, competing demands induced by economic growth and changes in life style patterns are aggravating further the pressure on land and water. Unsound interventions and fund driven imperfect technology application without holistic outlook cause cycle of floods and droughts and damage of groundwater. Moreover unchecked pollution and encroachments on water courses are seriously affecting availability and quality of water. Soil and water are finite elements and cannot be produced at will. Both are critical for our wealth and well-being. Soil and water resources conservation need priority attention. For sustained growth, our precious natural capital must be conserved and managed with utmost care. Otherwise our future will be in peril.

Application of technology of course is critical for achieving economic prosperity and welfare of the society, and for improving standard of living and quality of life of our people. There should not be any let up in ushering changes through developmental efforts; on the contrary there is tremendous need to accelerate technological changes since our rate of growth is rather slow. But this push forward should not in any way be allowed to damage the natural resources—land and water that are our basic life sustaining systems. How to march forward without harming our natural resources is a major challenge we face today? Watershed is an excellent unit for managing prudently the two natural resources, land and water. To get it done and sustain the health of our natural capital we must sharpen development objectives, design correct strategies and establish supportive policies and institutions.

REFINE OBJECTIVES

At present watershed development programmes have diverse objectives. Some address poverty and rural employment. Some focus exclusively on prevention of silting up of reservoirs. Some others address drought proofing and rainfed agriculture. Certain cases focus on soil and water conservation. Most programmes have multiple objectives. Almost all programmes, constrained by limited funds, are implemented in scattered small micro or mini watersheds with less than adequate impact. Having too many goals and a wide array of minuscule activities for achieving them dilutes efforts and leads to wastage of resources. Hence objective of watershed programme must be refined with a sharp focus on key issues.

A major resource critical for sustained growth and development is water, its availability in adequate quantity and quality. Water is a prime natural resource; its availability or supply is limited by nature. For the growth and development of agriculture, manufacture and service sectors we need increased supply of water. More than that for improving quality of life, water is indispensable. Every drop of water counts in an over populated and water deficient country like that of ours. Our effort must therefore first focus on water resource management, its conservation, allocation and judicious use to secure maximum gains.

Watershed is a geographic unit delimited by natural flow of water. It is ideal to use watershed development to primarily focus on water: its storage, quality, allocation, use, and disposal. Other considerations of development could then follow, preferably in a two-step sequential cascade. Water resource conservation and use to be effective must be preferably fully public funded and implemented in contiguous macro watersheds of catchments and sub-catchments of river basins. Integrated Catchment Management (ICM) must be the primary focal point of watershed development. Once a catchment has been treated for ICM, its mini or micro watersheds can be tackled with ease and effectiveness for achieving various other developmental goals including soil and moisture conservation, better land use, employment generation etc. These secondary objectives should be planned, in my view, by exclusive programmes designed according to location specific requirement by actively involving local people and without in any way clashing with ICM.

TWO-STEP SEQUENTIAL CASCADE

Why this two-step sequential approach? Reasons are three-fold. First, water is a finite common good, provided by nature; its conservation and use must be planned and implemented in a holistic manner to secure maximum gains to the society. Hydrology and geology must mainly influence decisions on water collection and storage. Its efficient conservation and use must be ensured through well-designed public policies and enlightened involvement of all stakeholders. Both for surface flows and ground water National and State Water Policies have laid down clear principles and guidelines for water collection, storage, allocation, use, and for safeguarding quality. Integrated Catchment (macro watershed) Management or ICM is a vital and integral component of River Basin Water Plan. Attempting to achieve goals of ICM through piecemeal disaggregated micro watershed treatment would be least effective, hurt larger interests of the river basin and result in wastage of resources.

Second, soil and moisture conservation and related tasks need a different scale and method of approach and so would merit separate handling. Individuals own land, except common land. Active interest of landowners, most of them small land-holders and their close involvement, is essential for success of soil and moisture conservation and better land and water use. Active and purposeful interaction of the service agency and landowners over time, from identification of problems through effective adoption of technologies, is critical for its success. For effectiveness it must be planned to cover micro or mini watersheds and implemented with application of cost-effective technologies. It would need capital investment and recurring costs. While capital cost, at best, can be shared between public and individuals, recurrent costs would have to be largely borne by individuals since they would be the sole beneficiaries of the investment.

Third, when other objectives like poverty alleviation, employment generation, livestock development etc, important though they may be, are packaged into a single watershed development programme, past implementation performance bears out that focus on a holistic outlook on water resource management invariably gets lost. Utilization of allocated funds assumes greatest importance, leading frequently to creation of needless and ineffective physical structures. These goals where needed could be combined with soil and moisture and land use programmes. So, water resource management in the catchments must be left to stand alone as an integral and major component of Basin Water Plan.

RECAST STRATEGY AND APPROACHES

To achieve the primary aim, development strategy must be recast. It must sharply focus on: Using river basin as unit for water planning, treating catchments / sub-catchments as integral part of the basin plan, strengthening R&D to provide cost- effective location specific support, improving knowledge and skills of involved human resource, promoting social action through involving user groups, and establishing institution and policy support for sustained management of water resources.

River basin plan

River Basin Water Plan must deal with both surface flows and groundwater. It must assess supply and demand over space and time by various user groups. It must assist in designing and implementing cost effective measures to conserve, allocate and use water resource most productively for the greatest benefit of the society. It should specially address quality, environmental and sustainability issues. Based on implementation experience the strategic water plan must be revised and refined at periodic intervals, perhaps once in three to five years. River basins for management purposes can be grouped into two broad segments

—catchments and command areas. Effective interventions must be planned for each segment according to need and resource potentials. Need for water for domestic use would be spread all over the basin; in resource planning domestic water need must be assigned first priority and provided with full allocation. Next, where and how to collect and store the balance of water and how to allocate it among competing demands and promote judicious use should receive careful thought and action.

Integrated catchment management

Interventions in catchments must consider the hydro-meteorology, and geomorphology characteristics of the watershed and focus on assisting in conserving water resources cost effectively. Interventions must not be stereotyped but should be designed to suit specific physical and socio-economic environment. Some important generic features of ICM worth mentioning are: Afforestation and improving the vegetative cover, strengthening and protecting existing surface water storage systems, creating new surface storage, and recharging groundwater with appropriate interventions, all to be crafted consistent with the water plan of the basin. Management aspects of catchments would include O&M of water resources with peoples' involvement, preventing encroachments and pollution, ensuring quality of water, monitoring silt-load, preventing environmental deterioration and promoting water use efficiency.

Research and development support

Effective R&D support is critical for designing location specific and cost effective interventions for treatment of catchments and command areas. At present interventions are planned with inadequate data and research support. Use of remote sensing technologies, and integration of knowledge of various disciplines like hydrology, geology, meteorology and economics etc are inadequate. The activities of research institutions functioning under different departments related to water resource operate in a fragmented compartmentalized manner with little coordination, direction or purposeful interaction. Operational research in command as well as catchments is conspicuous by its absence.

This situation leads to wastage of public resources by needless repetitive efforts and unproductive investments. Hence the functioning of the research system needs incisive review and improvement with a purpose of providing effective field support for water resource development through river basin planning approach. Development of innovative cost effective designs and interventions to improve the efficiency of storage and use of both surface flows and groundwater must be systematically encouraged. The net for capturing new technologies must be caste wide; experiences of other countries as well as traditional knowledge and practices must be collected systematically and tested for local application. Promising methods must be tested through ORPs in sizable extents involving the user groups and then promoted for adoption. All these would need a coordinated well-directed and efficient research effort; the present research system needs considerable improvement.

Strengthen data-base

For improving planning and intervention efficacy there is urgent need to improve the database. Various agencies are involved in collection and storage of data on land and water use. The methods of collecting data need constant improvement; accuracy, validity, timeliness and cost and benefit must be the governing factors. Establishing a central data center with networking facilities with data collection centers and user groups at State level would be of great value.

Human resource development

Human resource development is yet another aspect that needs considerable improvement to make the watershed development effective. Technology for better water conservation and use must be shared with staff and user groups through well-organized training programmes and through mass media. Effective linkages must be established among scientific institutions, training institutions, involved agencies and user groups. In all training programmes besides imparting technology the need for conserving water, its quality and improving its efficient use must be stressed.

Proposed organizational frame

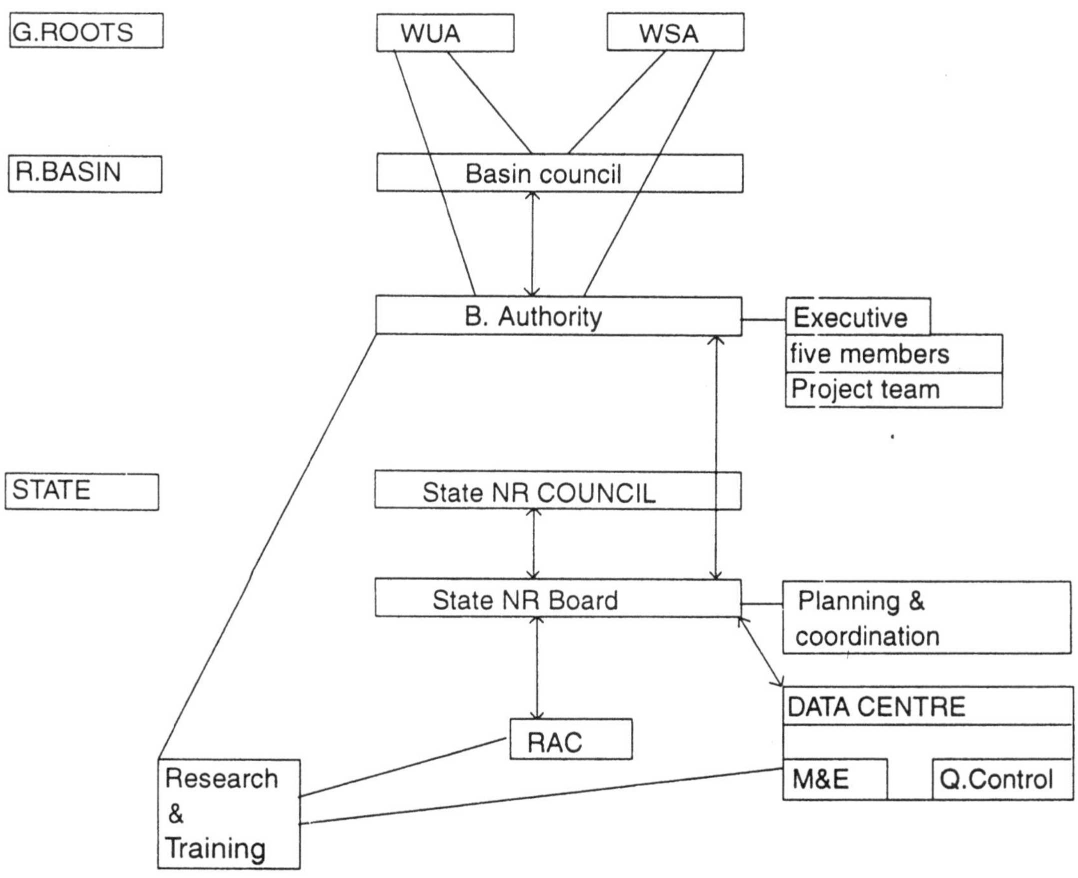

Institution development

Strengthening the data base, formulating research priorities, formulating river basin plan, providing direction, guidance and coordination among involved agencies; monitoring implementation performance, and ensuring funding support all would need the services of a central agency both at the state and river-basin level. Involvement of local user groups is critical for local planning, operation and maintenance of water resources and sustaining development efforts. Social action is the key for preservation and proper management of natural resources. For these purposes a well-knit supportive institutional framework must be established. A suggestive organizational model for this purpose is presented here.

Public policies

There is a general appreciation that people's involvement particularly at grassroots, is critical for sustaining natural resource conservation. But should this be done through establishing separate interest groups or should the Panchayats be empowered to look after natural resources with the assistance of local user groups? Views and experiences differ. National and State policies on water, both surface flows and groundwater, are either in place or under consideration. There are various enactments governing quality and pricing of water. Are these measures adequate to conserve and develop water resources, assure quality, effect planned allocation, promote efficient use and prevent pollution and encroachments? Are these adequately enforced? Or is there a case for enacting a separate all inclusive water law for effectively protecting and using the water resource? What measures would assist in improving the effectiveness of the system? These and similar issues on policy and institutional development deserve critical review and debating. Research institutions should address such socio-economic issues also and assist in developing suitable public policies.

CONCLUSION

The challenges we face are clear and not insurmountable. Let us plan, act and progress. Where there is will to act, where there is social cohesion in using technology for conserving and using natural capital prudently, economic prosperity and human well being are bound to blossom.

Watershed Management: Issues and Policies for 21st Century pp. 43-56

Watershed Development Research in Selected Watersheds of Tamil Nadu

K. Palanisami, D. Suresh Kumar and R. V. Selvi

INTRODUCTION

Impact of Green Revolution came about largely because of tremendous boost to irrigated agriculture and use of improved seed varieties. But the gains in agricultural production did not reach the vast rainfed areas of the country and development of dryland farming did not get the attention as it deserved though nearly 70 per cent of cultivated land in the country are under rainfed condition. As a result, though the objective of overall food availability in the country was achieved, the development process has created and aggravated serious but unintended agricultural, socio-economic and ecological imbalances. This called for special efforts to develop dryland agriculture to ensure equity and even distribution of gains of development. The only possible alternative to bring more area under cultivation is to go for watershed development since watershed has become an acceptable unit for planning soil and water conservation resources. To implement technologies related to watershed development programme, field tested recommendations are needed which can be generated from the applied research carried out in different agro-climatic zones of the country. In view of this, in 1983 model watershed projects were formulated and implemented with the involvement of ICAR as pilot projects. These projects generated valuable information that culminated in the launching of the National Watershed Development Programme for Rainfed Agriculture in the Seventh Five Year Plan. Based on the valuable lessons learnt from this experience, it has been well recognized that ending the neglect of vast rainfed areas is possible only through the principle of "integrated watershed management by adopting a holistic approach to development of rainfed farming systems". This new approach finds expression in the restructured National Watershed Development Programme for Rainfed Areas (NWDPRA) for implementation during the Eight Five Year Plan.

The National Watershed Development Programme for Rainfed Areas (NWDPRA) scheme, funded by the Government of India, is functioning in 84 watersheds covering six out of seven agro-climatic zones of Tamil Nadu. The State Department of Agriculture is implementing this programme in co-ordination with the Water Technology Centre, Tamil Nadu Agricultural University, Coimbatore. For effective implementation of this programme, 10 research stations/colleges viz., Water Technology Centre, Coimbatore; ARS, Bhavanisagar; RRS, Paiyur; RRS, Vriddhachalam; ARS, Virinjipuram; ADAC and RI, Trichy; AC and RI, Madurai; HC and RI, Periyakulam; ARS, Kovilpatti and TCRS, Yethapur were identified to carry out the research programmes in different watersheds. The project is being implemented with a total budget of Rs.84.0 lakhs for a period of five years from 1997-98 to 2001-2002.

OBJECTIVES

The National Watershed Development Programme for Rainfed Areas (NWDPRA) endeavours to achieve the twin objectives of sustainable production of biomass and restoration of ecological balances in the vast tracts of rainfed areas in the country. The specific objectives set out are outlined as follows:

- To determine the most suitable mix of agronomic (biological) and low cost engineering measures of controlling soil erosion and reduce surface runoff, water harvest and to conserve *in situ* soil moisture in a watershed.
- To evolve suitable soil fertility management technique through soil breeding and soil conservation, integrated nutrient management system involving organic and bio-fertilizers and reclamation of problem soils for increasing the productivity of crops in the watershed areas.
- To evolve suitable strategies for the alternate land use planning involving, agro-horti-forestry and silvi-pasture components for watershed areas.
- To evolve suitable farming systems under different conditions.
- To determine suitable cropping systems.

The progress of research carried out in different watersheds to improve the productivity is briefed hereunder.

IMPROVING GROUNDNUT YIELD THROUGH SOIL MOISTURE CONSERVATION USING LOCALLY AVAILABLE MATERIALS

The rainfed agriculture mainly depends upon the receipt of adequate rainfall, which should be evenly distributed. Even if the normal rainfall is received, the rainwater should be conserved most effectively. For this, different agronomical and engineering technologies are recommended. One of the technologies is the application of organic manures. The beneficial effects of organic manures are many, such as the improvement of physical properties of the soil like water holding capacity, aeration and drainage. It also improves the microbial activity of the soil. Further the fertility status of the soil is also improved because of the addition of major, secondary and micronutrients contained in them.

Table 1. Effect of soil moisture conservation materials on groundnut yield (ARS, Virinjipuram)

Treatments	Soil moisture (per cent)) 45 DAS	Soil moisture (per cent) 75 DAS	Pod yield (kg/ha)
Coir pith @ 12.5 t ha^{-1}	12.00	9.38	2029.80
FYM @ 12.5 t ha^{-1}	11.88	8.15	1865.80
Groundnut husk @ 5 t ha^{-1}	11.15	7.78	1715.80
Pressmud @ 12.5 t ha^{-1}	9.50	7.05	1604.30
Control	8.13	6.38	1542.30
Sed	0.20	0.15	56.80
CD (P = 0.05)	0.44	0.32	123.70

In rainfed agriculture, the organic matter addition is highly beneficial because it improves the water holding capacity of soil by increasing soil aggregation. Hence to study the effect of locally available organic materials like coirpith, FYM, groundnut husk and press mud on soil moisture conservation and yield of crops, experiments were carried out involving various crops.

Among the soil moisture conservation measures, the highest pod yield of 2029.80 kg ha^{-1} was obtained with the application of decomposed coir pith @ 12.5 t ha^{-1} against the control (1542.30 kg ha^{-1}). The next best treatment was application of FYM @ 12.5 t ha^{-1}. The highest pod yield from coir pith applied plot is may be due to higher percentage of available soil moisture on 45 and 75 DAS and more number of pods $plant^{-1}$.

In another study, the influence of the same organic materials on the tuber yield of tapioca both in the pure crop and tapioca + groundnut intercropping system and pod yield of groundnut were recorded and the data is given in Table 2.

Table 2. Tuber yield of tapioca (H226) and pod yield of groundnut (TMV 7) (TCRS, Yethapur)

Treatments	Groundnut pod yield (kg ha^{-1})	Tapioca yield (t ha^{-1})	
		Intercrop	Pure crop
Composed coir pith @ 10 t/ha	937	11.5	12.4
Pressmud @ 5 t/ha	910	9.7	10.5
FYM @ 10 t/ha	926	10.5	11.3
Groundnut husk @ 5 t/ha	878	8.7	9.4
Control	826	7.6	8.2
CD (P = 0.05)	5.0	0.28	0.17

Application of composted coirpith at 10t/ha recorded the highest tuber yield of tapioca both under pure and tapioca + groundnut intercropping system (12.4 t/ha and 11.5 t/ha respectively). The same treatment also resulted in the highest pod yield of groundnut (937 kg/ha) in the tapioca + groundnut intercropping system. The control treatment (no addition of organic manures) resulted in lowest tuber yield of tapioca (8.2 and 7.6 t/ha respectively in pure crop and intercropping system), as well as groundnut pod yield of 826 kg/ha in the tapioca + groundnut intercropping system.

MICRO WATER HARVESTING TECHNIQUES FOR PROMOTING GROWTH AND YIELD OF MANGO

The dry land farmers are traditionally cultivating annual crops like ragi, sorghum, cumbu, groundnut, gingelly, castor, pulses or cotton. The income obtained from these crops mainly depends on the suceess of the crops, which is decided by the quantity of rainfall received. Many times the dryland farmers are incurring loss in cultivation of such annual crops due to failure of monsoon. In recent times, an awareness has been created among the dryland farmers to change their cropping pattern by introducing fruit /tree crops. Many State / Central government schemes implemented in the dry land areas especially in watersheds motivated the farmers to take up cultivation of fruit crops. Further, these fruit crops can establish well in drought condition also. If these crops are maintained at their early stages of establishment by providing sufficient soil moisture they will fetch constant additional income for many years. The water received during the monsoon period should be effectively conserved by preventing the run off losses through formation of different systems for micro-water harvesting like pit system, crescent bund, compartmental bunding and saucer and basin.

The growth parameters like plant height (cm), stem girth (cm) and number of branches were measured three years after planting. Of the different micro water harvesting techniques tested for promoting growth of mango cv. Bangalora, compartmental bunding registered higher values of plant height (139.4 cm), stem girth (16.8 cm) and number of branches (10.6) followed by saucer and basin and pit methods. The lowest values were observed in control treatments. As regards Neelum variety, saucer and basin method of water harvesting recorded higher values of plant height (162 cm), stem girth (23.5 cm) and number of branches (12.1) followed by compartmental bunding, crescent and pit methods (Table 3).

Table 3. Mango growth parameters as influenced by treatments (RRS, Paiyur)

Treatments	Plant height (cm)		Stem girth (cm)		No. of branches	
	Bangalora	Neelum	Bangalora	Neelum	Bangalora	Neelum
Pit method	121.0	123.0	10.9	19.7	6.7	6.4
Crescent method	131.9	140.0	12.6	20.3	8.3	9.7
Compartmental bunding	139.4	144.6	16.8	21.7	10.6	10.4
Saucer and basin	125.1	162.0	16.7	23.5	9.0	12.1
Control	93.2	155.6	16.1	19.9	8.9	9.7
CD (P = 0.05)	8.49	5.67	3.00	2.40	2.13	1.56

MICRO WATER HARVESTING TECHNIQUES FOR PROMOTING GROWTH AND YIELD IN MORINGA

The trial on micro water harvesting techniques in moringa showed peculiar results as mentioned in Table 4. As far as the field survival is concerned there was no significant difference among the treatments and also between the treatments and the control. They were all at par with each other. Though the saucer and basin was at par (4.00) with the control (5.00) with respect to number of branches, it proved its significance in plant height (1.80 cm) and yield (167.33). From the experimental results the crescent bund treatment was found to be better performing and suitable for watershed areas (Table 4) as it performed better than other treatments.

Table 4. Field survival and biometrical characters of moringa in response to water harvesting techniques (HC &RI, Periyakulam)

Treatments	Field Survival (per cent)	Number of Branches	Yield (No. of Fruits/Tree)
Pit	98.33	3.66	173.33
Crescent bund	98.66	4.00	167.33
Compartmental bund	97.66	3.33	165.00
Saucer and Basin	98.00	4.00	176.00
Control	98.66	5.00	184.00
SED	0.62	0.76	7.43
CD (5 per cent)	1.42	1.75	17.14

MICRO WATER HARVESTING TECHNIQUES FOR PROMOTING GROWTH AND YIELD OF SAPOTA

Several water-harvesting techniques were implemented and the observations were made on the field survival and other biometrical characters of sapota. The data collected from the experimental plot were almost same showing least difference and are on par with each other. They performed equally with the control treatment at given. However, the character such as plant height and stem girth showed significant results compared to control. Among the four treatments, the crescent bund treatment was found to be good. The compartmental bunding treatment showed similar values (11.30cm) with the control (12.83 cm) in respect of plant spread (Table 5).

Table 5. Field survival and biometrical characters of sapota in response to harvesting techniques (HC &RI, Periyakulam)

Treatments	Field Survival (per cent)	Spread (cm)	Number of branches	Stem girth (cm)
Pit	98.00	11.23	4.00	156.56
Crescent bund	97.66	11.30	3.33	153.76
Compartmental bund	97.66	9.70	3.66	150.90
Saucer and Basin	97.33	11.66	3.33	154.16
Control	98.66	12.83	4.66	161.43
SED	0.80	0.54	0.59	1.10
CD(5 per cent)	1.85	1.24	1.37	2.54

MICRO WATER HARVESTING TECHNIQUES FOR PROMOTING GROWTH AND YIELD OF LIME

The water harvesting treatments viz., pit method, crescent bund, compartmental bunding, saucer basin and no conservation were imposed. The various micro water harvesting techniques showed significant difference in lime seedling height. The saucer basin method is superior over other methods of soil and moisture conservation. Compartmental bunding, saucer basin method and no conservation methods were at par with each other. The saucer basin method showed 9.5 per cent of increased height over no conservation. Four months after imposing the treatments, the height of lime seedlings in various treatments showed significant difference in height. Maximum mean seedling height (3.5 m) was observed in saucer basin method of water conservation followed by crescent bund and pit method. The increased height in saucer basin method over no conservation was 29.62 per cent. Pit method and compartmental bunding methods were at par with each other with reference to height of lime seedlings. From overall comparison, the saucer basin method of water harvesting technique is superior over other methods of moisture conservation method.

Table 6. No.of lime fruits/tree as affected by waterharvesting methods

Treatments	24 MAT	26 MAT
Pit	12.6	61
Crescent	14.6	59
Compartmental bunding	16.0	77
Saucer basin	20.0	92
Control	9.6	37
SED	1.0	0.8
CD (P = 0.05)	2.1	1.8

The fruit yield, two years after imposing treatment showed significant difference. Number of lime fruits/tree recorded was maximum in saucer basin method (20 No.of fruits/tree) of moisture conservation followed by compartmental bunding (16 No.of fruits/tree). The pit method and crescent bund methods were at par with each other. In saucer basin method 146 per cent increased fruits yield was recorded over control. Next to saucer basin method, compartmental method registered maximum fruit yield of 76.75 no of fruits / tree.

Table 7. Soil moisture content (per cent) as affected by water harvesting methods

Treatment	0-15	15-30	30-45
Pit	18.6	18.8	23.7
Crescent	19.4	19.7	23.5
Compartmental bunding	19.1	19.5	23.3
Saucer basin	19.8	19.9	24.2
Control	14.3	14.6	20.6

The soil moisture content in the top layer (0-15 cm) was maximum (19.8 per cent) in the saucer basin method whereas the moisture content of control treatment was lower (14.25 per cent). The same trend was observed in the 15-30 cm soil layer. In the sub-soil (30-45 cm) maximum moisture content of 24.2 per cent was recorded in saucer basin method. Top soil layer 0-15 cm, 15-30 cm and sub-soil layer 30-45 cm maximum mean moisture content was observed in saucer basin method 26.54 per cent, 27.64 per cent and 32.82 per cent respectively. In all the soil layers the control (No conservation) recorded 9.00 per cent to 9.64 per cent of moisture content. The saucer basin method registered maximum moisture content in the sub-soil layer (15.61 per cent).

IMPROVED NURSERY TECHNOLOGY FOR PRODUCTION OF ELITE TREE SEEDLINGS

The dry land farmers are generally cultivating annual crops like groundnut, gingelly, castor, finger millet, cholam, cumbu, cotton or pulses. Most of the times, crop failure has been witnessed due to the failure of monsoon resulting in heavy loss to the dryland farmers. Due to this fact the farmers should be motivated to change their cropping pattern by introducing fruit/tree crops.

If efforts are taken to maintain these crops in the critical stages of one or two years for their establishment, the farmers can realize a constant income from these crops for many years. The success of agro-forestry depends upon the use of healthy and vigorous seedlings so that the establishment percentage will be more, when planted in the rainfed areas. The use of biofertilizers found to be beneficial for producing elite tree seedlings by their favourable effect and root and shoot growth. Trials were carried out to study the effect of bio-fertilizers viz., *Azospirillum*, *Azotobacter*, *Phophobacteria* and VAM and their combinations on growth of tree seedlings.

Among the five-biofertilizer treatments, highest germination percentage (98per cent) was recorded in *Azospirillum* treated seedlings that was at par with *Azotobacter* and also *Phosphobacterium* treated seedlings. The *Azospirillum* treated neem seedlings recorded the highest mean plant height of 13.18 cm at 45 DAS that was at par with *Azotobacter* and

Phosphobacterium treatments. The root length of the neem seedlings was maximum in *Azospirillum* treated seedlings that was significantly different over the control.

Table 8. Effect of bio-fertilizers on neem seedling production (ARS, Virinjipuram)

Treatment	Germination (per cent)	Plant height (cm) 45 DAS	Root length (cm) 45 DAS
Seed treatment with *Azospirillum*	9.80	13.18	6.35
Seed treatment with *Azotobacter*	9.75	12.36	5.26
Seed treatment with *Phosphobacterium*	9.65	12.80	5.32
Seed treatment with *Rhizobium*	9.50	11.03	4.68
Seed treatment with VAM	9.00	11.52	4.52
Control	8.95	10.71	4.45
CD (P = 0.05)	0.59	2.27	0.73

Similar studies were carried out to produce elite seedlings of Tamarind and Teak. They also revealed the same results. Among the treatments tried, the treatment involving the combination of *Azosprillum*, *Phosphobacteria* and VAM recorded significantly the highest plant height and root length. On the other hand, the control treatment where no bio-fertilizer was applied resulted in the lowest plant height and root length.

IMPROVED SOWING TECHNOLOGY AND IPM ON GROUNDNUT

The germination of groundnut sown under rainfed condition depends mainly on the availability of soil moisture in addition to other factors responsible for germination. Several improved land preparation technologies for sowing of groundnut are available for conserving the soil moisture that is helpful to increase the germination of rainfed groundnut. Further, the groundnut crop is attacked by a number of pests resulting in reduced yield. IPM practices like summer ploughing, collection and destruction of larvae and pupae, using light traps, pheramone traps, growing castor in the border and need based pesticides application are recommended to control the pests

Table 9. Effect of sowing technology, IPM on growth parameters of groundnut (ARS, Virinjipuram)

Treatments	Plant population m^{-2}	Plant height (cm) 30 DAS	Plant height (cm) 60 DAS
Line sowing	30.60	15.40	25.40
Sowing behind the tractor plough	17.80	18.63	27.22
Sowing behind the country plough	14.40	18.43	28.30
S.Ed	2.20	0.24	0.26
CD (P = 0.05)	0.60	0.54	0.63
Quinolphos @ 750 ml ha^{-1}	20.70	17.19	26.76
Endosulfan @750 ml ha^{-1}	21.26	17.62	27.41
Dichlorvos @ 375 ml ha^{-1}	20.33	17.87	27.39
Dimecron @ 750 ml ha^{-1}	20.60	17.29	26.79
Monocrotophos @ 750 ml ha^{-1}	21.63	17.51	26.84
Neem oil spray @ 2 per cent	20.86	17.58	27.04
Control	21.27	17.15	26.57
CD (P = 0.05)	2.00	1.37	1.41

Among the treatments imposed, to study the effect of improved sowing technology and IPM on groundnut, the treatment with monocrotophos @ 750 ml ha^{-1} with line sowing recorded the highest pod yield of 787.50 kg ha^{-1} as against the control. The increase in yield of groundnut may be due to optimum plant population and effective control of spodoptera, leaf minor and green lacewing, through IPM practices.

SUITABLE LAND CONFIGURATION MEASURES FOR *IN-SITU* WATER HARVESTING FOR HIGHER PRODUCTION OF CROPS

To increase the moisture availability to the agricultural crops /tree crops in the individual farmers field / micro level basis and to increase the infiltration and percolation of rain water into the root profile, several *in-situ* soil and moisture conservation techniques may be introduced. One such experiment was conducted involving broad bed furrow, compartmental bunding and random tie ridge and the results are presented here below.

Table 10. Growth and yield of sorghum influenced by different soil conservation measures, (ARS, Bhavanisagar)

Treatment	Plant Height (cm)			Grain yield	Straw yield
	30 DAS	60 DAS	90 DAS	(kg/ha)	(kg/ha)
Broad bed furrow	44.6	110.6	136.7	1125.0	6483.0
Compartmental bunding	43.6	102.0	131.1	1060.0	5997.7
Random tie ridge	42.3	92.8	119.0	986.8	5585.2
Control	38.7	78.0	112.4	916.8	5188.0
CD (P = 0.05)	0.55	3.98	5.06	45.40	401.9

The treatment effect on plant height was significant at 30, 60 and 90 DAS. The highest plant height was recorded in treatments where broad bed furrow were formed. It was followed by compartmental bunding and random tie ridge treatments. The lowest plant height was noticed in control plot.The broad bed furrow treatment recorded the highest grain yield (1125.0 kg/ha) and straw yield (6483.2 kg/ha) respectively. This was followed by compartmental bunding and random tie ridge. The control treatment recorded the lowest grain (916.8kg/ha) and straw (5188.00kg/ha) yield, respectively.

Table 11. Soil moisture percent of sorghum crop influenced by different soil conservation measures (ARS, Bhavanisagar)

Treatments	60 DAS		90 DAS	
	0-15 cm	15-30 cm	0-15 cm	15-30 cm
Broad bed furrow	22.5	22.1	20.8	20.7
Compartmental bunding	21.8	21.1	20.3	20.1
Random tie ridge	20.7	20.3	19.5	19.2
Control	19.9	19.6	18.6	18.3
CD (P = 0.05)	0.64	0.61	0.46	0.54

In respect of the soil moisture percentage, at two stages (60 DAS and 90 DAS) and two depths of soil profile (0-15 cm and 15-30cm) the soil moisture is uniformly higher in broad

Watershed Association Meeting,
Banapuram Watershed
Salem (District)

PRA Exercises,
Eagapuram Watershed,
Salem (District)

Seedlings Distribution,
Karnandheri Watershed

Farm Surface Storage Tank,
Kattampatti,
Coimbatore (District)

Farm Pond, Kattampatti
Watershed,
Coimbatore (District)

Cablegation Irrigation System

Major Check Dam, Kattampatti Watershed, Coimbatore (District)

Percolation Pond, Kattampatti Watershed, Coimbatore. (District)

Vegetative Fencing with Agave, Kattampatti Watershed, Coimbatore (District)

Micro Sprinkler Irrigation

Drip Irrigation System

bed furrow. Next comes the compartmental bunding and random tie ridge. The lowest soil moisture percent is recorded in control plot.

IDENTIFICATION OF GRASS-LEGUME COMBINATION FOR EFFECTIVE SOIL MOISTURE CONSERVATION AND FODDER PRODUCTION

If the micro level fields in the watersheds are in patches, it is necessary that this land is treated with vegetative barriers / bunds on contour followed by planting and cultivation along the contours. A vegetative barrier on contour may be raised which could guide the farmers to take up all the farm activities like ploughing and planting along the contour. It is constructed at intervals of 20-30 m up to a slope of 5 per cent. The barrier will also obstruct the run off water and prevent erosion. Different grass and legume combinations were tried in the watersheds to acheive the twin objectives of soil moisture conservation and fodder production.

Table 12. Effect of grass-legume combination on soil moisture and green fodder yield (ARS, Virinjipuram)

Treatments	Soil moisture (per cent) Initial	Soil moisture (per cent) 45 DAS	Green fodder yield (t/ha)		
			Main crop	Inter crop	Main + inter crop
Fodder sorghum + Horsegram	9.32	11.38	18.76	9.61	28.37
Fodder sorghum + *Stylosanthus scabra*	8.43	8.24	19.28	1.92	21.20
Fodder maize + Horsegram	7.75	8.70	16.17	9.05	25.19
Fodder maize + *Stylosanthus scabra*	7.53	7.95	16.57	1.86	18.43
Deenanath + Horsegram	9.75	11.76	21.89	9.88	31.78
Deenanath + *Stylosanthus scabra*	8.68	10.30	22.31	2.15	24.51
S.Ed	0.94	0.49	0.25	0.17	0.25
CD(P = 0.05)	1.93	1.06	0.52	0.37	0.52

Among the grass-legume combination treatments, Deenanath + *Stylosanthus scabra* planted at 3 : 1 ratio registered the highest main crop (Deenanath) yield of 22.31 t ha^{-1} compared to all other combinations. But, the trial plot having combination of Deenanath + Horsegram recorded highest inter crop yield of 9.88 t ha^{-1}. However, the highest green fodder yield (main crop + inter crop) was obtained from the trial having combination of Deenanath + Horsegram (31.78 t ha^{-1}). Regarding available soil moisture, there was no significant differences among the different grass-legume combinations at initial stage. But, at 45 DAS Deenanath + Horsegram combination conserved highest available soil moisture (11.76 per cent) compared to other treatments.

In case of western zone, the highest plant height was recorded in Deenanath sp. followed by *Cenchrus ciliaris*. The lowest plant height was recorded in *Desmanthus* sp. On the contrary the biomass yield was higher in *Desmanthus* sp. and lower in *Stylosanthus scabra* when compared individually. Under intercropping situation, the highest biomass yield was obtained in Deenanath + *Desmanthus*, (1,940.7 kg/ha) followed by *Chloris gayana* + *Desmanthus* (10,887.7 kg/ha). The lowest biomass yield was obtained in Deenanath + *Stylosanthus scabra* (4842.00 kg/ha).Thus it can be concluded that best grass-legume combination for Vellakovil block of Bhavanisagar centre is Deenanath + *Desmanthus* and it can be recommended to the farmers for adoption.

Table 13. Effect of treatments on plant height and biomass (ARS, Bhavanisagar)

Treatments	Plant height (cm)			Biomass
	30 DAS	60 DAS	90 DAS	yield (kg/ha)
Cenchrus ciliaris	19.3	50.9	79.8	3149.7
Stylosanthus scabra	24.3	32.6	67.0	2553.6
Cenchrus ciliaris	21.8	50.6	83.1	3164.5
Desmanthus	17.8	31.7	58.1	9000.0
Chloris gayana	17.3	35.3	53.7	2700.5
Stylosanthus scabra	24.5	33.4	61.9	2477.7
Chloris gayana	17.6	37.2	61.0	2136.2
Desmanthus	18.7	33.8	52.6	8751.5
Deenanath	30.55	61.4	95.2	2415.0
Stylosanthus scabra	24.4	34.3	76.4	2427.0
Deenanath	30.2	65.7	109.6	2429.2
Desmanthus	19.3	33.9	60.1	9511.5

SUITABLE LAND CONFIGURATION MEASURES TO INCREASE THE YIELD OF CROPS

The study on suitable land configuration measures for *in situ* water harvesting and higher production revealed that under rainfed conditions broad bed furrow system recorded higher soil moisture percent which is higher by 2 per cent than compartmental bunding, 4 per cent than random tie ridging and 5 per cent over control. The broad bed furrow system of land configuration measures recorded higher groundnut yield than all other treatments.

Table 14. Effect of land configuration measures on groundnut yield

Treatments	Soil moisture (per cent)	Plant population m^{-2}	Pod yield (kg/ha)
Broad bed furrow	16.32	20.60	1411.00
Compartmental bunding	14.26	24.40	1350.20
Random tie ridge	12.60	24.20	1275.40
Control	11.14	23.40	1209.40
S.Ed	0.44	0.80	45.04
CD(P = 0.05)	0.96	1.74	98.14

Among the land configuration measures adopted in groundnut, the treatment of Broad Bed Furrow recorded the highest plant population, available soil moisture percentage on 60 DAS. The same treatment also registered the highest pod yield of 1411.00 kg ha^{-1} as against the control (1209.40 kg ha^{-1}).

Among the land configuration measures adopted in bajra viz., Broad Bed Furrow, Compartmental bunding and Random tie ridge for effective water harvesting, the treatment with compartmental bunding recorded highest available soil moisture of 9.22 per cent on 60 DAS and plant population of 12.45 nos. m^{-2}, sequently increased the yield of bajra (2433.30 kg ha^{-1}) compared to control (2036.70 kg ha^{-1}).

Table 15. Effect of land configuration measures on bajra

Treatments	Plant population m^{-2}	Soil moisture per cent 30 DAS	Soil moisture per cent 60 DAS	Plant height (cm)	Grain yield (kg/ha)
Broad bed furrow	11.77	12.68	8.62	92.93	2218.3
Compartmental bunding	12.45	15.52	9.22	104.21	2433.30
Random tie ridge	11.25	12.62	7.60	95.15	2163.30
Control	10.28	11.63	6.93	88.10	2036.70
S.Ed	2.20	0.22	0.12	1.44	34.90
CD(P = 0.05)	4.70	0.47	0.25	3.08	74.30

PERFORMANCE OF ARID ZONE FRUIT TREES UNDER DIFFERENT SYLVIPASTURE SYSTEMS

The results indicated that raising *Cenchrus ciliaris* + *Stylosasnthus scabra* as intercrops with tamarind and mango resulted in the highest green biomass production of 28.2 and 27.4 t/ha respectively. However, intercropping Deenanath + *Stylosanthus scabra* in sapota resulted in the lowest green biomass production of 19.1 t/ha.

Table 16. Green biomass production of grass and legume combination (G : L - 3 : 1)

Treatments	Grand total yield (t/ha)
Tamarind + *C.ciliaris* + *Stylo.scabra*	28.2
Tarmarind + Deenanath + *Stylo.scabra*	21.8
Mango + *C. ciliaris* + *Stylo. scabra*	27.4
Mango + Deenanath + *Stylo. scabra*	20.6
Sapota + *C. ciliaris* + *Stylo.scabra*	25.7
Sapota + Deenanath + *Stylo. scabra*	19.1

Among the different arid zone fruit trees tried, moringa recorded highest plant height when compared with other trees. In the development of improved nursery technology, the combination of 3 : 1 : 1 : 1 ratio of soil, sand, FYM and tank silt recorded higher germination and vigorous growth of neem seedlings.

Table 17. Establishment rate of different tree species

Treatments	3 MAP (percentage)	6 MAP (percentage)
Amla	100.00	99.10
Sapota	60.00	33.00
Mango	46.00	26.60
Curryleaf	40.00	25.00
Moringa	30.00	10.00
Sesbania	20.00	0.00
Deenanath	10.00	0.00

EVALUATION OF SPECIES FOR VEGETATIVE FENCING

Among the treatments, the establishment rate is higher in Kattamanakku (98.3 per cent) followed by Kodipoovarasu. The lowest rate (94.8) of establishment was recorded in *Agave* sp. The planting materials were collected from farmers field and planted with equal height of 60 cm. Among the treatments, the *kiluvai* plant species recorded highest plant height on 30 DAP (75.9 cm), 60 DAP (90.1 cm) and 90 DAP (103.3 cm). This was followed by kattamanakku and kodipoovarasu plant species. The lowest plant height was noticed in *Agave* species. Although the rate of initial establishment of *kiluvai* sp. is slightly lesser than kattamanakku and kodipoovarasu, the growth rate is higher in *kiluvai*. Hence, it can be concluded that *kiluvai* plant species was highly suitable for vegetative fencing in Anthiyur block.

Table 18. Treatment influence on plant height (cm)(ARS, Bhavanisagar)

Treatments	Establishment rate (per cent)	Plant height (cm).		
		30 DAP	60 DAP	90 DAP
Agave	94.8	65.2	70.8	79.4
Kiluvai	96.8	75.9	90.1	103.3
Kattamanakku	98.3	73.5	82.5	92.0
Kodipoovarasu	97.0	69.3	74.1	83.8
CD (P = 0.05)		1.36	1.74	1.74

Among the treatments, the establishment rate is higher in Kattamanakku (98.3 per cent) followed by Kodipoovarasu. The lowest rate (94.8) of establishment was recorded in *Agave* sp. The planting materials were collected from farmers field and planted with equal height of 60 cm. Among the treatments, the *kiluvai* plant species recorded highest plant height on 30 DAP (75.9 cm), 60 DAP (90.1 cm) and 90 DAP (103.3 cm). This was followed by kattamanakku and kodipoovarasu plant species. The lowest plant height was noticed in *Agave* species. Although the rate of initial establishment of *kiluvai* sp.is slightly lesser than kattamanakku and kodipoovarasu, the growth rate is higher in *kiluvai*. Hence it can be concluded that *kiluvai* plant species was highly suitable for vegetative fencing in Anthiyur block.

Table 19. Treatment influence on plant height (cm) (ARS, Bhavanisagar)

Treatments	Establishment rate (per cent)	Plant height (cm).		
		30 DAP	60 DAP	90 DAP
Agave	94.8	65.2	70.8	79.4
Kiluvai	96.8	75.9	90.1	103.3
Kattamanakku	98.3	73.5	82.5	92.0
Kodipoovarasu	97.0	69.3	74.1	83.8
CD (P = 0.05)	-	1.36	1.74	1.74

Among the treatments, the rate of establishment is higher in Kattamanakku (94.1per cent), followed by *killuvai*. The lowest rate of establishment was obtained in *kattunochi* sp. The planting materials were collected from the nearby areas and planted with equal height of 60 cm.

Among the treatments, the *kiluvai* plant species recorded the highest plant height of 82.0cm (30 DAP), 89.9 cm (60DAP), 93.0 cm (90 DAP) and 99.4 cm (120 DAP) respectively. This treatment was followed by *kattunochi* and kattamanaku. The lowest plant height was recorded in *Agave* species. Although the initial rate of establishment in *kiluvai* is slightly lower than the kattamanaku, the growth rate is found to be superior to other species. Hence it can be concluded that *kiluvai* can be recommended for vegetative fencing in Chennimalai block.

Table 20. Influence of treatments on establishment rate and plant height (cm)

Treatments	Establishment rate (per cent)	30 DAP	60 DAP	90 DAP	120 DAP
Agave	87.5	74.3	78.8	82.2	86.8
Kiluvai	89.3	82.0	89.9	93.0	99.4
Kattunochi	82.8	79.3	85.6	87.5	93.4
Kattamanakku	94.1	77.3	80.9	83.4	90.1
CD (P = 0.05)	-	1.67	2.48	1.81	2.07

The vegetative species planted to assess the suitability for fencing in dry lands revealed that the maximum establishment is in *Agave* followed by *Kiluvai*. Regarding the growth of the plant, *Kiluvai* grows faster than other species. Eleven tree species were selected and planted in areas polluted with tannery effluents. The plant height and number of branches produced by the different tree species showed that Neermathi, Naval, Neem and Pungam are in the order in producing taller plants and more branches indicating the suitability of these four species for growing in tannery polluted areas.

DOCUMENTATION OF TRADITIONAL PRACTICES OF FARMERS AS WELL AS LOCALLY AVAILABLE SOIL AND MOISTURE CONSERVATION FLORA

In rainfed agriculture, conservation of the rain water and prevention of soil erosion are considered as primary importance to get higher yields as well as to conserve the top soils. Even though modern technologies are available for this purpose like summer ploughing, application of organic manures, adopting broad bed furrow system, compartmental bunding, contour bunding, growing vegetative barriers like, *vettiver*, the farmers are traditionally using certain technologies for this purpose. Hence, a survey was undertaken with 100 respondents in each watershed to evaluate the traditional practices of farmers as well as locally available soil and moisture conservation flora for effective soil and moisture conservation. The survey indicated that there is an awareness among the farmers about soil erosion. Eighty per cent of the farmers are knowing that rain water is being lost from their fields as run off if no soil conservation measures is adopted and 65 per cent of the farmers are aware that run of water causes soil erosion in their fields.The survey conducted on traditional practices of farmers for soil and water conservation measures showed that the maximum number of farmers are adopting bunding (95per cent), summer ploughing across the slope (90 per cent) and application of FYM (75per cent). The other practices like contour bunding and placing granite bolders across the slope in erosion prone spots were followed by 50per cent and 40per cent of the farmers respectively. Planting of vegetative barriers is adopted by small number of farmers (planting *vettiver* along the contour is followed by 30per cent of the farmers and planting *agave* along the drainage channels is followed by 25 per cent of the farmers). There is no awareness among the farmers for application of coirpith for soil and water conservation. Only 10 per cent of the farmers were following the application of coirpith to the fields.

CONCLUSION

Among the soil moisture conservation materials tried, application of coir pith @ 10t ha^{-1} gave highest pod yield of groundnut besides increasing the soil moisture content on 45 and 75 days after sowing than control. Application of coir pith @ 10t ha^{-1} recorded the maximum yield in pure crop of tapioca as well as tapioca + groundnut intercropping compared to other treatments. The study on suitable land configuration measures for *in-situ* water harvesting, the broad bed furrow system registered higher crop yields followed by compartmental bunding. Four micro water-harvesting methods viz., pit method; cresent bund, compartmental bunding and basin were tried along with control. Among the water harvesting methods, tree seedlings grow faster under basin method when observed 70 days after planting followed by cresent bund and pit method. Among the micro water harvesting techniques tried for lime, saucer basin method was found to superior over other methods in conserving moisture from 0-15 cm. The study on improved sowing technology and integrated pest management revealed that the integrated approach of treating the seeds with thiram @ 4 g kg^{-1} of seed + raising bajra as intercrop 4 : 1 + spraying with monocrotophos 750 ml ha^{-1} recorded the highest yield in groundnut. Among the sowing technology, line sowing recorded the higher yield. Among the bio-fertilizers tried for the production of elite tree seedlings, *Azospirillum* treated seeds produced seedlings with vigorous growth than the other bio-fertilizers treatment and untreated seedlings.

Watershed Management: Issues and Policies for 21st Century pp. 57-64

Land and Water Management Strategies for Kalahandi District of Orissa

R. C. Srivastava, S. Mohanty, N. Sahoo, S. Pattnaik and Deepika Sahoo

INTRODUCTION

Kalahandi district of Orissa has become synonymous to hunger and poverty. In spite of plenty of natural resources, like land and water, the people are poor and dying of hunger. The agriculture in hilly parts of the district is at low subsistence level and they are able to produce sufficient only for about 6 months. The area is totally monocropped and shifting cultivation on marginal lands is prevalent. This has led to serious consequences of heavy erosion. As this area forms part of catchment of Upper Indravati Project, the heavy erosion will cause quick siltation of the reservoir. In view of it, it is necessary to manage the natural resources in such a way as to retard the erosion process by retiring marginal lands from cultivation, and increasing as well as stabilizing the productivity to meet the demand of population dependent upon it. For this a strategy is required for natural resource planning in the area. To develop such a strategy, a study was undertaken at Udri Gram Panchayat of Thumla Rampur block of Kalahandi district of Orissa. This paper presents results of this plan.

RESOURCE SCENARIO

Land resources

The land form of the area can be divided in five categories: Low land (Jhola), Medium land (Bhatta), I, Medium land II, Medium land III and Upland(Dangar). The characteristics of these land forms is presented in Table 1.

Water resources

Table 2 presents the rainfall data for the year 2000 which indicate that the total rainfall is around 3000 mm. This high rainfall leads to springs. In our study we found 3-5 springs with discharge varying between 2-10 lps in each village. Further the gully beds remain wet throughout the year due to subsurface flow. Satellite imageries have confirmed subsurface flow even during April. However, there is no significant exploitation of these water resource excepting 3 hand pumps for drinking water purposes.

Present food and nutrient availability status

The estimate of present availability of food and nutrients were made on the beds of present level of production and requirement on the basis of guidelines of ICMR. Based on this, it was found there is huge deficiency for all the items. The deficiency for energy is 58.93 per cent, for protein 34.19 per cent, for calcium 53.79 per cent, for iron 40.01 per cent, for cereal 59.66 per cent, for pulses 48.37 per cent, for vegetables 90 per cent and for oil and fats 77.82 per cent. This indicates that there is severe malnutrition.

Table 1. Characteristics of agriculture lands

S.No.	Type of land	Slope	Management practice	Crop rotation	Cropping intensity	Percent of total cultivated area
1.	Homestead	< 2per cent	Bunded	Vegetable-Vegetable	200 per cent	3.46 per cent
2.	Lowland	< 2per cent	Bunded Paddy terrace, Basically they are gully beds	Paddy-Fallow	100 per cent	9.60 per cent
3.	Medium land I	0-5per cent	Terraced	Paddy-Ragi-Fallow	33 per cent	23.27per cent
4.	Medium land II	5-10per cent	Terraced or unterraced	Ragi-Fallow	33 per cent	27.06 per cent
5.	Medium land III	10-20per cent	Unterraced	Ragi-Fallow	33 per cent	15.05 per cent
6.	Upland	>20per cent	Unterraced	Ragi/Echanochlra-Fallow	25 per cent	15.60 per cent

Table 2. Crop allocation under different irrigation scenario for maximization of net benefit.

Crop Activities/ irrigation in medium land	40 per cent irrigation in lowland			50 per cent irrigation in lowland			60 per cent irrigation in lowland		
	20 per cent	30 per cent	40 per cent	20 per cent	30 per cent	40 per cent	20 per cent	30 per cent	40 per cent
X_1	0	0	0	0	0	0	0	0	0
X_2	0	0	0	0	0	0	0	0	0
X_3	20	20	20	20	20	20	20	20	20
X_4	0	0	0	0	0	0	0	0	0
X_5	10.34	4.84	4.84	9.87	4.84	4.84	9.32	4.84	4.84
X_6	28.40	19	4.84	28.11	18.71	4.84	27.8	18.41	4.84
X_7	3.77	4.84	4.84	4.52	4.84	4.84	4.84	4.84	4.84
X_8	5.9	19.71	33.88	5.89	20.01	33.88	6.44	20.31	33.88
X_9	8	8	8	10	10	10	12	12	12
X_{10}	4.84	7.26	7.26	4.84	7.26	7.26	4.84	7.26	7.26
X_{11}	4.84	4.84	4.84	4.84	4.84	4.84	4.84	4.84	4.84
X_{12}	0	0	0	0	0	0	0	0	0
X_{13}	0	2.42	7.26	0	2.42	2.42	0	2.42	2.42
X_{14}	0	0	0	0	0	0	0	0	0

Present status of labour employment in agriculture

One of the major problem in this region is unemployment. Low level of industrialization makes people dependent upon agriculture for employment. With monocropping and poor agricultural base, the employment opportunities are very less. It was found that employment availability in agriculture is just 6.8 per cent of able bodied people (15-60 year age group). Even this availability is very skewed and available only during monsoon season.

MODEL FORMULATION FOR WATER RESOURCE DEVELOPMENT PLANNING

The objective of resource management planning can be either any of the following three or all of them are: (*i*) Maximization of net return in which total profit has to be maximized, (*ii*) Maximization of labour employment, (*iii*) Minimization of investment on water resource development to minimize the monetary resource requirement for providing sufficiency in food and nutrients.

These objectives functions will be subject to following constraints:

- Land availability constraints
- Crop rotation constraints
- Fortnightly labour availability constraints
- Food requirement constraints
- Nutrient requirement constraints
- Crop affinity constraints

For planning purposes, following crop activities have been considered

	Name	Notation
(*i*)	Direct sown rainfed rice in lowland	x_1
(*ii*)	Direct sown rainfed rice in medium land	x_2
(*iii*)	Transplanted rice in lowland	x_3
(*iv*)	Transplanted rice in medium land	x_4
(*v*)	Ragi	x_5
(*vi*)	Maize	x_6
(*vii*)	Monsoon season vegetables	x_7
(*viii*)	Cowpea	x_8
(*ix*)	Post monsoon season transplanted rice in lowlands	x_9
(*x*)	Potato	x_{10}
(*xi*)	Post monsoon vegetables	x_{11}
(*xii*)	Pea	x_{12}
(*xiii*)	Groundnut	x_{13}
(*xiv*)	Lentil	x_{14}

The objective function and constraints can be expressed mathematically as below:

Maximising the net benefit

$$Z_B = \sum_{I=1}^{N} B_i \cdot X_i$$

Maximization of labour employment

$$Z_B = \sum_{I=1}^{N} \sum_{I=1}^{26} L_{ij} \cdot X_i$$

Minimization of investment in water resource development

$$Z_B = \sum_{I=1}^{N} WC_i \cdot X_i$$

Subject to land constraints

$$\sum_{I=1}^{N} X_i \le A_L$$ for lowland crop activites

$$\sum_{I=1}^{N} X_i \le A_M$$ for medium land crop activites

Irrigated area constraints

(*i*) Monsoon season area constraint

$$X_1 + X_3 \le AL$$

$$X_2 + X_4 + X_5 + X_6 + X_7 + X_8 \le AM$$

(*ii*) Post monsoon season area constraint

$$X_9 \le K_L \cdot A_L$$

$$X_{10} + X_{11} + X_{12} + X_{13} + X_{14} \le K_m \cdot A_m$$

Labour availability constraint for fortnight

$$\sum L_{ij} \cdot X_i \le L_{aj}$$

Energy constraint

$$\sum E_i \cdot X_i \ge ER$$

Protein constraint

$$\sum P_i \cdot X_i \ge PR$$

Iron constraint

$$\sum IR_i \cdot X_i \ge IPR$$

Cereal constraints

$$\sum CR_i \cdot X_i \ge CRR$$

Pulses constraint

$$\sum PL_i \cdot X_i \ge PLR$$

Vegetable constraint

$$\sum VG_i \cdot X_i \ge VGR$$

Oil & fat constraint

$$\sum OL_i \cdot X_i \ge OLR$$

Crop area constraints

The area under vegetables have been restricted to 10 per cent of the total area in view of limited marketing infrastructure.

Crop affinity constraints

The area under maize and ragi have been kept to be at least 10 per cent of total area of medium land in view of affinity of the farmers with these crops.

As per irrigability conditions, it was decided to take five levels of irrigation (40 per cent, 50 per cent, 60 per cent 70 per cent and 80 per cent) development in lowlands and three levels of irrigation (20 per cent, 30 per cent and 40 per cent) development in medium land I. It was assumed that after irrigation development only these two categories of land will be put up under annual crops and rest of the lands will be retired to pastures, horticultural crops and agro forestry. The linear programming model was formulated on this basis and run for these combinations.

RESULTS AND DISCUSSION

The solution of the model showed that providing irrigation facilities will give a tremendous boost to benefits from agricultural production and mandays.

Maximization of net return

For this objective the net return increases linearly with increase in irrigated area under lowland (Fig.1). However, the increase in net return is not that sharp when irrigated medium land is increased from 20 per cent to 30 per cent. In case of labour employment, the increase in linear with respect to increase in irrigated area under lowland (Fig.1), but with increase

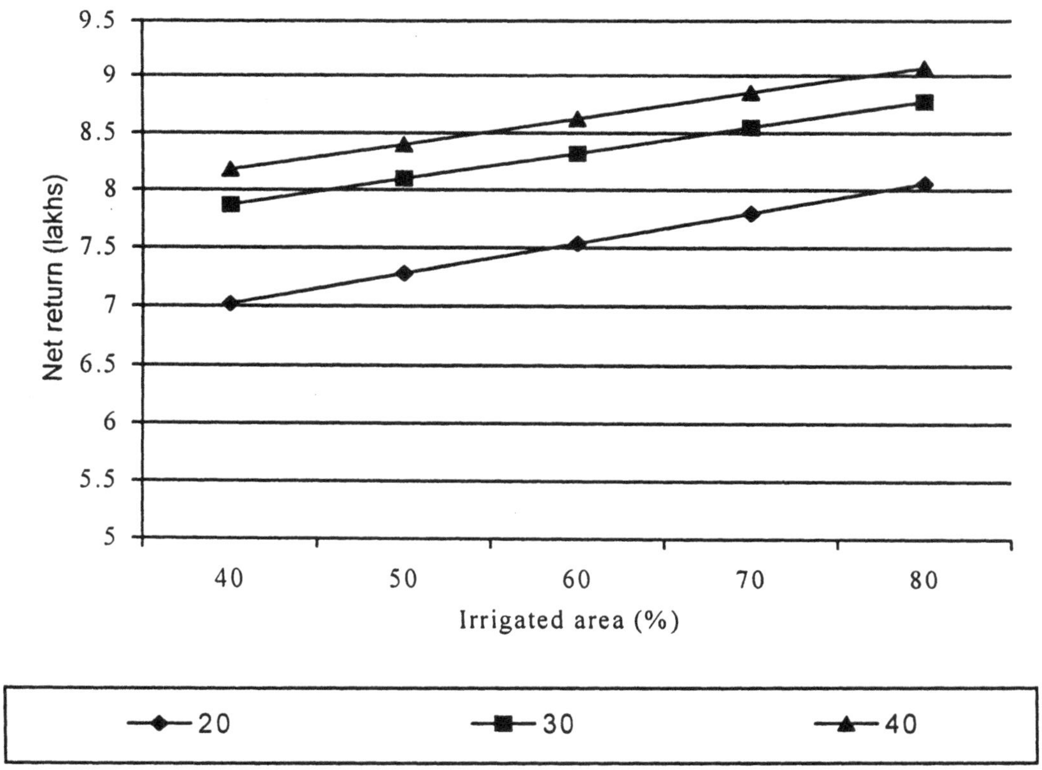

Fig. 1. Relationship between irrigated area and net return (Income Maximization)

in medium land irrigated area, the labour employment declines. This is due to constraints on vegetable crops which do not allow more area under vegetable crops during kharif as well as rabi. Fig 3 presents the relationship between irrigation water requirement and net return as well as labour employment. It is evident from the figure that both net return as well labour increases almost linearly with the amount of irrigation provided.

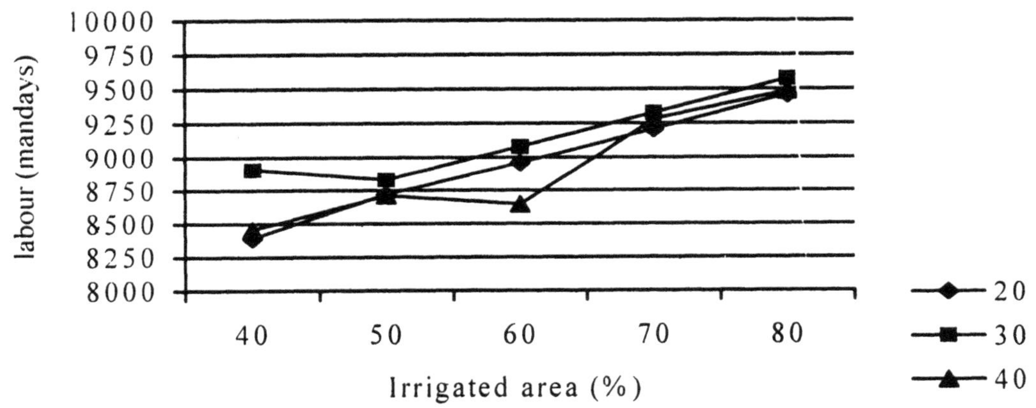

Fig. 2. Relationship between irrigated area and labour (Income Maximization)

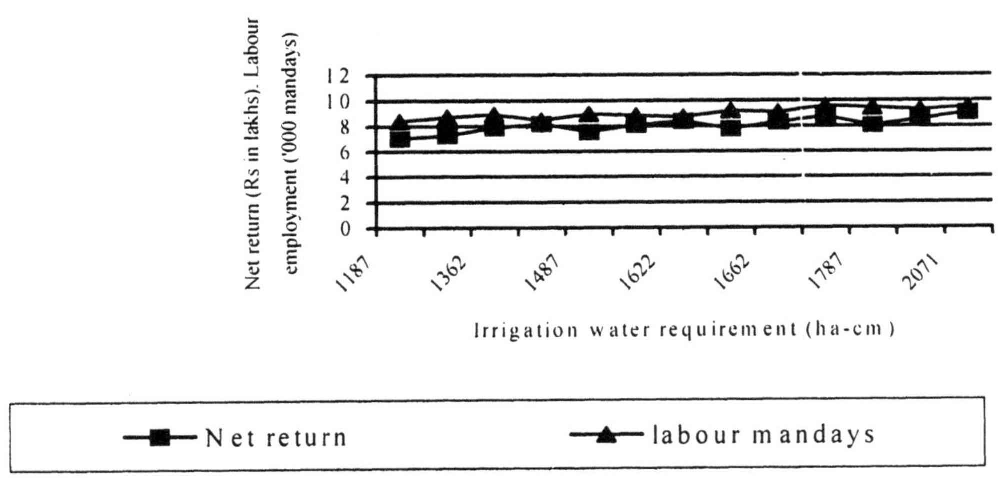

Fig. 3. Relationship between irrigation water and net return and labour employment

Maximization of labour

Figures 4 and 5 present relationship between labour employment and irrigated area of low land, and net return and irrigated lowland area. It is evident that both parameters are linearly increased with increase in irrigation in lowland. However, in this case the labour employment also increases in same fashion with increase in irrigated medium lands. Fig 6 presents relationship between irrigation water and labour and net return. It is evident that both increases almost linearly.

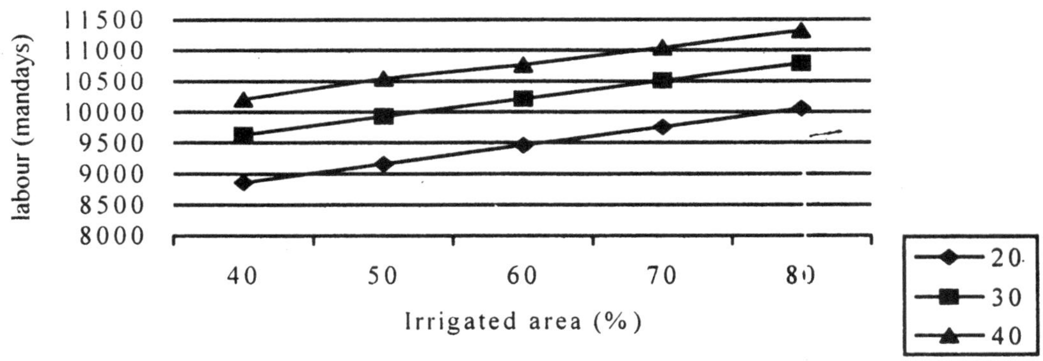

Fig. 4. Relationship between irrigated area and net return (Labour Maximization)

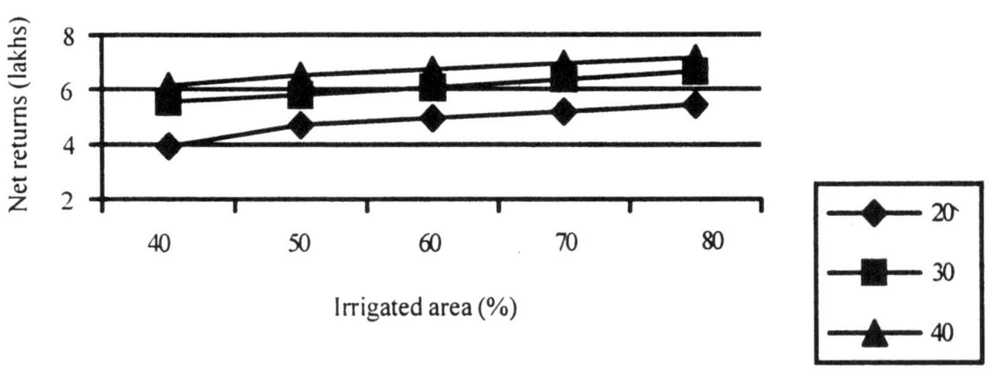

Fig. 5. Relationship between irrigated area and net return (Labour Maximization)

Minimization of water requirement

This model showed that with creation of 70.78 ha-cu of water, the food and nutrient efficiency can be achieved. However, net return will be just 3.32 lakh and labour employment will be just 5500 mandays.

Crop allocation

The crop allocation under objective of maximization of net return, maximization of labour and minimization of water requirement are presented in Table 2, 3 and 4 respectively. It is evident from the tables that whole of irrigated lowland goes to transplanted rice during *rabi*. It is natural as normally rice is most profitable crop in post monsoon season if water is available in plenty. In medium lands the vegetable area is at maximum allowable limit. Thus, if marketing infrastructure is developed, vegetable cultivation can be further enhanced which will increase the net return as well as the employment potential. It is evident from above that providing irrigation will bring a sea change in life of these tribal dominated area. Since water is available in plenty, what is required is sincere efforts to exploit them in an innovative manner.

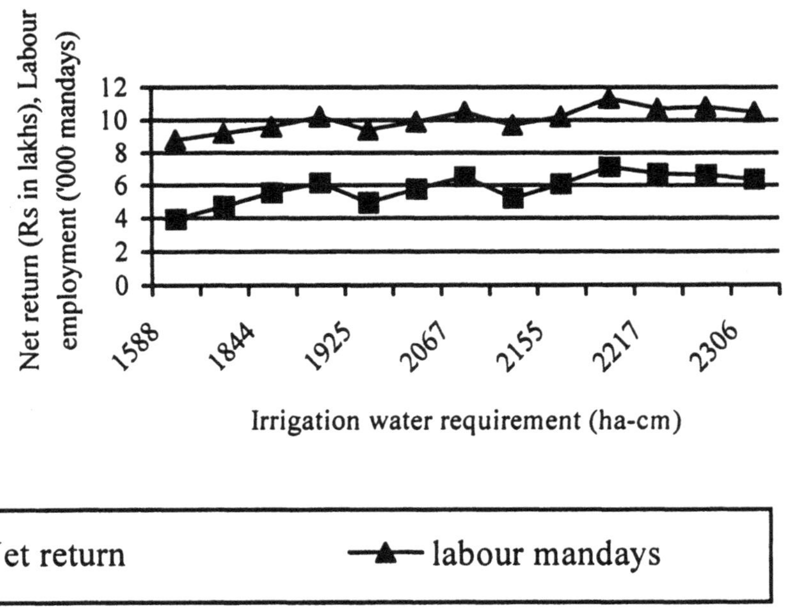

Fig. 6. Relationship between irrigation water and net return and labour employment

CONCLUSION

The irrigation development in tribal dominated areas of Orissa can stir activities which will lead to substantial change in net return and employment scenario. The water resource need to be identified and exploited properly. This should be backed up by suitable crop production package. It will also lead to retirement of marginal lands from crop cultivation.

Watershed Management: Issues and Policies for 21st Century pp. 65-69

Strategy for Reclamation of Eroded Land

K. R. Thiruvenkataswamy

INTRODUCTION

Nilgiri hills are situated at the north-western part of Tamil Nadu in peninsular India. The maximum altitude at Uthagamandalam is 2100 m, above mean sea level. The average annual rainfall is 131 cms, received mostly during pre-monsoon, south-west and north-east seasons. The topography is highly undulating with several hillocks and narrow valleys. The land slope varies from 0 to 40 percent and even exceeding this. The soil is lateritic in origin derived from charnokites - Nilgiri geneiss. The depth of soil ranges from 0.50 cms and that of the subsoil from 3 to 4m. The clay is ferruginous in character. The soil is acidic with pH ranging from 4.6 to 6.1. Ketty valley and Lovedale situated at the eastern and northern directions of Uthagamandalam are highly fertile areas where field crops such as potato, vegetables, wheat and Barely are raised.

The experimental areas received more than 450 m.m. of rainfall from January through September during the year 1991 to 1994. This amount rainfall was adequate to sustain the first crop of potato raised.

Table 1. Soil characteristics of Ketty Valley and Lovedale

Land slope per cent	Depth of top soil cms	Texture	pH	Total Nitrogen	Available P_2O_5	Available K_2O
12	83	Clayey	5.6	0.181	1.86	0.0486
16	76	Clayey	5.4	0.146	0.97	0.0267
20	65	Clayey silt	5.5	0.127	0.74	0.0207
24	53	Silty Clay	5.3	0.099	0.52	0.0190

Experiments on the mode of restoration of fertility on the severely eroded land were carried out for four years from 1991 to 1994. The details of experiments and the findings are described in this paper.

METHODOLOGY

Experiment - I. Productivity when the soil was removed to a fixed depth

The top soil was removed to a depth of 75 cms in a piece of land measuring 150 × 10 m. The slope of the land is 18 per cent. This area was divided into six plots of 20 × 5 m with sufficient interspaces. Potato variety great-scot was planted in these plots during the main season from March, through August-September. The following pattern of farmyard manure and Nanjanad mixture was adopted. (*i*) No farmyard manure and no Nanjanad mixture, (*ii*) 5 tonnes of farmyard manure along with normal dose of Nanjanad mixture per ha: (*iii*) 10 tonnes of farmyard manure + Nanjanad mixture, (*iv*) 15 tonnes of farmyard manure + Nanjanad mixture, (*v*) 20 tonnes of farmyard manure + Nanjanad mixture and (*vi*) 30 tonnes of farmyard manure + Nanjanad mixture. Excepting the above, all the other agronomic practices were identical in all the plots. The experiments were continued for four years in four different sites with similar slope, aspect and soil conditions.

Experiment II. Productivity when varied depth of soil was removed

Two sites with 14 and 24 per cent slope were selected and soil was removed to 25, 50 and 75 cms depth in plots of 20 × 5 m size. Potato variety great-scot was planted in these six plots during the main season namely March - September for four years. The crop received 30 tonnes of farmyard manure along with 2180 kgs Nanjanad mixture per ha per season. The other practices were common to all the fields throughout the period of experimentation (Nanjanad mixture supplies 95 kgs of nitrogen, 240 kgs of Phosphoric acid and 120 kgs of potash). A dose of 2180 kgs is applied per ha of potato crop.

The yield of potato obtained every year (averaged for four years) along with the other characteristics of the soil is furnished in Table 2.

Table 2. The potato yield and other characteristics of soil

Treatment	Yield of potato	pH	Organic carbon	Total nitrogen	Available ppm	
I	2.1	5.5	Nil	0.01	0.110	0.0201
II	4.6	5.6	0.14	0.02	0.430	0.0213
III	9.2.	5.9	0.19	0.07	0.810	0.0218
IV	14.6	6.2	0.31	0.12	0.130	0.0227
V	17.7	6.3	0.38	0.18	0.242	0.0234
VI	19.9	6.6	0.49	0.26	0.270	0.0242

RESULTS AND DISCUSSION

In treatment-I, the soil was almost sterile with hardly any biological activity, aeration and water holding capacity and hence it could not offer suitable conditions for root development and crop growth. Treatment-II had presented slightly better situation with organic matter, nitrogen, phosphoric acid and potash. Despite this, the sprouting was sparse and the number of leaves was few with pale green colour. Compared to this, sprouting was satisfactory in treatment-III. It was good in treatments-IV and V and the stand of the crop was quite superior to the rest with broad green leaves and the canopy could cover the interspace between the plants completely. The condition of the soil in treatment-V had become at par with the normal soil in the adjacent fields and as such, it had produced normal yield. The higher dose of organic matter together with Nanjanad mixture in treatment - VI had provided a far superior environment, which had produced the maximum yield. The rising trend of the yield had been steady from the beginning to the end exhibiting the unfailing chances of achieving increased yield.

The acidity had gradually decreased with every treatment. The organic carbon, nitrogen, available phosphoric acid and potash have improved progressively with every treatment.

From the above results, it is quite obvious that the subsoil can be transformed into a normal soil through proper management. Further it is apparent that the physical and chemical characteristics are replaceable features

The yield of potato in land with 14 per cent slope in the uneroded land was 22.0 tonnes/per hectare. This dropped to 18.2 tonnes when the loss of soil was 25 cms accounting for 15.2 kgs per mm loss of soil. For a loss of 50 cm of soil, the decline in yield was 16.0 kgs per mm. of soil lost and for 75 cms loss of soil, the reduction in yield for one mm of soil lost was 14.3 kgs. The yield of crop in uneroded land with 24 per cent slope was 15.5 tonnes. With the loss of 25cm of soil, the yield was 11.2 tonnes indicating a drop of 17.2 kgs, per mm of soil lost. For a loss of 50 cm of soil the extent of drop in yield was 13.6 kgs. In the plot where 75 cm of soil was removed, the drop in yield was 11.6 kgs per mm of soil lost. The decline in yield for every mm loss of soil in land with 14 per cent slope was 15.2, 16.0, 14.3 kgs for 25, 50 and 75 cms of soil removed respectively. It was 17.2, 13.6 and 11.6 kgs per mm. of similar loss of soil in land of 24 per cent slope.

Execution of anti erosion measures ensures prevention of soil erosion, which vary with the rainfall, topography, soil and cultural practices. In Nilgiris lands steeper than 30 per cent can be kept under perennial vegetation or provided with trenches for plantation crops: those with slope between 30 and 16 percent can be bench terraced for cultivation of field crops like wheat, potato etc, and lands having slope below 16 and above 10 per cent can be protected with contour stone walls or trenches for raising crops. Contour cultivation is essential for uniform distribution of moisture along with the above structures.

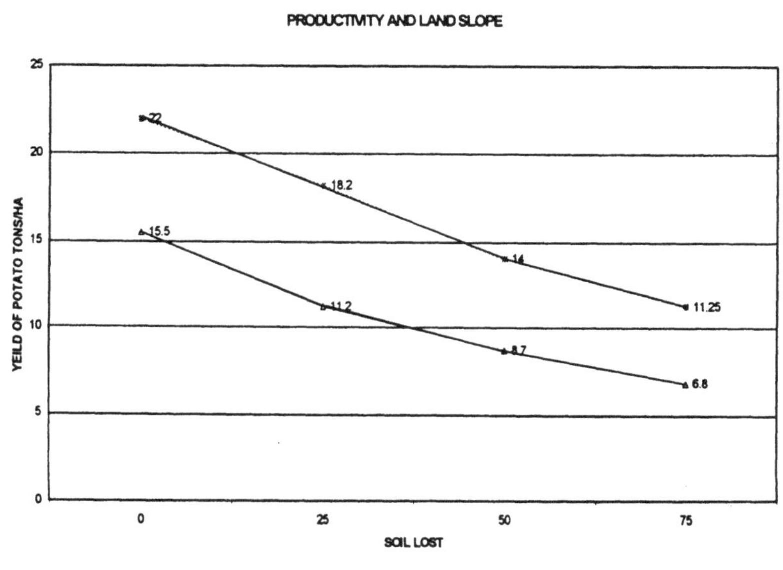

Fig. 1. Productivity and land slope when the depth of soil was 25,50,and 75 cms in land having 14 per cent slope and 24per cent slope

Soil is a renewable resource as it is formed from weathering of the parent material. The factors involved are (1) the rate of annual soil erosion, (2) Chemical activity involved in soil formation due to weathering and (3) intensity of weathering. Values for each of the above factors can be assigned with a thorough information on the extent of weathering of the various parent materials. Accurate accounting of the magnitude of weathering of the various types of parent materials and soil formation under natural conditions and cultivated state are still lacking. Frequent turning of the soil in clean cultivated land exposes fresh surface for weathering, which is totally absent under perennial vegetation. Hence, the rate of soil formation is comparatively quicker in cultivable lands notwithstanding its vulnerability to erosion.

The finer fractions of the soil settle between the interspaces of the coarser ones during splash erosion due to the impact of rain drop and consequently, a crusty surface with slower infiltration is produced. Receipt of further raindrops commence soil movement which is supported by land slope. Under this situation, removal of the soil constituents is selective in fashion. Desurfacing involves removal of the entire slice of soil uniformly over the surface. Although this is a quick method to quantify the response of the denuded land, seldom be it a substitute for natural erosion.

ECONOMIC ANALYSIS OF RECLAMATION

Application of 30 tonnes of farmyard manure in company with 2180 kgs of Nanjanad mixture has been able to produce 20 tonnes of potato per ha comparable to uneroded land. Let Fl be the investment on reclamation of eroded land and F be that for uneroded land on cultivation. The difference between Fl and F represents the cost of reclamation of the degraded land. The investment to produce 20 tonnes from the eroded land is more than that on the uneroded piece of land. It is essential that the investment on reclamation is restricted to a fair level.

The following tables will indicate the cost effectiveness of reclamation .

Table 3. Cost effectiveness of treatment on productivity

Application of manure and fertiliser	Additional cost treatment-normal Rs./ha.	Rate of incremental yield to additional cost
Normal practice	–	–
Treatment-I	-5000	–
Treatment-II	-8000	–
Treatment-III	-3000 -	–
Treatment-IV	+25UO	0.16
Treatment-V	+5500	0.15
Treatment-VI	+7500	0.19

Table 4. Cost effectiveness of the treatments

Treatment	Additional cost Rs./ha. Treatment Kg/ha	Incremental yield treatment normal	Ratio of incremental yield to additional cost. kg/ Rupee
I	– 5000	–10000	–
II	–500	– 8000	–
III	+ 200	– 3000	–
IV	+ 400	+2500	6.25
V	+ 800	+5500	6.88
VI	+1400	+ 7500	5.36

CONCLUSION

Reclamation of severely eroded land is a feasible proposition through application of additional dose of organic manures and fertilisers in the lateritic soils of Nilgiris. The study provides/ unmistakable evidence that the loss of soil-fertility can be compensated through higher dose of organic matter and fertiliser although at enhanced investment. It is highly essential to keep the investment within economic limit. Since cultivation has to be continued to meet the needs of the people, a permissible rate of soil erosion has to be set out: such a limit shall be the maximum rate of erosion which would enable achievement of optimum productivity on a sustained basis with the minimum investment. The above study has enabled arriving at an appropriate method of reclamation of land ravaged by erosion. Our prime objective should be to prevent soil erosion rather than reclamation after allowing the lands to be denuded of the soil. The productivity of land primarily depends on the scientific management of soil and water and this is the sheet anchor of prosperity of humanity at large.

References

Aina, P. O., R. Lai and G. S. Taylor (1977). Soil and Crop Management in Relation to Soil Erosion in the Rain Forest of Western Nigeria. *In*: Soil Erosion Prediction and Control. *Soil Cons., Soc., Am.*, Ankery. Iowa.

Dhruvanarayana, V. V. and G. Sastry (1977). Soil Conservation in India, *In*: Soil Erosion and Prediction and Control. *Soil Cons., Soc., Am.*, Ankery, Iowa.

Das, D. C. (1977). Soil Conservation and Management in Developing Countries, Food and Agrl. Org. United Nations, Rome, Italy.

Department of Agriculture (1976). Annual Research Report Soil Cons, Sect., Lucknow, India.

Ministry of Agriculture and Irrigation (1976). Report of the National Commission on Agriculture, New Delhi.

Lal, R. (1981). Soil Erosion Problems on Alfisols in Western Nigeria. VI. Effects of Erosion An Experimental Plots *Geoderma* **25**: 215-300.

Ram Babu and Associates (1979). Rainfall Erosion Potential and ISO-Erosion Map of India, Bull No. 2 CS + W CR & TI - Dehradun, India.

Sastry, G. and V. V. Dhruvanarayana (1982). Hydrologic Responses of Small Watersheds to Different Land Uses. Int. Symp. Hydrological Basins and their Use in Water Resources Planning, Bern, Switzerland, September 21-23.

Skidmore, E-L. (1982). Soil Loss Tolerance Determinants of Soil Loss Tolerance. Publ. No. 45 *Am. Soc. Agron.*, Madison, Wisc.

Stamey, W. L. and R. M. Smith (1964). A Conservation Definition of Erosion Tolerance *Soil Sci.* **97**: 1983-1986.

Tejwani, K. G., S. K Gupta and H. N. Mathur (1976). Soil Conservation Research in India 1956-71. *Indian Coun. Agr. Res.* New Delhi.

Thiruvenkataswamy, K. R. (1976). Pattern of Soil Erosion and Spacing of Contour Bunds. Compilation of Research Data.

Thiruvenkataswany, K. R. (1987). Engineering and Economic Aspects of Soil and Water Conservation. *J. Soil Wat. Conserv. India.* **32:** 97-107.

Watershed Management: Issues and Policies for 21st Century pp. 70-72

Rainfall Analysis and Cropping System in a Watershed Area: A Case Study

R. Kulandaivelu

INTRODUCTION

Semi arid regions today face more difficult problems than ever before. The potential of semi arid lands can be best developed by concepts and methods specially suited to dry regions. Sound watershed management in such areas leads to improved soil moisture regime by rain water harvesting. In order to select a suitable crop in a watershed area, a knowledge of rainfall climatology and moisture adequacies are important. In this paper a case study was taken up for Bhavanisagar watershed area and the results are reported here.

DESCRIPTION OF THE AREA

Bhavanisagar watershed area is located at 11°8' North latitude and 77°28' East longitude in Erode district of Tamil Nadu. The soils are red loamy type, shallow in depth with poor water holding capacity. The mean annual rainfall of this area is 640 mm received in 44 rainy days. Fifty per cent of the rainfall is recorded in the North East Monsoon period (October-December), 35 per cent in South West Monsoon (June-September) and the rest during summer (March-May). Millets and groundnut are the important crops grown under rainfed condition.

RAINFALL ANALYSIS

Water is the most limiting factor in dryland agriculture and improved crop production should be aimed at optimizing its use through adoption of efficient land and water management practices. The daily rainfall data of Bhavanisagar Centre was analysed for its distribution within a season and the frequency of occurrence of dry spells of different durations would enable to select optimum time of carrying out tillage operations, time of sowing crops, length of growing season and suitability or otherwise of successful cultivation of a given crop or the amount of risk involved so that the farmers of that area can be advised properly.

SUMMER TILLAGE

The mean annual rainfall of Bhavanisagar is 640 mm + 178 mm. Most of the annual rainfall is received during September to November. For sowing of groundnut, the farmers used to take up summer tillage and keep the land ready for taking up sowing in June-July. The feasibility of carrying out summer tillage, however, is dependent on the pattern of rainfall received during the summer months. The probabilities of receiving atleast 10 mm rainfall in three consecutive days (sufficient to soften the surface soil) during 1st March to 31st May have been calculated (Table 1). Results show that during 1st March to 31st May the probabilities of at least two opportunities for summer tillage are high at Bhavanisagar. Atleast one tillage could be given in May with a probability of 92 per cent.

Table 1. Probabilities (per cent) of summer tillage one or two during March to May

Month	March	April	May	March-May
One tillage	46	85	92	100
Two tillage	15	61	77	92

SOWING OF GROUNDNUT

The probabilities of receiving atleast 30 mm rain in three consecutive days in June, July, August, September and October are 8,0,0, 61 and 61 per cent (Table 2) indicating that groundnut crop can not be sown upto end of August and any crop that can be sown can be thought of only in September. The maximum period upto which groundnut is generally sown is middle of August. Hence, it is not advisable to take up sowing of groundnut in this watershed area.

Table 2. Probabilities (per cent) of receiving 30 mm rain during June to October

Month	June	July	Aug	Sep	Oct	June-Oct
Average rainfall (mm)	26	17	24	118	112	297
30 mm (per cent)	8	0	0	61	61	100

DRY SPELLS

To confirm this recommendation, a dry spell analysis was carried out to find out the probability of raising the groundnut crop, between July and October, cutting across the drought. The probabilities of dry spells of 14 and 21 day durations during the rainy months are calculated. (Table 3). The dry spell is defined when the total rain during 14 and 21 day periods was ≤ 20 and ≤ 30 mm respectively. These values are chosen on the assumption, that the rainfall below these limits would not be sufficient to meet the 33 per cent evapotranspiration needs of the crop. The probabilities of getting dry spells of 14 day period of two, three and four dry spells are 100 per cent and that of 21 days period are 100, 77 and 15 per cent. Hence, even if the groundnut crop is sown in the above period, it is sure of meeting with a dry spell and the crop can not survive.

Table 3. Probabilities (per cent) of ≤ 20 mm and ≤ 30 mm rainfall in a 14 and 21 day period dry spells (July-September)

Event	≤ 20 mm in a 14 day period	≤ 30 mm in a 21 day period
Two dry spells	100	100
Three dry spells	100	77
Four dry spells	100	15

From the above case study, it is clear that groundnut crop is not suitable for Bhavanisagar watershed area since, the rainfall quantity and distribution pattern is not conducive for the growth of the crop, even though the soil is ideal. Based on the rainfall distribution pattern sorghum crop is found suitable for sowing in this area during September and the crop can come to maturity without meeting any moisture stress.

CONCLUSION

The probability analysis of rainfall pattern of Bhavanisagar watershed area indicates that groundnut crop cannot be raised successfully under rainfed condition. Sorghum crop is found to be more suitable to the rainfall pattern which can be raised successfully by taking up sowing in September.

References

Kulandaivelu, R. (1984). Probability Analysis of Rainfall and Evolving Cropping System for Coimbatore. *Mausam* **35**(3): 257-258.

Kulandaivelu, R., N. Kempuchetty, and Y. B. Morachan (1980). Rainfall Pattern and Cropping System in Coimbatore. *Madras Agric. J.* **67**(3): 171-176.

Watershed Management: Issues and Policies for 21st Century pp. 73-87

Geographical Information System Based Decision Support for Annur Sub-watershed Planning

K. Palanisami

IMPERATIVE OF WATERSHED PLANNING

In the southern states and Tamil Nadu in particular, farmers are adopting rainfed agriculture (54 per cent). Well-irrigation is practiced in most of the districts mainly to supplement the crop water need in places where there are no major dams or tanks for irrigation. Total number of wells in the state has increased from 1.18 in 1968 to 1.84 millions in 1995 indicating the importance of well-irrigation. However, due to the hard rock strata, the recharges to the wells are poor and the area irrigated by wells is declining over years. Construction of water storage structure such as tanks, percolation ponds, check dams etc., for irrigation or groundwater recharge has not been fully exploited in many districts and villages of Tamil Nadu. The reason for this is the lack of a good decision support system to analyse the possibility of maximising the benefits from available resources and presenting it to the farming community and district and village officials for acceptance and subsequent implementation of the scheme. Sound planning before the construction of irrigation structure such as tanks and percolation ponds will ascertain the cost and benefits to the farmers using these structures for irrigation or indirect benefits such as ground water recharge.

DECISION SUPPORT WITH GEOGRAPHICAL INFORMATION SYSTEMS (GIS)

The bridging of the resource gaps among various sections of the society can be done only with the help of sound planning and with the introduction of computerised database management. The efficient management and further use of the same can benefit our society. The collection, storing and manipulation of geographic data are essential at a regional or national level for planners. This calls for the use of advanced tools such as Geographical Information Systems, which is used for the storing, manipulation, display and analysis of both spatial and non-spatial data. The tool can provide valuable information for decision making and can be further used to develop user-defined set of rules and thus create a decision support system. The user can pose a set of different question to the system for which he can expect replies that match his needs.

The computerising of large voluminous cartographic and demographic data which until now are available in the village and town offices in the traditional methods of record keeping will ease the development activity. One of the scientific approaches in the management of available resources and in the location of additional resources is the economic and financial evaluation of the investment that could be used in the development of a decision support system for the planners. As a particular case in this study it is used to help the location of soil conservation structure in rainfed agriculture and thus eliminate wasteful expenditure on additional facilities. This can benefit the farming community in particular in the southern districts of Tamil Nadu where agriculture is largely restricted to rainfall, well, and tank irrigated areas. Keeping this as the main objective, the study focuses on the development of a decision support system for the Annur Sub-watershed in Coimbatore district of Tamil Nadu.

OBJECTIVES OF THE STUDY

- Development of a database for the study villages in Annur subwatershed for soil, crop, irrigation, economic, social and education categories.
- Locate additional soil conservation structures, irrigation wells in the study village for augmenting the available water supply.
- Incorporation of pond equations using data collected for the study villages for assessing the possibility for additional structures.
- Financial and socio-economic analysis of investment options.
- Development of a decision support system for watershed development

LOCATION OF STUDY VILLAGES

The Annur Sub-watershed of the Kallar sub-catchment comprises of five villages viz. Vadavalli, Pogalur, Karekoundanpalayam, Vadakallur and Kuppanur. The subwatershed fall between 770 01' and 770 05' latitude and 11014' and 11019' longitude. The villages are located near Annur, which is approximately 40 kilometers from Coimbatore on the road from Coimbatore to Satyamangalam. The villages are at short distance between 10 kilometers from Annur. The major occupation of the villagers is largely farming restricted to rainfed agriculture and in a small level to irrigated agriculture.

DATA COLLECTION

All data pertaining to crop, irrigation facilities for the study area was collected from the village offices located in the respective study village. Data from Season and Crop Reports (1995) were also utilised. Soil data was collected from the Joint Director of Agriculture, Soil Survey Manual (1995) and gully particulars, percolation pond details were collected from the Department of Agricultural Engineering (1995). The data was collected for the study for the cropping seasons of the year 1996 and stored in Excel spreadsheets. The Geographical Information software used in the study is MAPINFO version 4.0 (1995). It is vector-based software with limited map analysis capabilities. The software was used for the storage of the collected data and putting the collected data in a geographic context. GIS was used in the creation of thematic maps for all the observed parameters.

CREATION OF THEMATIC MAPS

The GIS was used to create the thematic maps that serve as ready reference for the user interested in any of the stored information. Thematic maps were created for all soil parameters, crop, irrigation and social data. The creation of the maps was done using Geographical Information Systems software.

Land capability classification

One of the major geographical analyses carried out was the creation of a land capability class, used to determine the use to which the given land can be put for maximum productivity. It is based on a set of rules to be applied to the study area and extracting the results. This part of applying the rules was accomplished by extracting the required data for soil such as slope, erosion, soil texture and soil depth from the available database and using programming in BASIC to apply the rules. Once the rules were applied the results were stored in a separate database and later incorporated into GIS.

Grouping of soil to indicate the land suited to agriculture and non agricultural purpose and kind of soil conservation measures to improve the productivity is called as Land Capability Classification (LCC). In LCC emphasis is given to inherent soil properties, morphological and chemical properties which are of permanent nature. Second order of importance is given to the external features like slope, erosion and environmental features like climate. The three categories of LCC adopted are Land Capability Class (LCC), Land Capability Subclass (LCSC), and Land Capability Unit (LCU). In this analysis the factors considered for LCC are Soil Texture, Depth of Soil, Slope of Soil and Erosion Class.

The above classification scheme was used to classify land in the study villages and the resulting Land Capability Classification map was obtained. GIS database was used in an overlay analysis and based on this the land capability classes for all the fields in the villages were identified. The overlay analysis matches information in the slope, depth, texture and erosion columns in the database and assigns the LCC to each field or survey number in the study village. The results of the analysis with all the fields grouped by their land capability classes was later created as thematic maps using MAPINFO GIS for all study villages. (For subsequent maps also only the figures relating to Vadavalli village are presented as illustrations and to conserve space). The results of the land capability classes were used in the recommendation of suitable crops for the fields, their adaptability and alternates.

Crop classification

Crop data obtained for the various farms in the village from the village office was categorised into 13 major crop categories. The major groups are banana, paddy, cotton, groundnut, cholam, pulses, turmeric, gingelly, tobacco, tomato, chillies, and sugarcane and farms with no crop. Crop type was assigned to the farm based on the maximum crop area of any crop in the farm.

Farm holding

Based on the area of the holdings, the farms were classified as marginal (with area less than 1 ha), small (with area between 1 and 2 ha) and large farms (with area greater than 2 ha).

Irrigated area

A map showing the per cent irrigated area was developed using the criterion developed from the field survey, that a well can approximately irrigate about 2 ha. Using this norm the percentage of irrigated area for the farms in the village was determined and grouped under 1-25 per cent 26-50 per cent, 51-75 per cent and greater than 75 per cent categories.

Employment map

Thematic maps were generated from the employment details collected for all farms in the village. The major classification was agricultural occupation and non-agricultural occupation.

Education status map

The level of education of farmers was grouped under illiterate, primary school, middle school, high school, intermediate and graduate. A map was generated for this classification of employment in the village.

Caste breakup in the village

The major sections as applicable to government norms such as backward class, most backward class, scheduled caste and scheduled tribes and other castes were grouped. The major castes were used in the generation of thematic maps for different castes in the village.

Economic status map

The economic status of the farmers was divided into three groups. Farmers earning less than Rs 1250 per annum were below poverty, farmers earning atleast Rs 1250 were low, Rs 1250 to 5000 were medium and greater than Rs 5000 as high. A map was generated for these variables to reflect the economic status of the villagers in the study village.

SELECTION OF PERCOLATION PONDS

Having developed the thematic maps, the first step in developing the decision support system was to decide the developmental activities to be undertaken in the geographical location to assess their impact and viability. Development of watershed being the prime focus of this study, the logical option was the construction of percolation ponds. The decision making process then involved the location of the pond in the area and assessing its economic viability with the help of the GIS.

Percolation ponds are essentially small concrete masonry structure with an earthen bunding to store water. The surplus discharge weir at the downstream end of the pond can safely discharge the excess water. Storage capacity of the ponds typically ranged from 15,000 to 30,000 cubic meters. The catchment area was in the range of a few square kilometers. The structure could be planned in any area that accumulates sizable flow during rainfall, along gullies, ravines, and streams or in upland areas where there is a minor gully or rivulet. The area can also be delineated using a field survey typically within one sq. km. The ponds selected for the present study were located where a pond once existed and is not in use because of silting and consequent abandoning of the structure. Some of them were located in the gullies using the gully maps obtained from the Agricultural Engineering Department and from the Village Administrative Officer's record. The gullies are small to medium type with a bed width of about 18 m and the depth ranging from 3-9 m. The details of the cost of the ponds proposed in each village along with the storage capacity are given in Table 1.

The zone of influence of the ponds was fixed from previous studies on percolation ponds where a 700 m downstream influence was observed as the pond benefited area (Palanisami, 1991). Since upstream of pond will also be benefited an arbitrary distance of 150 meter in

Table 1. Ponds location cost, catchment area and capacity in study villages

Karegoundanpalayam

	Location Survey No.	Catchment Area sq.km	Cost of Pond (Rs)	Capacity Cu.Meters
Pond 1	810	5	120000	9000
Pond 2	785	0.6	120000	
Pond 3	151		120000	
Pond 4	630	1.5	74579	5000
Pond 5	229		120000	
Pond 6	724	0.7	120000	
Pond 7	396		188000	9000
Pond 8	613		120000	

Kuppanur

	Location Survey No.	Catchment Area sq.km	Cost of Pond (Rs)	Capacity Cu. Meters
Pond 1	459		120000	
Pond2	252		120000	
Pond 3	362	10	120000	30000
Pond 4	103		120000	
Pond 5	63		120000	
Pond 6	801		120000	
Pond 7	769		120000	

Pogalur

	Location Survey No.	Catchment Area sq.km	Cost of Pond (Rs)	Capacity Cu.Meters
Pond 1	405	25	199302	35000
Pond2	509	1.5	120000	
Pond 3	280	0.6	120000	3000
Pond 4	207		120000	
Pond 5	583	1.1	120000	6000
Pond 6	746	4.5	120000	3000

Vadakallur

	Location Survey No.	Catchment Area sq.km	Cost of Pond (Rs)	Capacity Cu.Meters
Pond 1	373		120000	
Pond2	154	1	120000	6000
Pond 3	300	0.5	120000	3000
Pond 4	311		120000	
Pond 5	500	0.45	120000	3000
Pond 6	429		120000	

Vadavalli

	Location Survey No.	Catchment Area sq.km	Cost of Pond (Rs)	Capacity Cu.Meters
Pond 1	86	2.1	192808	35000
Pond2	151	15.1	120000	40000
Pond 3	684	1	176986	16000
Pond 4	603		120000	
Pond 5	442	2.1	120000	
Pond 6	506	1.5	120000	10000

the upstream direction was also assumed to be benefited by the percolation ponds. A buffer zone of approximately 700-m for each pond location was generated using geographical analysis function of MAPINFO GIS. From this buffer zone the upstream benefited farms and down stream benefited farms were extracted. The recharge pattern was assumed to be semicircular in shape.

To select the location of adjacent percolation ponds the criteria that the recharge area of no two ponds should intersect was followed. Based on this criteria 6 to 8 ponds were selected with their benefit areas (crop fields) delineated using buffer functions. A geographical analysis was performed to extract only the farms in the benefited zones of each pond with other details such as the total farm area, number of wells (dug and dug-cum bore well), crop grown, and acreage in the benefited farms and stored in table format.

EFFECT OF PERCOLATION PONDS USING POND EQUATIONS (PONDEQ)

The benefits accrue to the farms in the beneficial zones of the percolation ponds through groundwater recharge that is utilised through wells. Such benefits were demarked using the zone of influence of the wells from studies completed through pond equations (PONDEQ). Using the groundwater recharge functions, additional area covered as well as the additional benefits due to the structures in the village were quantified. Any change in the existing cropping pattern was also documented.

Based on the studies of 10 percolation ponds in Coimbatore district, relationship of pond water level, groundwater recharge and benefit to farmers were determined using simultaneous equations model (Palanisami, 1991). A system of three recursive equations were specified to model the gross income of the farmers. To start with, the water level in ponds is specified to be determined by the nature of physical location of the pond, rainfall and catchment area of the ponds. This pond water level in turn, along with rainfall, distance of the well from the pond and well density is specified to determine the groundwater recharge levels. Finally, the recharged groundwater along with cropping intensity and depth of the well is specified to determine the gross income of the farmers.

Two stage least square (2SLS) method was used to estimate the equations. The estimated pond equations are:

Pond water level function

$$PW = 0.231 + \underset{(0.002)}{0.004}\,{}^{*}RF + \underset{(0.021)}{0.126}\,{}^{*}PL - \underset{(0.139)}{0.0427}\,CA \qquad R^2 = 0.27;\ N=10$$

Well recharge function

$$GR = 1.425 + \underset{(0.009)}{0.021}\,{}^{**}PW + \underset{(0.034)}{0.054}\,RF - \underset{(0.04)}{0.021}\,DS - \underset{(0.017)}{0.036}\,{}^{*}WD \qquad R^2 = 0.23\ N=46$$

Benefit function

$$GI = 876.47 + \underset{(3.51)}{6.94}\,{}^{*}GR + \underset{(0.09)}{0.21}\,{}^{*}CI + \underset{(0.017)}{0.14}\,DW \qquad R^2 = 0.34;\ N = 46$$

(Figures in parantheses are standard errors)

** Significant at 1 per cent level ; * Significant at 5 per cent level

Where

PW	=	Water level in the ponds in m
PL	=	Physical location of the pond (1 = good, 0 = otherwise)
CA	=	Catchment area of the pond Sq. km.
GR	=	Groundwater recharge in m
RF	=	Rainfall in mm
DS	=	Distance of well from the percolation ponds in m
WD	=	Well density (distance between two successive wells) in km
GI	=	Gross income of the farmer in '00 Rupees
CI	=	Cropping intensity in per cent
DW	=	Depth of well in m

The pond equations were applied to rainfall conditions in Coimbatore. Rainfall ranging from 400 to 650 mm per annum were used in the equation for certain catchments in the study villages. The areas of the catchments were recorded from the Agricultural Engineering Department. The location of pond was chosen to be good. The distance between percolation pond and the wells was about 200m. The well density obtained from previous studies in Coimbatore district was found to be 175 m. Based on these conditions the pond equation were used to obtain the pond water level and the groundwater recharge in the pond benefited. It was found that the recharge was high, ranging from 1 to several meters, which is adequate to raise a second crop after the rainy season. Based on this assumption it was decided to locate as many percolation ponds in the village as possible without overlapping benefit areas or water storage areas of the ponds. The ponds were located in all possible locations so as to benefit as many farmers as possible. Their costs and benefits were analysed through the decision support system developed for the purpose.

BENEFITS DUE TO PERCOLATION PONDS

The crop related benefits due to additional area irrigated by wells and the additional employment related benefits were the two main criteria adopted to analyse the benefits due to the percolation ponds.

Crop related benefits

The pond functions were used to determine if there was sufficient ground water recharge due to the percolation ponds. An average annual rainfall of 650-mm was used. It was found using the function that there would be sufficient groundwater recharge for the farmers to irrigate the crops in the second season (post-rainy season). This result was used in the study to arrive the additional cropping area in the post rainy season.

The difference in the actual farm area and the cropped area is the direct additional area available for cultivation because of percolation ponds. In the post rainy season typically January to April, the farmers owning wells can raise a crop in their farms. Hence only the second season crop benefits were used for calculating the benefits due to ponds, as most farmers in the neighbouring pond benefited villages, interviewed during field visits, also reported that the water level increased only during the rainy season.

Development of alternate cropping

Soil survey manual guidelines were used in determining the alternate crops suitable for the study watersheds. The manual lists the recommeded crops in 4 categories at high medium and low levels of suitability and unsuitability based on the soil properties tested by the soil survey and land use organisation. Based on the recommendation the most suitable to moderately suitable crops for the soil series present in the village was selected for alternate crops.

The farmers continued to grow the same crop in the season or the previous season as the case may be as most of the farmers in the field survey reported that they would not change the crop pattern suddenly due to market constraints. Comparison of the existing crop pattern and the crops suitable based on the land capability classification has indicated that most farmers had deviated from the suggested crops. The deviation was mostly in sugarcane, groundnut, vegetables and sorghum (as an illustration and to conserve space the above information for Vadavalli village is presented in Table 2).

Table 2. Crop acreage under existing and alternate cropping patterns for benefited farms (area in acres)

Vadavalli

Crop	Pond 1	Pond 2	Pond 3	Pond 4	Pond 5	Pond 6
Sorghum	16.18	7.64	8.65	15.41	32.03	26.09
Groundnut	(17.5)	(20.2)	(15.5)	3(29.7)	(38.8)	(18.0)
Pulses	7.52	1.21	3.25	2.15	6.76	3.49
Vegetable/Fruits	3.12	4.15		3.92	2.3(5.8)	5.7(23.6)
Sugarcane	1.1(13.1)	4.81	3.63	4.83	3.51	6.30
Coconut		2.37				
Cotton						
Others	2.66					

(Figures within parenthesis are crop acreage for alternate crops)

Data from the field survey on the crop yield per hectare, value of the produce using existing market rates and cost of cultivation were used in obtaining the value of the crop benefits in both seasons. The details of the yield, cost and returns are given in Table 3.

Table 3. Production cost and returns from major crops produced in the study villages

Crop	Yield/ha (Quintals)	Irrigated Value (Rs/ha)	Operating cost (Rs/ha)	Net Income(Rs/ha)	Labour Required in Mandays /ha
Sorghum	12	5400	1961	3439	37
Groundnut	20.2	21210	10296	10914	85
Pulses11	18700	3961	14739	32	
Tomato	198	54450	18172	36278	330
Sugarcane	1100	66000	29342	36658	172
Cotton11.115	31122	14575	16547	214	
Banana	2541.5*	139782.5	56397	83385.5	300

* Number of bunches

The benefits obtained due to additional wells in the second season was calculated for each pond for all the five villages. On an average, the additional area benefited in each pond was ranging from 10 to 40 ha. The details of the existing wells and benefited area under each pond are given in Table 4. Based on the rainfall pattern and the pond water level, it was observed that only five out of 10 years, the pond could help the wells in terms of additional recharge. This means, the benefits due to ponds were assumed to occur in every alternate year only.

Table 4. Summary of area benefited and well details

	Pond 1	Pond 2	Pond 3	Pond 4	Pond 5	Pond 6	Pond 7	Pond 8
Karegoundanpalayam								
Area benefited (ha)	19.9	23.8	31.6	28.5	22.6	22.9	23.8	7.7
Additional wells	2	7	6	7	5	3	3	4
Existing wells	7	26	21	14	13	13	12	12
Kuppanur								
Area benefited (ha)	28.2	49.1	46.4	50.7	55.6	28.6	36.2	
Additional wells	0	0	1	0	0	2	1	
Existing wells	8	23	23	25	18	19	18	
Pogalur								
Area benefited (ha)	19.7	36.2	24.8	28.0	41.0	41.0		
Additional wells	1	0	1	2	0	0		
Existing wells	16	24	12	18	21	31		
Vadakallur								
Area benefited (ha)	22.1	32.4	10.3	25.8	36.2	37.5		
Additional wells	3	0	0	3	0	3		
Existing wells	16	27	3	15	20	21		
Vadavalli								
Area benefited (ha)	30.6	20.2	15.5	29.7	44.6	41.6		
Additional wells	1	0	0	1	0	0		
Existing wells	17	6	6	15	15	16		

As a rule for construction of new wells it was fixed that the distance between no two wells should be less than 150 m. This eliminates the interference in the well draw down due to neighbouring wells. Using this criterion, all potential farms (farms with no existing wells) were extracted from within the pond-benefited zone. Typically additional wells due to a pond ranged from zero to 7.

Employment benefit

Additional mandays of labour generated due to ponds were calculated from the standard values for mandays/season for various holding size of farms for major crops. The product of additional area for each season in the crop and the average value for mandays/season calculated from the mandays for different farm holding size was used in the calculation of additional employment potential. The benefits for each pond in the village are presented for the two crop seasons as well as with additional wells in the second season.

ECONOMIC ANALYSIS OF THE POND INVESTMENT

Since the crop benefits and employment benefits address the productivity and social aspects respectively both the benefits as such could not be combined, as the wages for additional employment were already accounted in the cost of cultivation of the crops. In the present analysis only the crop benefits are considered in working out the economics of the irrigation structures. The measures of evaluation include Net Present Value (NPV), Benefit Cost Ratio, and Internal Rate of Return.The crop benefits with percolation ponds as well as benefits due to additional wells were calculated using the above benefit calculation methods.

The internal rate of return (IRR) has been ranging from 12 per cent to 72 per cent among the ponds indicating that some ponds are not at all financially feasible for investment (Table 5). The reason was due to the location of the ponds with respect to benefited farms as well as the pond catchment. Out of the eight ponds proposed in the Karigoundanpalayam village, only four are financially viable, as the IRR is higher than the commercial bank rate of 14.5 per cent. However, when alternate crops are suggested for cultivation, the crop benefits due to ponds recharges also changed. Even with the alternate crops, out of the eight ponds, only four are financially viable. It is observed that the order of the viable ponds had changed between the existing and alternate crops. In the case of Kuppanur village, out of the seven ponds proposed, six are financially viable and when alternate crops are included, all the seven ponds are financially viable. In the case of Pogalur village, out of six ponds proposed, five are financially viable with existing crops and all the six are viable when alternate crops are included. For Vadakalur village, out of the six ponds, five are financially viable with existing crops and only four ponds are found to be financially viable with alternate crops. In the case of Vadavalli village, out of the six ponds, five are financially viable with existing crops and only three ponds with alternate crops.

It is interesting to note that though most of the farms are deviating from the proposed crops, the benefits from them are more in some villages and comparatively less in others. The major reason was the marketing of the output as well as the small size of the farms wherein they couldn't invest more on crop cultivation. For example, in Karegoundanpalayam village, pond 1 has higher IRR (68 per cent) with the alternate crops compared to only 15 per cent with the existing crops. In the case of pond 4, the IRR is high with the existing crops (72 per cent) compared to 26 per cent with the alternate crops.

In the case of Kuppanur village, except for pond 1, IRR could not improve with alternate crops. For Pogalur village, the IRR is marginally high in all the ponds with alternate crops. The IRRs are not much significant between the ponds with existing and alternate crops. In the case of Vadavalli village, except the three ponds, the proposed crops could not increase the IRRs. The inference from the results is that even though it is possible to change the crops according to the soil series, the financial returns are not much attractive because of the above reasons.

Given the financial aspects of the ponds, it is important to rank the ponds for investment purposes (Tables 6 and 7). Among the eight ponds proposed for Karegoundanpalayam village, pond numbers are ranked in that order: 4, 6, 3, 8, 1, 7, 2 and 5. In the case of Kuppanur, the ranking of the ponds is: 4, 2, 3, 6, 7, 1 and 5. In the case of Pogalur, the ranking is : 6, 2, 5, 4, 1, and 3. For Vadakalur, the ranking is: 6, 2, 5, 1,4, and 3 and for Vadavalli village, the pond ranking is: 6, 4, 2, 5, 1 and 3. However, when the alternate crops as per the land capability class is recommended, the benefits had changed much and the ranking of the pond investment are varied considerably. For example, in the case of Karegoundanpalayam village, pond number 1 has been ranked first followed by ponds 6 and 3. Similarly, in all the other villages also, the ranking had been varied because of the changes in the crop income due to change in the crops to be grown.

Table 5. Financial analysis for existing and alternative crops without (A) and with (B) wells

Details	Existing crops				Alternate crops			
	A		B		A		B	
	IRR	B/C	IRR	B/C	IRR	B/C	IRR	B/C
Karegoundanpalayam								
Pond 1	0.15	1.02	0.01	0.56	0.68	2.26	0.57	2.54
Pond 2	−0.12	0.28	−0.07	0.36	−0.09	0.34	−0.07	0.38
Pond 3	0.41	1.95	0.05	0.70	0.56	2.20	0.14	0.97
Pond 4	0.72	3.03	0.06	0.74	0.26	1.76	−0.06	0.39
Pond 5	−0.14	0.25	−0.03	0.48	−0.04	0.46	0.00	0.54
Pond 6	0.65	2.82	0.16	1.07	0.57	2.26	0.34	1.72
Pond 7	0.06	0.73	0.30	1.54	0.15	1.02	0.01	0.57
Pond 8	0.22	1.27	0.07	0.75	−0.03	0.48	−0.15	0.23
Kuppanur								
Pond1	0.40	1.93	0.25	1.39	0.70	2.10	0.74	2.20
Pond2	0.58	2.00			0.44	1.50		
Pond3	0.56	1.98	0.71	3.10	0.81	2.20	0.49	1.70
Pond4	0.60	2.10			0.32	1.30		
Pond5	0.13	0.96			0.43	1.60		
Pond6	0.55	2.10	0.28	1.48	0.34	1.40	0.40	1.60
Pond7	0.47	1.95	0.66	2.89	0.43	1.70	0.25	1.50
Pogalur								
Pond1	0.28	1.46	0.17	1.10	0.30	1.80	0.36	1.78
Pond 2	0.47	2.20			0.52	2.40		
Pond 3	−0.09	0.35	-0.14	0.24	0.25	1.35	0.05	0.69
Pond 4	0.35	1.73	0.10	0.85	0.45	2.20	0.46	2.15
Pond 5	0.42	2.00			0.60	2.80		
Pond 6	0.53	2.67			0.53	2.20		
Vadakalur								
Pond 1	0.40	1.91	0.08	0.79	0.37	2.20	0.34	1.89
Pond 2	0.43	2.03			0.51	2.50		
Pond 3	0.04	0.68			−0.11	0.31		
Pond 4	0.20	1.19	−0.01	0.51	−0.05	0.42	−0.01	0.51
Pond 5	0.40	1.93			0.41	2.30		
Pond 6	0.57	2.50	0.13	0.95	0.58	2.80	0.42	2.34
Vadavalli								
Pond1	0.19	1.15	0.12	0.91	0.51	2.68	0.43	2.04
Pond 2	0.39	1.60			0.02	0.60		
Pond 3	0.10	0.85			−0.11	0.31		
Pond 4	0.41	2.10	0.45	2.12	0.08	0.78	0.00	0.56
Pond 5	0.34	1.50			0.59	2.89		
Pond 6	0.49	2.30			0.61	2.98		

Table 6. Ranking of ponds based on financial analysis for existing crops

	Karegoundan-palayam	Kuppanur	Pogalur	Vadakalur	Vadavalli
Pond 1	5	6	5	4	5
Pond 2	7	2	2	2	3
Pond 3	3	3	6	6	6
Pond 4	1	1	4	5	2
Pond 5	8	7	3	3	4
Pond 6	2	4	1	1	1
Pond 7	6	5			
Pond 8	4				

The other benefit due to ponds will be investment in new wells by the farmers. It is equally important to compare the ponds with additional wells also. However, it was observed that only in few ponds the additional wells proved to be financially viable. For example, in Karegoundanpalayam village, only in two ponds both with the existing as well as alternate crop pattern, additional well investment is financially viable. In the case of Kuppanur village, four out of seven ponds are attractive for additional well investment with existing as well as alternate crops. For Pogalur village, additional wells are possible only in three ponds that too with alternate crops. For Vadakalur village, only three out of six ponds are attractive for well investment that too with alternate crops. In the case of Vadavalli, only one out of six ponds is attractive for new well investment. The inference is that additional wells are possible only in locations, where number of existing wells is limited and where the zone of influence is high enough to allow more benefited area under the pond command. Since the Annur sub-watershed falls on a hard rock zone, where groundwater exploitation is more than 100 per cent, the additional wells are feasible only in selected locations. Also each well will cost about 70-80 per cent of the pond investment of Rs 1.2 lakhs. Given the high investment cost, farmers will be reluctant to invest on new wells.

Table 7. Ranking of ponds based on financial analysis for alternate crops

	Karegoundan-palayam	Kuppanur	Pogalur	Vadakalur	Vadavalli
Pond 1	1	2	5	4	3
Pond 2	8	5	2	2	5
Pond 3	3	1	6	6	6
Pond 4	4	7	3	5	4
Pond 5	7	4	1	3	2
Pond 6	2	6	4	1	1
Pond 7	5	3			
Pond 8	6				

ANALYSIS OF SOCIAL PARAMETERS DUE TO PERCOLATION PONDS

It is equally important to compare the benefits due to percolation ponds in terms of social parameters such as education, income level, type of farmers (marginal, small and large) and social status (backward, SC/ST etc.,). In addition, employment generation was also incorporated as one of the social parameters, as the beneficiaries are not necessarily from the farm households. The factor share would highlight who benefits from the pond which is a public investment. The social background such as economic status of the farmers, their profession, education and their caste affiliations for the pond-benefited areas were analysed. The data were collected and stored in the database for the respective villages. Geographical analysis/data query function was used to extract the social details from the database for the pond benefited areas and presented here for each village on a pond wise basis for the predominant crop season that would be benefited due to the ponds (post-rainy season).

The ponds were ranked according to the social status data obtained for the villages. The needy class in each category was assigned a high score of 10 followed by the medium priority farms, which were assigned a value of 5, and the low priority farms were assigned a value of 2. Accordingly the data on social status was used to obtain the scores in the high, medium and low priority farms by multiplying the assigned scores with the number of farms in each category. A weighted average of the scores were obtained as the ratio of the sum of high, medium and low priority farm scores and the total number of farms benefited for each pond. The weighted average can now be used to interpret the importance of the ponds and they can now be ranked based on the decreasing order of priority. Thus in Karekoundanpalayam village, the pond ranked based on social status of the farmers are 6, 7 and 3. Similarly, for Kuppanur village the ponds ranked based on weighted average are 3, 4 and 2. For Pogalur village the ponds are 6, 2 and 5, and for Vadakalur village the ponds are 2, 3 and 6. In the case of Vadavalli village, the ponds are 2, 6 and 3. The ponds ranked 1, 2 and 3 based on financial analysis.

It is interesting to note that the ranking of the ponds based on the social parameters had also differed much when the employment benefits were not included. This is mainly because the social parameters other than the employment mainly relate to the beneficiary farmers of the pond in that village, whereas the employment parameter mainly denote the benefits that occur to the non-well owners, viz., the labourers. Hence it is relevant to consider the pond ranking including the benefits that occur to the labaour class in the village. This is important as in any factor of share analysis, the benefits occurring to the labour class is also given importance while interpreting the income distributional consequences of the public investment. Further, if the investment authorities are still interested to analyse the ranking based on individual parameters such as social status, economic status etc., for pond investment, the results will be highly useful for such comparisons.

DEVELOPMENT OF THE DECISION SUPPORT SYSTEM

Lack of proper planning is likely to result in reduced returns and under utilization of resources. Sound planning supported by a decision support system for regional planning will also help the village and district officials in implementation of the irrigation structures in the districts and in the villages. Considering the need for a good decision support system for planning on a regional scale, the study analyzed the economic aspects of the construction of irrigation structures to augment the water supply for irrigation and for percolation ponds for increased crop production and employment generation in the watershed programmes.

A system programming for decision support for Annur sub-watershed has been developed using Geographical Information System. Data from the GIS are retrieved and plugged in an

Excel sheet where the decision analysis is being done. The sets of input data taken from the GIS database are:

- Area of individual crop
- Yield for individual crop
- Area benefited by additional ponds

The other sets of input data are

- Price for individual crop
- Cost of cultivation for individual crop
- Discount rate for finding the benefit cost ratio

Provisions are made in the Excel sheet for easy data entry for individual pond. Analysis is being made for two scenarios viz., with additional wells and for alternative cropping pattern. The number of additional feasible wells is derived from GIS. The alternative cropping pattern is also derived from the land capability classification (LCC) from within GIS. Benefit cost ratio and Internal Rate of Return were calculated using the standard for the discounted net returns. Based on the BC ratio and IRR values the ponds were ranked for making investment policy decisions.

Using the above analyses a decision-maker can arrive at a solution for watershed development based on the following financial priorities. If the funding for investment is coming from commercial banks or lending agencies then the decision is made based on the financial measures such as BC ratio and IRR. Similarly, social index analysis is also done for the individual ponds using five parameters viz., caste, economic status, education, farm holding and employment.

CONCLUSION

Tamil Nadu has about 84 watersheds and development works are urgently needed in most of them. Exploitation of ground water resources is increasing in many locations without making any efforts for their sustainability. Identification of those locations that need recharge measures and location of percolation ponds were done using the GIS in the selected villages of Annur sub-watershed. The land use and alternative land use with additional wells recharged by the percolation ponds were compared. Financial analyses using B/C ratio, and IRR were undertaken to assess the benefits of the percolation ponds both at farm and watershed levels after construction of additional wells.

Some ponds were not found to be financially viable for investment due to the location of the ponds with respect to benefited farms as well as the pond catchment. When alternate crops were suggested for cultivation, the crop benefits changed. The order of the viable ponds also changed between the existing and alternate crops. Though it is possible to change the crops according to the soil series, the financial returns are not much attractive because crop suitability alone could not decide the crop profitability.

The other effects due to ponds is investment in new wells by the farmers. However, it was observed that only in few ponds the additional wells proved to be financially viable implying that additional wells are possible only in locations where number of existing wells is limited and where the zone of influence is high enough to allow more area to be benefited under the pond command. In the hard rock dominated Annur sub-watershed, where groundwater has almost been fully exploited, additional wells are feasible only in selected locations. Also investment on wells is a costlly proposition as each will cost about 70-80 per cent of the pond investment of Rs 1.2 lakhs. This discouraged additional investment on wells by the farmers.

Regarding social aspects, the extent and type of beneficiaries, whether small, marginal or large farmers as well as whether backward, SC/ST and other groups, after the construction of additional wells and percolation ponds and adoption of alternative land use were also studied. Further employment generation to the labour class of the village was also quantified and included in arriving an index of social parameters for each percolation pond. If the funding is from the government agencies or community donors then the decision is to be made based on the social index analysis. Here again, the scores for various categories can be reassigned based on the priorities of the region.

References

Government of Tamil Nadu, Department of Agricultural Engineering (1995). Guidelines for construction of percolation ponds, Coimbatore.

Government of Tamil Nadu, Director of Statistics (1996). Season and crop report of Tamil Nadu, Chennai.

MapInfo Corporation (1995). MAPINFO Professional V. 4.0. Reference Manual, Troy, New York.

Palanisami. K. (1991). Economic evaluation of percolation ponds in Tamil Nadu State. *Indian Journal of Soil Conservation.* **19**(1&2) 45-52.

Soil Survey and Land use Organisation, Department of Agriculture, Tamil Nadu (1995). Soil Resource of Annur Block, Avinashi Taluk, Coimbatore.

Watershed Management: Issues and Policies for 21st Century pp. 88-96

Prioritization of Micro Watershed for Management of Bijapur District of Karnataka

K. Rajakesavan, M. A. Ashwin and R. Shailaja

INTRODUCTION

A watershed consists of all land and water area, which contribute to the runoff at a common point. It is a natural geographical unit with a certain extent of homogeneity and uniformity. Further, it is a limited, convenient, clearly defined and unambiguous topographic unit, whose size can be altered on the basis of stream ordering, to suit the needs of the project or the availability of the resources. Also, it has been observed that any sort of work or practice in the upstream area has a bearing on the downstream area. The objectives of the watershed approach for development stretches beyond those of the conventional models—like land and water conservation or productivity increase. While taking all these into consideration, it aims at the greater intent of restoration of the ecological balance of the region to the maximum extent possible.

Over the years, the concept of watershed management further evolved, the most significant transformation happening with it becoming more people-oriented. This was borne out of the conviction that despite the massive inflow of funding for these projects and their innate potential to create better livelihood opportunities for the people of the area. Some of the main reasons attributed to the low performance of these projects were, planning from an outsider's point of view and poor post-project management. It was rightly thought that, involvement of the beneficiaries, as one of the major stakeholders would help solve these problems and in turn make these projects both, viable and sustainable. Thus was ushered in the era of participatory watershed management.

PROJECT BACKGROUND

A substantial part of the land in Karnataka comes under rainfed cultivation and does not get assured rainfall. It is estimated that 75 per cent of the area is drought prone. Surface water bodies are degrading due to various activities. So it's very important to understand the watershed and manage them systematically. We discovered that there are some important criteria missing in the watershed programmes, they are:

- There are many watershed programmes in the State managed by various governmental agencies such as DDP and other externally funded projects have not included the information on socio economic condition of people in the watershed area. Unfortunately in any one of these projects, planning has not considered systematic inclusion of both technical and socio-economic information. The reason could be the laborious, time-consuming process involved in this exercise. Hence the projects are not sustainable.

- Socio-economic condition is the essential components of the development of watershed development and planning, if it has to be a success and sustainable.
- To some extent technical data on soil, land use, hydrogeomorphology and information on rainfall, temperatures are available. But information like socio-economic condition of any given area is largely lacking. Geographic Information System (GIS), which reduces the labour, and time consuming factors, are involved in the analysis.
- Existing technical data along with socio-economic data linked and integrated through GIS, will be very useful in the process of efficient planning and management of the natural resources in a watershed.
- Also delineation of micro watershed having all the information will be useful for the watershed treatment, as the area (500-1000ha) is small and easy to operate with in 1- 2 years.
- Watershed Planning Support System (WPSS)—a query system developed through GIS will be useful for the implementers, farmers and decision makers analyze the data and farmers for sustainable water management for the watershed area. It can shine as quick references to take decision. This prompted us to take up the project.

Objectives

- Develop a simple package of geo-reference maps and attribute data for a portion of the Doddahalla watershed area in Bijapur District.
- Develop a farmer information system and a query tool, which would be useful in providing synthesized information to the people involved in the watershed projects.
- Prioritize the micro watersheds in the project area using all the information generated.
- Following prioritization, a GPS survey would be conducted for one micro watershed, for which specific treatments would be suggested. In addition, a general management plan would be arrived at for the entire watershed.

The GIS based query tool would not only aid the immediate implementation of any watershed project in the project area, but would also act as a prototype, which could be used by other watershed practitioners after modifying it to suit their needs and interests. Geographic Information System for watershed management is well-established, effective tool for resource planning, conservation, development, management and utilization. The availability of information regarding the project area is one of the key factors that help the watershed practitioner prioritize his/her goals and objectives.

Global Positioning System is a tool, which seems to be very useful for planning the watershed in better manner especially at micro level. Since the cadastral maps are about 100 years old there may be considerable changes in the landforms, which are quite important for planning a watershed. We have just made a beginning to find out the utility of the Global Positioning System, which is an emerging technique, in the micro watershed planning. We may need to do some more exercise, before we recommend the use of this technique. However, this considerably changes record in the land fragment, slop pattern and suggests that it is a wonderful exercise for planning micro watersheds.

The Watershed Planning Support System (WPSS) seamlessly links different tables through a series of user-friendly graphic menus. The tool streamlines the whole process of data input development and output analysis, and thereby greatly speeds up a complex query process. In the process, the tool offers various flexible options for user-defined applications. This tool can be used for several watershed analyses in-house.

Project area

The area selected for the project is the Doddahalla Watershed in Bijapur District of Northern Karnataka. The district is located in the Northern Dry Zone in between the longitudes of 750 40' and 780 0' and the latitudes 160 45' and 170 05'.The approximately 30,000-hectare area considered for the project, is a natural extension of the watershed areas, which have already been treated. To the west is the area treated under the Indo Swiss Participatory Watershed Development, Karnataka. The area that lies southwest to the project has been treated under a DDP watershed project. Thus the GIS based query tool developed as a part of the project and other information generated will definitely help any immediate treatments implementation plans in the area. Also, it will be a prototype, which can be used with reference to other watershed areas with certain modifications done depending on the needs and interests of the institutions involved.

Suitability of the area

- Reports of the Drought Monitoring Cell, Karnataka, as well as that of the Integrated Mission for Sustainable Development (IMSD) project of National Remote Sensing Agency (NRSA) and Department of Space (DOS) classify Bijapur as a chronically drought prone district.
- Slightly more than 80 per cent of the total population of Bijapur lives in the rural areas (1991 Census record). Industries do not have a strong presence in the district, which is reflected in the 1999 data, according to which there are only 398 of them employing only 6712 persons. Thus, it can be inferred that most people are predominantly dependent on agriculture.
- The land use pattern of the district for the year 1997-98 suggests that the net cultivated area occupies 72.8 per cent of the total geographical area. But only 11.96 per cent of the cultivated area is under irrigation.
- The agriculture in the district is mainly rainfed. Further, the leaving of 13.76 per cent of the lands fallow.Overall background of the district can be mainly attributed to low or erratic distribution of rainfall.

Physiography

Physiographically the study area belongs to the northern maidan regions. It falls in the dry semi arid tracts of Deccan plateau. The terrain is undulating, but without conspicuous mounts or deep valleys. Although it forms a part of the central upland, the Doddahalla watershed is considered as a separate geographical region. The elevation ranges from 460 to 640 meters above Mean Sea Level. Generally the slope is from north to south. The drainage system generally shows dendritic to subdendritic patterns. Drainage texture is coarse to medium and drainage frequency, high. Doddahalla drains the seasonal water off to the Bheema River.

Climate

The area experiences dry semi arid climate. The 30 years cumulative weather data for the period 1950 to 1980 collected from the Indian Meteorological Department, show the highest temperature of 41.9° C in the month of May, and the least of 12.2° C in December. The area experiences variable rainfall since it is an inland area. Maximum rainfall is dependent on the southwest monsoon with a little contribution from the northeast monsoon. The annual average rainfall in the watershed area is 560mm; it indicates that the average number of rainy days is 46.4. Most of the area experience strong winds throughout the year and direction changes from season to season. Wind speed ranges between 3.9 to 14.3km /hr.

Geology

Almost the entire study area is comprised of basic lava, commonly known as Deccan traps. These are mainly of massive basalts and horizontal beds overlying the earlier formation. Taking vesicular basalt zones as the top of individual flow, three flows have been identified (given by DMG). Secondary minerals like zeolites, calcite, and quartz usually fill up the vesicular basalt. In fact the vesicular basalts and some red boulders on boundary at the ridge portion of the project areas are usually good repositories of groundwater.

Methodology

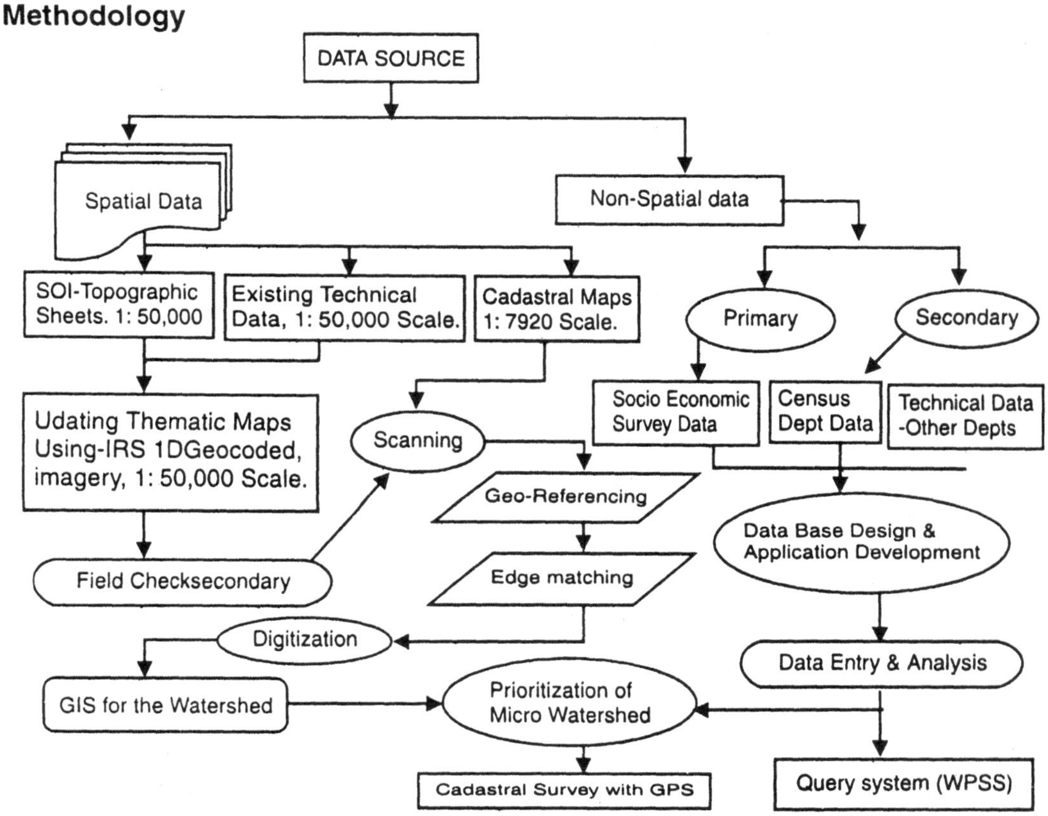

Socio-economic survey

The objective of the survey was to collect primary data, which could be mined to arrive at information that would reflect upon both the social and economic status of the farmers as well as the landless people of the project area. Two NGOs at Bijapur were identified for carrying out the survey work.

A detailed survey questionnaire with 24 tables, which focused on documenting various issues like the family structure and details, literacy rates, land holding information, the cropping pattern, livestock data, fodder and fuel requirement, annual income, expenditure pattern, migration habits if any, and the perception of the people towards watershed projects was prepared with active inputs from the partner NGOs. A guidelines booklet which was easy both to read and comprehend, to aid the NGO personnel who would be involved with surveying was also prepared.

It was decided that though the watershed boundary differed with village boundaries, within a village, the surveying could not be limited to only those households whose lands fell in the watershed area. So a criterion was arrived at whereby, complete household survey of a village would be taken up only if at least 50 per cent of its land area fell within the watershed boundary. Cadastral maps were acquired from the Department of Land Records, Bangalore, Bijapur and Indi, and the project area boundary was roughly demarcated on them, using drainage streams, roads etc. as reference points. A visual estimate was done to arrive at the approximate percentage of each of the village areas falling within the watershed boundary.

PREPARATION OF SPATIAL DATA

The spatial database developed consist of the following thematic digitized coverages

- Contours
- Drainage
- Slope (derived)
- Hydrogeology (from Satellite imagery)
- Land use and land cover (from satellite imagery)
- Soil series
- Village boundaries
- Transport
- Micro watershed boundaries

These maps are based on 1:50,000 Survey of India toposheets. All the thematic coverages are projected and transformed to the real world coordinates ie geo-referenced. From the cadastral maps the following geo-referenced thematic coverages were generated.

- Revenue plot boundaries for each village
- Transport
- Settlement

The spatial data prepared were converted to Arc View compatible format and the respective attribute data were attached to the thematic maps. To facilitate the end user four thematic projects were created and packaged using Arc Explorer. Arc Explorer is a geographic data explorer developed by ESRI, and distributed free of cost. It is designed to view, query and share geographic data. Using this, one can measure distances on the map, find features, identify and query geographic and attribute data, create maps using classifications, symbols, and labeling, pan and zoom through multiple map layers, change legends, overview maps, save and retrieve projects, and print map. The four thematic projects created were natural resources information, village information, cadastral information and GPS survey theme. The following screen shots show the capability of GIS in aiding the decision maker to prioritize the watershed in a more scientific and quick manner. The guidelines broadly followed for watershed prioritization were complied and step-by-step way to arrive at a final decision through GIS was formulated.

SWAT MODEL FOR DODDAHALLA WATERSHED

Soil and Water Analyzing Tool (SWAT) is a distributed hydrological model. GIS is used to generate the basic input data needed to run the SWAT model on the study area. Digital Elevation Model (DEM) was generated using the digitized contours. DEM along with the digitized drainage was used to automatically delineate the topographically defined sub basins. The sub basin and channel characteristics were derived using GIS functions. Sub basin theme was overlaid with the land use and soil layer to derive the Hydrological Responses Units (HRU). Model inputs including soil, weather, groundwater and management are required for each HRU. These input are automatically derived using the GIS interface. The study area is a part of the bigger watershed of size of about 1024 sq km. SWAT run was made on this bigger watershed. Since the model is distributed, the output can be generated at the sub basin level. The entire area was divided into 50 sub basins and 175 HRUs. The daily weather data for Bijapur station was used for rainfall and temperature. Model outputs include all the water balance component at various levels ie, basin, sub basin or watershed level and at intervals of daily, monthly or annual.

- Surface runoff
- Evapo-transpiration
- Lateral flow
- Recharge
- Percolation
- Sediment yield

The following figures show some of the pre-processing steps before running the model and the average water yield and sediment yield maps after running the model from 1969 to 1990.

DESIGNING DATABASE

Designing database is about discovering and completely defining the application's data characteristics and processes. Data design is a process of gradual refinement, from the coarse "What data does the application require?" to the precise data structures and processes that provide it. With a good data design, the application is fast, easily maintained, and can gracefully accept future data enhancements. The following are the concept that are used for designing the database.

Database for socio-economic survey data

Since the collection of the secondary data from various sources was a long drawn process, initially a database was designed only to accommodate the socio economic data. It was thought that the database for storing secondary data could be a part of the query tool application, which would be built in due course of the project.

Data modeling

- Identifying the data and associated processes (such as the farmers in the field need to see the availability of the close seed society, health centre, education centre etc.,)
- Defining the data (such as data types, sizes, and defaults)
- Ensuring data integrity (by using business rules and validation checks)
- Defining the operational processes (such as security reviews and backups)
- Choosing a data storage technology (such as relational, hierarchical, or indexed)

All the tables and fields in the application are designed for storing data for various purposes abiding the concept of database design. The idea behind designing the database was to develop an application which would facilitate data entry, storage and simple querying. The application developed using Visual Basic Version 6.0 front end and with Microsoft Data Access Object 3.5 (DAO) engine. Developing of the GIS and the query application, this involved the linking of the spatial and the non-spatial data using the software ARC Explorer. It was planned to develop an application, the entire programming to be done in Visual Basic, as an integral part of the GIS, which would allow data entry, storage and querying of both the primary and secondary data with reference to the spatial data.

PRIORITIZATION OF MICRO WATERSHEDS

Watershed prioritization guidelines follow physical criteria and socio economic criteria. Under the physical criteria watershed having the following characteristics are to be taken up first

- Head water
- Erodability, land degradation, water scarcity problem
- Rainfall less than 750 mm
- Net cultivated area is not more than 20 per cent.
- Proportion of irrigated area should not exceed the state average or 30 per cent
- Cropping pattern - no long duration crop or water intensive crops
- Villages fully falling in watershed

Under the socio economic criteria

- Economically weaker
- SC./ST population
- Uniform land holdings and many others

CADASTRAL SURVEY AND GLOBAL POSITIONING SYSTEM (GPS)

The cadastral survey of a three prioritized microwatersheds was planned using the GPS. The work included conducting topographical sight survey on 1 : 1000 and 1 : 5000 scales with spot levels at regular intervals and contouring at 1 m intervals. It also included super imposing of the revenue survey numbers in the topographical/contour survey drawings and the establishment of references/permanent bench marks in the site. Marking of the outer perimeter boundary, this was demarcated on the site with reference to the village boundaries, ridges and the road.

PLANIMETRIC CONTROL

Planimetric control (to measure the area of plan of any shape very accurately) was carried out using total stations by precise and scientific methods, based on procedures evolved by Survey of India. The method adopted ensured self-checks at every stage of work in respect of direction and length. Magnetic North was established at the site area and girding established North-South from East of Magnetic North. Thereafter, the centerlines E 30000 along East-West direction and N 20000 along North-South direction were established as reference base line.

THEODOLITE TRAVERSE

The horizontal angles were observed on two zeros (0-180) using Electronic Total Stations. Precision targets with optical centering and leveling devices were used at the stations observed. The distances were measured twice (back and forward) by using electronic distance meter in built in total station, which will give the measurements to the millimeter accuracy. All traverse lines were commenced and closed on traverse station of adjoining lines. The accuracy was determined by noting the closing error in the Easting and Northing co-ordinates.

HEIGHT CONTROL

Fly leveling, using a Wild Automatic Level provided this. Bench Mark value of 552.30 situated on stone plinth about 45 cm above ground level and 2.7 m North of South East corner of South Verandah of Public Works Department Inspection Bungalow at Kamatagi was taken as reference bench mark. Bench marks were established on all the permanent structures and traverse points along the project site. Spot levels at regular interval were carried out between the benchmarks established.

CONTOURING

Contouring at 1.0 m interval was carried out at the site by interpolating spot levels and taking into consideration the ground features. The land use in the area was documented. Also recorded were the exact location of the wells, hand pumps etc. along with the depth of water. Further, with the active help of villagers, the fragmentation of the land that had taken place over the years was recorded. The drawings were prepared in the scale of 1:1000 and 1:5000 by specialized CAD operators using Auto Civil Software.

RESULTS AND DISCUSSION

General overview of the socio-economic data analysis.

Demography

- Compared to the 1991 census, there is an increase of 27.62 and 20.66 per cent in the household and population, respectively; women comprise 45.21 per cent of the population, compared to the 48.39 per cent for men and 39 per cent of the population belongs to SC/ST categories.
- Area literacy is 52.29 per cent, less when compared to the 66 per cent district average and male and female literacy ratio is 2.2 : 1

- 49.21 per cent of the working population is dependent on agriculture, indicating that the major source of the area is agriculture. Women in agriculture are mostly represented as agricultural labourers rather than as farmers.
- The landless, small and marginal farmer households represent the larger section of the population. However, their ownership and access to the resources is limited.
- Any project to be implemented will thus have to consider the interests of the weaker section of the population. The decreasing female population trend and low women farmer population reflect the need for gender sensitization.

AGRICULTURE

52 and 18 per cent of the total land area are under rainfed and irrigated agriculture, respectively. Open wells are the main source of irrigation. But water level in summer is very low.

- Though private ownership of lands is less (2 per cent), vast tracts of wastelands are found, which fall under the responsibility of the government departments.
- Cereals, oilseeds and pulses are being cultivated in large areas. Cultivation of horticultural crops is negligible. Only 2.61 per cent of the lands are cultivated in summer.
- Local seed usage is mostly seen in crops like sorghum, groundnut and green gram, which contribute to human consumption and also fodder production. Large areas of other crops like bajra, maize and cotton are under hybrid seed cultivation.
- The cost of cultivation incurred is found to be less when local seeds are used. Productivity of jowar and wheat is lesser than the taluk averages; it is in contrary with bajra.
- It is observed that the rate of water use is more than the rate of its collection and storage. Planning water-harvesting structures within the vicinity of the wells might help increase the ground water levels.
- Wasteland development can be taken up through afforestation in and around various villages. But for this, proper linkage creation is to be made, between the people and the government departments like the Forest or Revenue.
- The households dependent on sheep rearing for their livelihoods must be taken into confidence while planning any wasteland development.

POVERTY AND MIGRATION

- 8.07 per cent of the households are below poverty line. 593 families migrate mostly in summer in search of better work and wages.

Watershed Management: Issues and Policies for 21st Century pp. 97-100

Land Use Planning and Vegetation Dynamics of Janahi Nala Watershed of Orissa through Remote Sensing: A Case Study

K. Kannan, R. C. Srivastava, Shishir Rout, N. Sahoo and G. P. Reddy

INTRODUCTION

Land use, land cover and land capability surveys are basic for making rational use of land in a watershed. The five basic surveys namely: soil survey, slope analysis, land classification and land suitability classification, present land use survey and land adjustment survey should be undertaken in a watershed before watershed development planning (Sheng, 1990). Remote sensing methods are becoming increasingly important for mapping land use and land cover for its faster, less expense and elimination of problems of surface access that often hamper ground surveys. But the disadvantage of this system is that most images lack the horizontal perspective which is valuable for identifying many categories of land use such as specific crop. Ground survey information is generally needed to identify a particular land use within the agriculture (level II) like cropland, pastureland and specific crops within the crop, and canopy density and their management (level III). A study was undertaken in Janahi Nala watershed in eastern Orissa to compare land use map taken from remote sensing map with ground survey and to develop land use planning for the watershed.

MATERIALS AND METHODS

The Janahi Nala watershed which drains into the Baitraini river is located in the Keonjar district of Orissa. Remote sensing map of the watershed was obtained from IRS 1C-Pan and LISS-III map. At the beginning, close checks were made to establish the relationship between photo images and actual ground condition. Tideman (1996) formula was employed for present land use and vegetation dynamics appraisal. Canopy density of forest tree species was calculated by adding all the basal areas of a particular species within the quadrat and dividing it by the area of the quadrat and expressed as dense (D), thin (T), degraded (Dg) and scrub (Sc) in the map.

RESULTS AND DISCUSSION

Basic information

The watershed is situated on the either side of the national highway number 6 and lying between 21°40' - 25°45'N latitude and 85°48'-85°50' E longitude. Dykes are the long sheet of rocks which are 10-15 metre height supporting mixed vegetation. The valleys are 20-100 metre wide. Rainwater collected in the watershed drains through these valleys. The total rainfall of the watershed is 1400 mm/annum, which is mainly due to southwest monsoon and confined to five months from June to October. The slope of the watershed varies between two and twelve per cent. The water table level fluctuates between 1-5 metre. The land distribution pattern and irrigated area under three villages which come under the watershed is given in Tables 1 and 2.

Table 1. Land distribution pattern in Janahi Nala watershed in hectare

Items	Kheritangri	Dabarchuan	Naugaon
Cultivable land	388.97 (46.50 per cent)	137.61 (41.90 per cent)	28.04 (16.35 per cent)
Cultivable waste	51.63 (6.16 per cent)	19.97 (6.07 per cent)	16.75 (9.77 per cent)
Pasture land	30.53 (3.64 per cent)	5.91 (1.80 per cent)	8.488 (4.95 per cent)
Forest land	82.56 (9.85 per cent)	55.07 (16.75 per cent)	35.38 (20.63 per cent)
Others	283.79 (33.88 per cent)	110.16 (33.13per cent)	82.80 (48.29 per cent)
Total	837.41 (100 per cent)	328.72 (100.00)	171.47 (100.00 per cent)

Table 2. Land use pattern in Janahi Nala watershed, Keonjhar in hectares

Items	Kheritangri	Dabarchuan	Naugaon
Holdings	333	146	117
Irrigated land	41.20	28.63	–
Dry land	347.77	108.99	28.04
Total	388.97	137.61	28.04

Present land use system and vegetation dynamics

The present land use as seen from the remote sensing map (level I) and ground truth (level II & III) is given in the Table 3. The detail vegetation dynamics and its maintenance are given here.

Table 3. Present land use as seen from remote sensing and ground truth

IRS (level I)	Ground truth (Level II)	Ground truth (level III)
Agriculture	Field crop	Direct seeded rice with no definite spacing and current fallow in-between.
Agriculture	Orchard	Mango (local variety with poor maintenance)
Agriculture	Pasture	Native grass with low productivity
Forest land	Pure forest	*Shorea robusta* (Thin)
Forest land	Mixed forest	Dense, mixed species on rocky soil with moderate to deep slope and uncultivable waste land in between
Forest land	Plantation forest	Eucalyptus, *Terminalia arjuna*
Water	Natural ponds Man made water harvesting structures	–
Small streams(nala)	Low land crop on flowing water	Rice
Barren land	Exposed rocks	
	Deep gullies	Deep gullies filled with sand

Agriculture

Rice-fallow rotation is followed in more than 95 per cent area followed by rice-vegetables in isolated pockets where water is available in ponds (Table 4). The cropping intensity is only around 100 per cent. Frequent crop failure is common in upland and the yield is also very less (1.0-2.0 t/ha) because of lower level of input use. Beusening (cross ploughing of rice fields under standing water at 30-40 days of sowing) is a unique practice in this region to control weed and redistribute the plants. The slope of the arable land varies from 1 to 4 per cent slope which needs only biological measures like, mulch tillage and stubble mulch for erosion control. The problems found in agriculture are low productivity, uncertainty of rainfall, mono-cropping, lack of water for taking up second crop.

Table 4. Cropping system followed in the watershed

Name of the crop	Rice
Area	1100 acre
Variety	Local, swarna and lalat
Preparatory tillage practices	Time: May Method: three cross ploughing with country plough followed by broadcasting of seed.
Sowing	Seed rate: 40-45 kg/acre Time: I week of June(Broadcasting) Source: own seed, agricultural department
Manures and fertilizer	1.5t FYM/acre no chemical fertilizer method: incorporation during last ploughing
Intercultural operation	Time: August Method: Beushening(cross ploughing with country plough in standing paddy crop followed by gap filling).
Plant protection measures	No chemical spraying
Irrigation	Rainfed
Harvest	Time: November-December
Yield	1-2t/ha

Forest

Forest is sub-tropical type with fair management. Pure and mixed forest is found in the watershed (Table 5). *Shorea robusta* is the principal species in pure forest and *Diospyros melanoxylon* in mixed forest. Ground cover and litter are medium in both the cases. The forest department has taken up plantation of *Terminalia arjuna* and Eucalyptus in small area. Crop density at different site is found dense to degraded. The problems found in the forest land are erosion (deep gullies) in limited pockets and illegal deforestation.

Agro-forestry and orchard

Agro-forestry is not being practiced in the village. There are only four hectares of orchard (mango) in the watershed. The age of the trees are more than 20 years and the stocking, condition and yield (50 kg/tree) are poor.

Table 5. Trees species and their occurrence

Tree species	Relative dominance	Relative frequency	Relative density
Diospyros melanoxylon (Kendu)	23.57	25.0	25.0
Syzygium cumini (Jamun)	31.42	25.0	25.50
Dalbergia sissoo (Sisoo)	22.00	16.3	14.20
Shorea robusta (Sal)	23.57	33.3	21.40

Grazing lands and grass lands

i. Grassland type : Natural pasture
ii. Botanical composition : *Digitaria, Panicum repens, Chloris barbata, Dactyloptenium*
iii. Condition : fair
iv. Management level : poor

MANAGEMENT STRATEGIES

Crop productivity improvement

1. Introduction of new high yielding varieties, more input use and green manuring
2. Encouraging transplanting in low lands
3. Introduction of short duration high yielding varieties like vandana in upland
4. Early sowing of short duration rice varieties in kharif followed by second crop like safflower and sesamum in rabi in medium land and black gram in low land on whole village basis
5. Reclamation of acidic soil by application of lime
6. Renovation of old orchard crop
7. Reseeding of grass land with legume fodder like *Stylosanthus* and *Desmodium*
8. Introduction of cross bred cattle, fish production and agro-forestry

Water conservation and use

1. Rain water harvesting in water harvesting structures for ground water recharge
2. Ground water exploitation through open well for taking up high value rabi crop like potato, ladies finger and tomato
3. Biological measure like stubble mulching, strip cropping in arable land.
4. Gully control in forest land
5. *In situ* soil moisture conservation

References

Sheng, T. C. (1990). Watershed Management Field Manual. Watershed Survey and Planning, FAO Conservation Guide. Food and Agriculture Organization of United Nations, Rome p. 170.

Tideman, E. M. (1996). Watershed Management Guidelines for Indian conditions. Omega Scientific Publishers, New Delhi. p. 372.

Watershed Management: Issues and Policies for 21st Century pp. 101-108

Socio-Economic Concerns in Watershed Management and Planning: Practical Approaches

G. P. Reddy, H. N. Verma, R. C. Srivastava, K. Kannan

INTRODUCTION

Watershed management planning in developing countries is usually hampered by limited manpower resources and time. Therefore, the work should be carried out as practical and economic as possible. The final products should be oriented to problems and their management rather than theories and academic studies. As watershed problems grow with population and time their management tasks are almost endless. Planners must keep in mind that planning is a continuous process and there is no single plan or programme, regardless how sound it is that can solve the problems forever and completely. When major problems are tackled and data bases are established, such plans should be considered useful and beneficial. Too ambitious a plan causes non-approval and non-operation. Also, a small success is better than a big failure. Government goals may not always coincide with the interests of farmers or watershed inhabitants.

CHALLENGES TO WATERSHED PLANNERS

To start any watershed planning, planners should be prepared themselves to face many challenges. They also need to determine what principles should be best observed and what courses taken. The following examples include some major challenges frequently confronting the planners in the developing countries;

- Political views may differ considerably from those of technical persons on types of management work, priority areas and timing.
- An ideal physical plan or a most effective work may not always be socially receptive by local communities or farmers.
- Sometimes, intangible benefits of a watershed programme may out-weigh tangible benefits whereas the former can not be easily assessed in monetary terms.
- Planning from bottom up may cost lots of time and energy in watersheds which populated with illiterate farmers, yet, involvement of local communities in planning process in a necessity.

There are no universal answers to these challenges. Each of them should be carefully examined in the context of the given conditions and the existing environment. As stated in a world Bank paper, watershed project design is inevitably a compromise. Selecting a practical approach to obtain early agreement of the plan and initiating operation for benefiting both the people and watershed are much more fruitful than arguing a perfect answer or remaining undecided.

QUICK IDENTIFICATION OF PROBLEM AREAS

National reconnaissance and watershed classification

At the national level, a quick reconnaissance type of survey, assisted by remote sensing techniques, is often sufficient for identifying major watershed problems and areas. For instance, in Jamaica such a survey was practically completed in three-month period for the whole island of 11, 300 sq. km. mainly by one technical person. It is not advisable for less-developed countries to spend many years on a detailed survey.

A quick classification of a nation's watersheds can also be made by using existing data and knowledge. They can be classified into reservoir watersheds, forestry and wild land watersheds, agricultural watersheds, municipal watersheds etc. for ready pinpointing of their major problems and for establishing management policies and priorities.

Problem oriented surveys

At the local or watershed level, further surveys or investigations are needed to obtain basic information for formulation of management plans. These surveys should be problem oriented, in a way similar to a doctor's approach to a patient. Healthy patients or watersheds should not be ignored but they should be put on routine care or maintenance while special attention or urgent treatment is given to serious ones. The "right medicine for the right illness" will not only save money and time but also lives.

Watershed problems can generally be grouped into categories as follows:

- Physiographical problems (e.g. steep slopes, heavy rains, excessive run-off, problem soils, etc.)
- Resources use problems (e.g. shifting cultivation, forest destruction, fire, over-grazing, uncontrolled mining, poor road construction, etc.)
- End problems (e.g. erosion, sedimentation, flood, water pollution, water shortage, etc.)
- Socio-economic and other problems (e.g. illiteracy, low acceptance of innovation, labour shortage, land tenure, poor infrastructures, etc,)
- Detailed surveys or investigations should be centred on the major watershed problems identified during preparatory missions or preliminary investigations.

Considering management possibilities

A practical watershed plan should consider, at the very beginning, the management possibilities and suitable strategies. For example, some foreign experts of outsiders, when faced with local watershed problems, may simply suggest to move the people out of the watersheds. It may sound attractive but they don't realize the difficulties and expenses of resettlement, not to mention creating social problems. Solving a problem in one place by creating more problems at other places is not a wise policy to pursue. Management possibilities which need to be taken into good considerations are suggested as follows:

- **Technological aspects:** Whether the present technology is sufficient for coping with major problems in the watersheds or more local trials are still needed. What will be the effect of the transfer of exotic technology, etc.

- **Available resources:** What are the possible sources for obtaining resources, monetary and the like, for the proposed watershed programme ? The magnitude of the programme should not exceed the expected resources.
- **Institutional ability:** The strengthening of any present institution is quite limited because it needs time to train technical staff. The programme should grow only as fast as trained and experienced staff are available.
- **People's acceptance:** Although this will depend on a continuous effort of education and extension, extreme care should be given to those management practices which are against local culture, religion or tradition. People's acceptance is a key to the success of most watershed programmes, but it is always a slow process.
- **Expected outputs:** What really can be accomplished at the end is often an ambiguous and neglected subject. Exaggeration of results may attract funding but it eventually will cause despair, to many funding agencies. The planner should be frank at the planning stage and realize that there are limits of management practices.

Watershed management and people

The focal point of any programme is the betterment of the people. Thus, the wider socio-economic concerns of any package of technology ought to include the size of population, population of ethnic groups or communities, composition and structure of households, and basic and secondary needs of the people. High population growth and the failure of the governmental programmes to keep pace with population growth lead to such problem as unemployment, underemployment, inequalities in income and assets, and over exploitation of natural resources. Poor people have to live more on natural resources, and they get caught up in a vicious cycle of removing more from fewer resources. The best example is Jhum cultivation. At present no in-depth survey in this regard is being done before planning a project, which could assist in locating persons and also might act as a catalyst in the programme and encourage people to participate.

Land capability and social needs

Improved land use is not to be pursued independently and relentlessly regardless of all other factors. Worldwide, there is a starting correlation between poverty and abusive land use. But the poor are not always to blame. Their immediate problems are often too pressing. Therefore, to solve the gigantic problem of soil and water conservation, it is imperative to match the land's needs with the needs of the land holder. For this to occur, an indepth survey of available natural resources as well as socio-economic factors in needed.

More emphasis must be put on the tenure system, leasing system, insecurity and capability of resources. Policies and programmes should answer such questions as these: which crops will grow in which areas? Where will new arable land be developed. What agro-forestry practices, if any, are feasible? It is not enough to impose these changes on the people who must practice them.

PLANNING MECHANISMS

Extent of coordination

The extent of coordination among various agencies depends on the degree of integration of a watershed programme. Generally speaking, because of the composite nature of watersheds, no single agency could plan or manage a watershed fully. Coordination is almost a mist. Nevertheless, the word "coordination" is more often spoken than acted upon in the developing countries due to lack of qualified staff and resources in each agency. Only when there are surplus staff and its main task is being taken care of, will an institution render assistance to the others. Usually, the least efficient agency limits the whole progress of work.

There, only the essential agencies should be coordinated into the programme. A watershed is land mass involving all human activities. If all activities are to be integrated, they will need no less a complex government bureaucracy to manage them. Sometimes, it is difficult even to call a meeting. Each country has its different institutional set-up and its unique environment. Essential agencies vary from one country to another. Experience in many countries has shown that involving seven or eight agencies may already feel bulky and beyond that it became impractical. The overall responsibility usually rests with one organization, the chief organizer, such as the forestry or watershed conservation agency. After the coordinating agencies are decided upon, the ways and means of coordination should be well spelled out at the early stage of planning to avoid duplication or confusion.

Involvement of local communities

As mentioned before, this is a necessary but difficult task. If there are local bodies such as local governments, farmers' associations, private interest groups and specially organized districts, their appropriate representatives can be involved in watershed planning. Care should be given not to involve those who are too strongly into party politics that to involve them would cause trouble or make others non-cooperative.

Another technique is to directly pull farmers or inhabitants by specially designed surveys. In developing countries farmers are mostly illiterate, conservative and hard to find in day-time. They may not like to be bothered by many repeated socio-economic surveys of this nature. Their answers may not be very useful unless the questionnaires are properly designed and tested.

Organizing proper bodies for planning

First, there should be one agency designated to initiate the work. Usually this agency should have its chief responsibility in managing watersheds or have its own interest in watershed products (i.e. irrigation, electricity authority or municipal governments). A steering committee should be organized involving representatives of essential agencies and local communities with the organizer as the convener. The committee thereafter monitors the processes of design, survey, analysis and final reporting. If required, the committee will be involved in implementation and evaluation.

Below the committee, a number of field survey teams are usually needed. A team can have members from different agencies but its leader should be drawn from an agency related to the survey subject. For instance, the leader of a soil conservation agency, although members may come from extension, crops and soil department and farmer's association, etc. The leader will not only report their progress to the steering committee but also serve as a bridge from his mother organization to the committee. Periodically, all the teams need to meet and report their progress and problems to the steering committee. Any inter-team matters should also be settled in the committee meetings. Funds for planning can be provided mainly by the chief organizing agency because they have more interest than others or it may be shared among coordinated agencies.

At the end of the planning, each agency should agree to an overall plan consisting of work schedule, staff needs and budget. When no external aid is available each should share the cost and programme in their regular budget year by year, according to the overall plan. The programme will eventually be carried out by each responsible agency in the watershed with mutually understood objectives and overall supervision. Sometimes, a liaison office is needed in the field to represent the steering committee.

This kind of joint and coordinated planning and decentralized implementation proves much more effective in many developing countries. It is better than piecemeal approaches and staff-borrowing type of undertakings. As long as each agency considers it as its own responsibility and is not merely to do others favour, the work will be done more efficiently and fruitfully.

DATA COLLECTION

Preparatory work

There is a general tendency in survey and planning to collect more data than necessary in one area and insufficient information at the other. Therefore, before data collection, preparatory work should be done to determine what is really needed, how it can be collected and where to get it. Valuable existing data, maps and reports should not be overlooked in order to save time, money and efforts.The data to be collected should be relevant, to the point and useful for the final analysis and report. Survey forms or tables should be concise, practical and well designed and checked in the field before use. In the case of land capability and land use planning, for example, the smallest unit should be compatible to small farms and the capability class should promote safe land use rather than eliminate their use totally.

Data should be collected or produced in an orderly manner so that all for the necessary planning work can be completed in a specified time period.

Data requirements and collecting techniques

Data required for watershed planning vary according to management objectives, watershed problems and given conditions. Generally, the following categories of information are needed:

- Physiographic data including location, elevation, sub-watersheds, soils, geology, land forms, slopes, drainage patterns, etc.
- Land use and cover types including forest, grass/range land cultivated land, orchards, wildlife reservations, recreation areas, urban, water area, worded areas, land capabilities, etc.
- Climate and hydrology including precipitation, wind, evaporation, temperature, stream flows, floods, erosion, sediment, etc.
- Socio-economic data including demography, land tenure, farming systems, education, infrastructures, human resources, farm enterprises, rural employment, production, income, marketing, transportation, credits, labours, etc.

- Institutional and cultural data including policy and administration, extension services, farmer's organizations, community and private groups, traditions, religions, cultural practices, acceptance to innovations, group actions, etc.
- Management oriented data including watershed problems, environment impacts, land management techniques, treatment needs, unit cost, sectoral cost, cash flows, work schedule, financial arrangements, expected benefits and results, etc

To do the above surveys and analysis, multi-disciplinary teams of professionals and technical persons are often needed. However, in a small developing country where professionals are scarce, a body consisting of an engineer, an economist, a sociologist, and a forester or agronomist, together with some field assistants may do a satisfactory job. The general techniques employed for such field oriented surveys includes simple statistics and sampling, air photo interpretations, mapping and design of questionnaires, etc., most of which can be taught and learned by sub-professionals or technical assistants.

Establishing data base

In some advanced developing countries, computers can be employed for storing most of the basic data for furture use. For instance, periodic surveys on land use, farm incomes or erosion/sediment data will provide clear pictures of changes. The effect of watershed management work over time can thus be evaluated.

FEASIBILITIES, ALTERNATIVES AND RECOMMENDATIONS

Assessment of feasibilities

Assessment of project feasibilities by economic means is one of the measures often employed in planning. However, economic viability is not the only criterion for decision making. Watershed projects, unlike simple engineering works, have their special characteristics which are not easily assessed by monetary means, i.e .long term and inter-generation benefits, externalities, spatial distribution of costs and benefits intangible and indirect benefits, etc.

Besides the usual cost and benefit analysis with and without project and the calculation of internal rates of returns etc., the social values or benefits should also be added for overall assessments of watershed projects. For developing countries, social benefits derived from increased employment, equitable distribution of wealth, improving the farming environment and farmer's well being and providing better infrastructures etc. should be weighed heavily in the final assessment.

Selection of alternatives and strategies

There are many ways to achieve management goals. During the planning stage, possible alternatives should be explored so that the best alternative can be chosen. For instance, if the main objective is to restore vegetation on the steep slopes for protection, expensive reforestation may not be the best answer. Alternatives could include protection of the existing land for natural regeneration/succession or supplementary planning or seeding, etc. Some technological alternatives need field trials which can be built into the project at the beginning of the implementation stage. Alternatives in cost and effect, in incentive needs vs. education, in group actions vs. individual assistance, among others, should also be well considered and selected. For watershed project to be successful, strategies for effective extension for streamline operations, cost control, inducing farmer's participation and for work maintenance, etc. should be well chosen before initiation of implementation. Many constraints in physical, technical, administrative and budgetary aspects should therefore be clearly identified and resolved in the planning stage.

Making realistic recommendations

The watershed planners, being not always the future project executors, tend to make recommendations good in theories but questionable in practices. Examples are not lacking such as to move large farming communities out of watersheds, to complete treating a huge watershed in 3 or 4 years, or to ask government to increase staff out of proportion to its existing structure, etc. Planning should by no means be only for planning's sake. To sell the plan to either the government or to any international aid agency, recommendations have to be relevant, responsible and realistic.

Monitoring and evaluation

Monitoring and evaluation are critically important in realising the objectives of developmental programmes and projects, particularly watershed projects having multiple objectives. In India, integrated rural development projects, including soil and water conservation and watershed progammes are being monitored at four levels. Central, state, district and block. The Ministry of Rural Development and/or the Ministry of Environment and Science reports from state level. Rural development aggregate or couple this reports. This monitoring is done mainly on a secondary data base. In most cases, the question of what should be monitored and at what stage is not clear to the field staff. These reports rarely are of any use in terms of suggesting mid term corrections. Often, monitoring gets underway after, rather than before a project has begun. Monitoring of social cultural factors and of infrastructure becomes increasingly important but neglected either because of rigid, target oriented progammes and lack of proper guidance. In fact, there are no separate monitoring and evaluation cells at district level. Monitoring of project effects after withdrawal of a project is rarely done, which could be useful for future projects.

Evaluation of a process of determining systematically and objectively the relevance, efficiency, effectiveness and impact of activities in light of their objectives. It is an organizational process for improving activities still in process and for adding management in future planning and decision making. But in country, most of the evaluation studies are confined to achievement or fulfilment of targets and efficiency of implementation. Evaluation of these programmes must include the process of percolation spill over effects, lacunae in the programme, and the phenomenon of non participation on the part of local people in the programme. Lack of standard or well developed evaluation methodologies for evaluation of watershed programmes has clearly hampered progress. This indicates that there should be a team of social scientists at least at the district level that can develop a suitable methodology for evaluation of programmes and up date data banks. At the district level involvement of social scientist at all stages of a project or programme will not only increase programme efficiency but also indicate needed improvements.

CONCLUSION

Watershed management technologies must be compatible with socio-economic concerns in a developing society. Efficient and properly linking technology and various socio-economic aspects is critically important. Watershed technologies will gain acceptance among the rural masses only when these socio-economic considerations are adequately addressed. A strong data base is needed at least at the district level as effective coordination between various governmental and non-governmental agencies; other establishments of planning, monitoring and evaluation cell at the district level; surveys and their documentation before planning; development of alternative set of technologies and the active involvement of local people in preparation, execution, and maintenance of structures in a watershed. Otherwise, huge amounts of scarce resources could be wasted on ill-conceived, poorly designed, and badly executed watershed programmes.

References

Gregersen, H. M. and K. Books (1978). Economic analysis of watershed projects. Paper presented at the 12th Session of the Working Party on the Management of Mountain Watersheds, Rome.

Food and Agriculture Organisation, United Nations (1986). Guidelines for Watershed Management, Rome, Italy.

International Centre for Integrated Mountain Development (1990). Guidelines of Monitoring and Evaluation in Watershed Management, Kathmandu, Nepal.

Sheng, T. C. (1981). Physical planning for watersheds. FAO/UNDP: Project Working Paper 17.

Srivastava, R. C., Nand Sahoo, G. P. Reddy (2001) Eds. Training Manual for Watershed Management Technology for Eastern Region. Bhubaneswar.

Watershed Management: Issues and Policies for 21st Century pp. 109-114

Socio-Economic and Institutional Issues in Watershed Development Project

A. Oppilamani

INTRODUCTION

The demand for basic human necessities, viz., food, fuel, fodder and fibre are constantly on the increase. The two districts, Tirunelveli and Thoothukudi located at the southern tip of the Indian sub continent are exposed to suffer from variable, low and erratic rainfall as well as severe wind erosion coupled with edaphic factors. Parts of the region are subjected to suffer from moving sand dunes. Also due to strong human and livestock pressure and continuous and frequent drought, the land degradation process in the two districts is further aggravated. Therefore to make the land productive and sustainable, wind velocity is to be reduced through area specific and appropriate measures of land and water resources management. With this in background, Danida aided Comprehensive Watershed Development Project, Tirunelveli, Phase I and II have been implemented in the two districts since 1990-91.

The main objective of the project is to enable land users in priority villages to practise dryland agriculture through conservation and use of natural resources on a sustainable manner. Though the project focuses on immediate objectives and development objectives, the ultimate aim is to identify and cater to the felt need of the rural community.

This is taken up in the following stages.

- To prioritise the needs and requirements by identifying the problems.
- To solve the problems through introducing appropriate technologies and sustainable practices.
- It is positively expected that in the long run it will help in the upliftment of socio - economic status and betterment of the environment of the rural community.

MAJOR COMPONENTS OF THE PROJECT

In advocating sustainable land use systems alongwith conservation of natural resources, the main activities focused and implemented by the project are:

- Tree planting intervention.
- Sinking of borewell.
- Assistance to landless.
- Land development and moisture conservation.
- Capacity building of beneficiaries and staff.
- Village development associations.

The main focus of this paper is on the community involvement, socio-economic and institutional issues in (Watershed development) this project.

COMMUNITY INVOLVEMENT

It is an experienced and very well established truth from the implementation of Comprehensive Watershed Development Project that any project ventured for socio-economic development of people can be made popular, successful and sustainable among the beneficiaries only through proper and whole hearted participation of the beneficiaries themselves from the planning stage of the project itself. Hence acceptance by the beneficiaries of any development project is an important factor that results in willingness and active participation of beneficiaries in the implementation, monitoring, maintenance and so on to achieve sustainable benefits as per the objectives of the project.

This project has emphasised the above factor and the community involvement is achieved through people's/beneficiaries participation. In any such project the primary target group consists of all rural farming households in the priority villages especially marginal and small farmers, women and SC/ST farmers. This project specifically includes the landless families also to receive direct and indirect benefits from the project.

The Community involvement is carried out in the following five stages.

- Rapport building stage.
- Group formation stage.
- Planning stage.
- Implementation stage.
- Monitoring and evaluation stage.

Rapport building stage

At this stage, the project staff conduct a number of village level visits to exchange, inform and get feed back from the beneficiaries as well as the whole rural community about the necessity to work as a group, the objectives of the proposed project, direct and indirect benefits of the project, merits and demerits of the project, immediate and long term development of the village, the package-programme activities that are to be executed by the project, the participation and contribution of the project, institutes concerned and also the participation and contribution from the rural community to their own development through the project. These frequent village visits have to be definitely combined with required number of informal village level meetings so that the bilateral information both passed onto and collected from the community will be complete and sufficient to proceed further with the project in full co-operation of the rural community.

Group formation stage

Once the community feels the need and usefulness of working as a group, they get motivated to organise themselves into groups. Draft byelaws have to be given to these groups that can be modified by the group to suit their own needs. As per byelaws Office Bearers are to be elected. This stage gets over when the Associations establish proper records such as books of accounts, related registers and records and opening of a Savings Account in a nearest scheduled bank. The Associations from then on start their full-fledged functioning by regular periodical meetings to discuss their common development issues.

Planning stage

Planning starts from the analysis of the situation. The group at this stage takes up socio-economic survey through Participatory Rural Appraisal and assess the actual beneficiaries both landowners and landless, requirements, needs and problems relating to the project from which basic facts are derived for preparation of participatory development plan. Facilitated by the officials, the development plan is prepared by the Village Development Association with all proposed work components, action plan with time schedule, responsibilities of Village Development Associations and estimated budget.

Implementation stage

Village Development Associations play a vital role during this stage. They will be the contact points for the project officials for implementing the project components as per the participatory watershed plan. Through regular interaction with the beneficiaries by the staff, the problems and constraints faced by them are identified and solutions made then and there and the beneficiaries themselves under the technical guidance of the staff execute the project successfully.

Evaluation stage

With the basic information about the village collected by the Village Development Association through Participatory Rural Appraisal, the Village Development Associations have to conduct an evaluation every year and identify the significant development changes both positive and negative and assess the socio-economic status of its members. Through regular monitoring and proper self-evaluation, the Village Development Associations will grow to be self-reliant and achieve sustainability by repeating the process or recycling the stages continuously.

SOCIO-ECONOMIC ISSUES

Recent trends in Rural Development have established that the success of a project largely and solely depends on the effective participation of the beneficiaries as a group. Unless the rural people are given the means to get involved fully in development programmes, they will continue to be excluded from its benefits. Formation of rural community into a single group is itself a great achievement in respect of social harmony. With social-harmony achieved among a rural community, remaining process of introduction and implementation of any development activity in future and their sustainability are assured.

Social issues

In a development programme like watershed management the social issues involved are categorised as follows.

- Motivating and mobilizing of local support.
- Establishing links with outside institutions viz., implementing agency, various government schemes programmed by different departments, finance institutions, social institutions etc.

- Discussing merits and demerits of new development programmes.
- Carrying out the programmes effectively and efficiently.
- Keeping on to the time schedule prescribed for the programme.
- Identifying not only beneficiaries but also volunteers, coordinators, leaders, labourers, etc.
- Suggesting and obtaining appropriate feed-back from the community to the implementing agency.
- Meeting periodically to reflect on the progress and discuss on current vital issues.
- Initiating social binding on protection of assets both individual and common.
- Encouraging the public to raise home gardens, medicinal plants, etc.
- Ensuring timely grant of loans and subsidies from government and other donor institutions.
- Ensuring proper distribution of loans among the members.
- Forming of subgroups for women.
- Ensuring collection and timely payback of loans from the members to the donor agencies.
- Arranging tools and facilities for the members to enrich their knowledge, to cater their basic amenities in respect of public health, animal health, agricultural practices, education etc.
- Ensuring timely supply of necessary inputs for the above.
- Encouraging for taking up developmental measures and replicating such activities on their own in a sustainable manner.
- Encouraging co-operative activities in agricultural production and live-stock produce.
- Propagating gender sensitization.

Economic issues

The economic issues involved are:

- Sharing of benefits among the community.
- Raising of funds to strengthen the Association by way of subscription, donation, institutional finance, etc.
- Utilising the resources of their own village.
- Raising of nurseries following modern techniques.
- Hiring of implements and timely supplying of the same to the community in respect of agricultural practice, education, information technology, etc.
- Marketing the agriculture and other produce of the community/society at a profitable value.
- Utilising the funds for regenerative purpose as revolving funds distributed as loans to meet out agricultural, medical and educational requirements of individuals.
- Linking with commercial banks to avail credit facilities at low interest rate.

INSTITUTIONAL ISSUES

Watershed management in India at this stage of about 50 years of development appears both fantastic and frightening. It is fantastic because Government, Non-Governmental

Organisations and bilateral donor agencies are increasing priority and providing astronomical funding support. It is frightening because is somehow getting mystified. Watershed contains a number of biophysical resources viz., soil and water, alongwith vegetation in the form of trees, grasses and crops and also provides live-stock production. Hence watershed management is essentially a resource-based approach to livelihood enhancement, which should support the livelihood of poorest groups and are institutionally sustainable. The recent spurt in the programme is a policy response to the increasing environmental crisis and non-sustainability of agriculture especially in the dryland/semi-arid regions. The process is interactive and continuous hence requires fine-tuning between physical and social engineering.

The process is wide ranged from social to physical engineering and starts at micro-watershed level. Therefore it aims to establish an enabling environment for the integrated use, regulation and treatment of land and water resources of a watershed based eco-system to accomplish resource conservation and biomass production objective. If watershed development has to become a people's movement, technologies would have to be simple, low cost and should be based on vegetative measures which are self representative.

Therefore this needs a multidisciplinary approach of involving various Institutions viz., Government departments concerned, Donors, Non Governmental Organisations and Financial Institutions like banks, etc. The same approach is followed in the Comprehensive Watershed Development Project, Tirunelveli, Phase II. The primary responsibility of these institutions is to provide technical and financial support and participate in people's effort to upgrade and consume their natural resources.

Responsibilities/role of institutions

- To Provide financial resources.
- To provide technical guidance.
- To prepare the Project Plan with details of physical and financial outlay in the context of Watershed Development/Management.
- To obtain training in the relevant fields of social and physical engineering aspects.
- To impart training in turn to the rural community in the relevant fields.
- To transfer technology.
- To provide necessary infrastructure facilities.
- To co-ordinate among the line departments and Non Governmental Organisations and other Technical, Social and Financial Institutions.
- To form various committees at different levels from the grass root level of Village Development Association to the apex body of the Government.
- To analyse, compile, document and disseminate the experiences and lessons learnt so that such documentation will be useful to other similar development projects.
- The beneficiary Institution i.e., Village Development Association to participate from planning stage itself.

CONCLUSION

A Watershed Development Project carried out with the community involvement dealing with their socio-economic problems through a multidisciplinary system will be definitely a successful project achieving its immediate and development objectives. The participatory approach to watershed development would secure the commitment of people to execute, monitor and maintain the project after its completion thereby establishing sustainability. In any case, one should be careful of the fact that approaches and guidelines should not become substitutes for the ultimate goal of ensuring livelihood support to most, if not to all of the households within the watershed.

Watershed Management: Issues and Policies for 21st Century pp. 115-122

Watershed Plus Activities: An Intervention Approach for Increasing Watershed Productivity

P. Nanda, R. B. Singandhupe, R. C. Srivastava, N. Sahoo, R. K. Mohanty and M. Das

INTRODUCTION

The watershed development programmes have been designed for integrated development of area comprising natural, animal and human resources. However, most of the projects result in sub optimal utilization of resources due to weak linkages and lopsided planning and execution. As a result, the productivity potentials of the developed watersheds are not optimally utilized. In Orissa out of 5.9 million hectares of net cropped area, around 34 per cent is irrigated. Most of the areas of plateau district of Orissa are rainfed. The rainfed agriculture in this large area of Orissa has led to serious socio economic consequences. The area is predominantly mono-cropped and the productivity of rice in the plateau district of Orissa is around 1 tonne per hectare. Due to poor agricultural performance in this region, the large segment of this population is below poverty line for which the per capita income is very low. Under different schemes of Central and state governments, watersheds are increasingly being adopted as unit of development in the district. Due to integrative nature of watersheds with respect to natural and human resources development, the investment impact is supposed to be wide based in the district. The district of Keonjhar has been in the forefront of watershed development in the state due to topographical and administrative advantages.

OBJECTIVES

As a case study for technology evaluation and intervention in a developed watershed, following objectives were set for the study.

- To study the efficacy of watershed technologies through farmer participative method and to suggest location specific suitable technologies for intervention by watershed functionaries. To study the effectiveness of technological intervention in utilization of common and owned resources with special emphasis on water utilization from water harvesting structures.
- To assess the impact of technological interventions on socio-economic condition of the beneficiaries.

BRIEF REVIEW OF THE KULADERA NALLA WATERSHED

A developed watershed named Kuladera Nalla in Keonjhar district (saturated watershed as per implementing agency's parlance) was selected for study. The mini watershed on

Kuladera Nalla encompasses the villages: Jhadbelda and Nelung in sadar block of Kendujhar district. The upper-most ridge of the Nalla was taken into consideration for treatment by Soil Conservation Department under EAS scheme during 1995-96 that comprises of Nelung and Jhadbelda. The topographical expanse of the project lies in between 21°-45' N to 21°- 50' N latitude and 85°-15' E to 85°- 13' E. The upper reaches of the Nalla in consideration forms a part of the tributary network of Keonjhari nalla sub-watershed, which in turn feeds to river Baitarani.

RESOURCE APPRAISAL IN THE WATERSHED

Human resources

The total population of the watershed area was 2009 with 371 household families, the demographic data of which is produced in Table 7. Most of the population i.e. 92 per cent depends upon agriculture for their livelihood. Further 238 of the farmers are marginal having holding size of less than 1.0 ha. The sex ratio in the watershed was found out 1027 females per 1000 males. This sex ratio is in higher side in comparison to state average of 972 as per 2001 census and district sex ratio of 977.

Livestock

The total livestock figure is given in Table 6. There are 1097 animal and 1067 birds. As mentioned earlier there is no scientific management of the pastureland for the large cattle population. So excessive and over grazing of common property resources accelerates land degradation in non- arable lands.

Plantation

Silvi – pasture plantation was introduced in plantation crop as well as the proprietary boundary of the farmers to meet the fodder demand of the cattle in the area. Besides the above, pasture development by covering productive grass species were also taken up. The planting materials and techniques will be available to people on free of cost to take up fodder development in their tenanted land also.

Natural resources

The geographical area of the project comes to 821.59 hectares, out of which the treatable area is 762.16 hectares, rest being the area under community places like road, village site etc.

Table 1. Treatment information on Kuladera Nalla mini watershed

Measures taken	Area reclaimed in ha.		
	Private wasteland	Govt. wasteland	Total
Gully & ravine reclamation	4.00	17.00	21.00
Moisture conservation devices	6.00	23.00	29.00
Total	10.00	40.00	50.00

Table 2. Changes made in land use pattern in arable sector

Type of land	Pre-project cropping pattern	Post-project cropping pattern
Up land	Mostly fallow or sporadic coverage of up land paddy.	Sabai , Niger, Arhar, Maize, Horsegram,
Medium land	Paddy (local)	H.Y.V Paddy, Vegetable
Low land	Paddy (local) (single crop)	Paddy (Double crop) (H.Y.V)

SOIL PROFILE OF PROJECT AREA

Twenty soil samples viz. 1 to 12 from eastern part (Nelung) and 13 to 22 from western part (Jharbelda) was collected. Soils were collected from two depths i.e. 1-0-0.15m and 0.15-0.30m, processed and analyzed for estimating different important soil chemical properties. Soils are in general moderately acidic to slightly alkaline in reaction, non saline in nature and low in organic carbon, nitrogen and phosphorous (P) content. Soil in the hill top is more acidic (1.17 times) than the soil at hill bottom. It indicates erosion of the soil is still prevailing and thus the soil at hill top is devoid of basic metal ion. Organic carbon content of the soil at 0-0.15 m depth is lower than the soil at 0.15-0.30m depth which is further 4.35 at hill top and 2.36 times lower at mid hill soil than the organic carbon content at bottom hill. Available N also showed the similar pattern as soil organic carbon. In contrast available(P) status of the soil at mid hill is bit higher than the average (P) content of soil at hill top and bottom-hill. The exchangeable K was distributed at differential manner in soil at various situations.

DRAINAGE LINE TREATMENT

The rills, gullies and streamlets were treated to carry the run-off with non-erosive velocity and free of silt load. This was done with plugging the gullies with loose boulder structures and good number of drop and sunken structures. The sides of the gullies have been protected by vegetation as well as stone patching.

FAILURE OF LINKAGE

From the field studies and evaluation, it is evident that the visibility of treated watershed got blurred due to non-utilization of post project potentials of the developed resources like land and water. Arable lands have been made available due to rehabilitation of eroded lands, *in situ* soil moisture is abundant in the down stream of the water harvesting structure, but the farmers are of the opinion that they are not able to take up a second crop as there is not water. Basically the tribal farmers do not go for a second crop in large scale as traditionally they practice a single crop in kharif. Interaction with the farmers and survey by the centre in the Kuladera nalla watershed in Keonjhar district of Orissa revealed that the farmers were not trained to utilize the developed resources of the watershed and the maintenance of post project watershed capital base was not emphasized.

Viable and workable common property management strategy is lacking in the watershed villages as green pasturelands and forestlands are left unattended even though a great potential is existing in fodder availability and minor forest produces. Indiscriminate cattle grazing is still prevalent in the developed watershed area. The water harvesting ponds having large potential for supplemental irrigation and pisciculture remains unutilized. In the absence of animal husbandry and fishery programme in the area by line departments the potential created in the form of pasturelands and ponds remain unproductive. As there is no management of pasture lands repeated cattle and ruminants grazing have exposed them to erosion, which may turn alarming in near future. The biomass yield potentials remain unutilized.

Socio-economic status in the watershed area

Table 3. Literacy

Name of the village	Total	Total	Male	Female
Neulang	1340	387	282	105
Jhadbelda	669	215	147	68
Total	2009	602	429	173

Table 4. Profession

Name of the village	Cultivators	Agril. labourers	Indus labourers	Other workers	Marginal workers	Non workers
Nelung	99	160	9	186	328	544
Jhadbelda	103	36	3	22	154	351
Total	202	196	12	208	482	895

Table 5. Land holding

Name of the village	No of farmers	Holding size					
		Below 1.00 ha		Below 2.00 ha		Above 2.00 ha.	
		No.	Area	No.	Area	No.	Area
Nelung	220	128	85.26	52	83.19	40	91.57
Jhadbelda	176	110	40.73	40	60.14	26	61.51
Total	396	238	125.99	92	143.33	66	153.08

Table 6. Present land use statistics (area in hectares)

Land use pattern	Village wise distribution		
	Nelung	Jhadbelda	Total
Agricultural land			
Upland	187.52	121.69	309.21
Medium land	37.50	24.43	61.93
Low land	25.00	16.26	41.26
Total	25.002	162.38	412.40
Non- Arable land			
Village forest	49.75	40.04	89.79
Culturable wasteland	117.18	104.50	221.68
Patit	16.67	11.11	27.78
Gocahr	7.27	3.24	10.51
Public places	1.15	1.42	2.57

Table 6 (*contd. on* page 119)

Table 6 (*contd. from* page 118)

Homested	2.09	2.83	4.92
Ponds/nalla/road	10.20	13.46	23.66
Rockout crops and hills	28.14	0.14	28.28
Total	232.45	176.74	409.19
Grand Total	482.47	339.12	821.59

Table 7. Livestock resources

Name of the village	Livestock population						
	Cow	Calf	Bullock	Buffalo	Goat	Sheep	Birds
Nelung	101	28	147	32	296	26	520
Jhadbelda	106	32	152	42	313	29	547
Total	207	60	299	74	609	55	1067

Table 8. Infrastructure

Name of the village	Road	School	College	Post office	Vet centre	Bank/co-operative society	Hospital	Market
Nelung	5 kms from Keonjhar on NH-6	Primary-2 M.E School-1	At Jhadbelda 1 km from the village	One post office in the villages from the village	At Keonjhar 5 kms from the village	At Keonjhar 5 kms from the village	Head Qrs Hospital 5 kms from the village	Keonjhar
Jhadbelda	7 kms from Keonjhar on NH-6	Primar-1 M.E. School-1	1 college at Jhadbelda	At Nelung 1 km from the village	At Keonjhar 5 kms from the village	At Keonjhar 5 kms from the village	Head Qrs Hospital 5 kms	Keonjhar

Table 9. Crops grown during 1999-2000 (area covered in hect, Yield in Qtls)

Name of the village	Name of the crop	Area covered	Yield obtained
Nelung	Paddy	195	3300.00
	Maize	22	160.00
	Niger	21	82.00
	Ragi	6	37.00
	Arhar	21	92.00
	Black gram	24	82.00
	Horse gram	23	80.00
Jhadbelda	Paddy	161	2800.00
	Maize	19	102.00
	Niger	21	86.00
	Ragi	9	26.00
	Arhar	23	96.00
	Black gram	22	86.00
	Horse gram	12	46.00

Assessment of impact of dugwells on the cropping pattern: Production, consumption and marketed surplus of the farmers at the project site

After surveying 25 farmers having dugwell at their backyard, it was found that there was a differential impact on their land utilization, consumption pattern, cropping pattern and income as compared to their living standard before watershed development programme. From the survey, it was found that the major crops practiced in the backyard by the farmers were onion, garlic, potato, brinjal, tomato, chilli etc. The survey was conducted in the month of Feb-March when the Rabi crop was at the harvesting stage. Among the 25 dugwells, the average depth of well was 32.04 ft whereas the average depth of water present in the dugwell was 8.26 ft in the month of March. Water level data in the well indicated that upto March light duty crops could be grown with three to four supplemental irrigation. The farmers produced on an average 285 kg of vegetables from the average backyard farm size of 21.42 dcml using water from the dugwell. After meeting their domestic consumption, they were able to generate some marketable surplus and sold the same in the local market which fetched them on an average of Rs 1152/ha in the season. The non-food consumption expenditure on non food items of the households having dugwell was on an average of Rs 700 per month which was almost equal to their consumption before having dugwell. The expenditure on non food item is a function of agricultural income and income elasticity of the items that did not change in the present scenario for the surveyed households.

The survey indicated that the non food consumption expenditure was almost equal in pre and post watershed project scenario. However, household food security increased due to increase in cropping intensity wherever dugwells were present to harvest water from water harvesting ponds through percolation. Though the marketed surplus of vegetables was not satisfactory, meeting the domestic demand for vegetables in the Rabi season could be realized through backyard vegetable production utilizing the water from dugwells. The survey covered households on a mini plot basis having assured source of irrigation in the form of dugwell in the backyard of the house. It is significant that, if water is harvested in the dugwells through increased infiltration rate, the possibility of non-paddy crop with available water resources can increase the economic condition of the people.

Intervention design

After study of resource potentials and their utilization in the selected watershed, it was inferred that the developed potentials of the watershed resources are suboptimally utilized. Absence of proper working links of extension departments and non involvement of people in the programme, from then beginning, were major shortfalls in the programme that was otherwise effective, if conservation of natural resources and agri-horticultural aspects are considered. In view of the diagnosis, an intervention decision was made by the project team to demonstrate the resource use potentials through people's participation in on farm and off farm activities.

AGRONOMIC INTERVENTION

Kharif and *Rabi* intervention for 2000

The farmers were practising mono-cropping of paddy in all the landforms i.e. up, medium and low. Possibility of non paddy crops in rainfed uplands with good economic returns was not given due importance by the farmers. They used to direct broadcast the crops in uplands that get destroyed in the event of dry spells that are common in the monsoon and post monsoon periods. In order to demonstrate the farmers the better land utilization technology and to increase economic return per unit of holding, the team under the leadership of the agronomist designed an intervention in the Kharif and Rabi-2000 for Jhadbelda village coming under Kuladera nalla watershed. The project team interacted with state agriculture department, ITDA, horticulture department for arranging substantial input supply for upland crops like groundnut, arhar, sesamum and maize. The WTCER arranged critical inputs from the project fund and supplied to 30 farmers for 16 acres of land. Groundnut-arhar intercropping, groundnut solo, arhar solo and maize solo germinated profusely and crop health was vigorous in the middle of September. The crop production was satisfactory and gave a good economic return to the farmers. The average productivity of groundnut was 1.5 ton / ha with a low of 0.8 ton/ha for 2 farmers. The productivity of groundnut is far above the average yield rate of the district. With supplementation of improved crop production strategies and use of water from water harvesting structure inspite of erratic rainfall, the productivity of Kharif paddy gave an average yield of 1.8 tons per hectare. The district average of Kharif paddy was 1.3t/ha. The 30 farmers unitedly collected funds among themselves for supporting the WTCER efforts and state department efforts in terms of input supply. During Rābi, 4 acres of land was put under gram and pea for utilization of *in situ* moisture for a second crop. The production of pea and gram gave a yield of 0.9 t/ha and 0.7 t/ha respectively.

Kharif intervention 2001

To break the prevailing monocrop of paddy and to introduce non-paddy crops as monocrops and intercrops was the main motive of Kharif intervention of 2001. The project is interacted with the farmers of Jhadbelda and convinced them to have intercrops and monocrops of non-paddy crops in upland soils. For this, 22 farmers were selected with total land holding of 8.5 acres of land and seeds of four different crops i.e. biri, arhar, groundnut and paddy along with other inputs like fertilizer, pesticides etc. were supplied to the farmers by WTCER. Due to heavy rain, only biri+groundnut intercrop sustained along with solo crops of biri, groundnut and arhar.The intercropping of groundnut and biri was done in 5 acres of land and solo crops of biri, groundnut and arhar were taken in the rest of 3.5 acres of land. Five different varieties of groundnut i.e. JL-24, AK-12-24, ICGS-11, GG-2, K-134 were raised for varietal demonstrations in 1.5 acres of land. Various improved agronomic practices like sowing across the slope, earthling up in the groundnut, intercultural operations, line sowing, etc. were practised in this current kharif. The yields of these crops are yet to receive. Fishery was also one of the important aspect of this season. One pond, which was maintained by four farmers, was selected and 3000 fingerlings were supplied to them by WTCER. All important activities of pisciculture was maintained carefully by the instructions of the specialist.

CONCLUSION

To bridge the missing links and to scientifically utilize watershed natural resources, WTCER attempted to develop the skill of farmers and demonstrate through participative effort the rational utilization of land, labour, and water. The agronomic intervention 2000 and 2001 sensitized the farmers to realize the watershed resource potentials and the improved agronomic practices like sowing across the slope, earthing up in groundnut, intercultural operation, linesowing, intercropping etc. The watershed capital base utilization for higher production and productivity, departmental links resulted in improved economic condition and resource conservation in the developed watershed. The watershed plus activities like link up of natural resource conservation programmes, socio-economic programmes and demonstrative activities can go a long way in increasing overall watershed productivity.

Watershed Management: Issues and Policies for 21st Century pp. 123-127

Constraints Analysis of Beneficiaries of Integrated Watershed Development Programme

S. Ganapathysankaran, M. Manoharan and M. Ramasubramanian

INTRODUCTION

In order to help the dryland farmers in maintaining soil health and the stability and sustainability of agricultural production, Government of India with World Bank financial support has introduced IWDP. Integrated watershed management focuses on combining improved farming practices with soil and water conservation and appropriate land use. Watershed approach refers to both the types of farming i.e. irrigated and rainfed (dryland) and widely differs in their approach but the concept remains more or less the same. It also includes both arable and non-arable land and therefore, needs equal emphasis in terms of improvement for maintaining ecological balance and sustainable development. The concept of integrated treatment of all lands on a watershed basis to improve the moisture retention capacity of soil and minimize the soil erosion by effectively checking the flow of losses run-off rain water is important specially in dryland/rainfed areas. The specific objectives of the watershed programme include promotion of soil and water conservation, optimal use of land and water resources (Singh, 1993).

Venkateshwarlu and Fletcher (1985) defined watershed in physical terms to the area lying above a given drainage point. It may cover less than a hectare or more than thousands of hectares depending upon the point of reference. In functional terms, the watershed programme tends to become synonyms with area development as it increases in size. The planning commission recognised the importance of agriculture in India and introduced the National Watershed Development Project for Rainfed Agriculture (NWDPRA) during the sixth five-year plan period. This project continued during the seventh and eighth five-year plans also. During the eighth plan period, several changes have been introduced in the watershed development project, including the mode of central assistance to the states for implementing the project. During eighth plan period watershed development programme has been extended to all the blocks having less than 30 per cent assured irrigation.

While commending the importance of watershed M. S. Swaminathan as quoted by Sangawan (1994) said "for the future welfare of dryland farmers and for the conservation and development of our land and water resources, scientific management of watersheds need to become a mass movement." There is need to develop suitable and efficient farm implements and tools or deep placement of fertilizers, suitable water harvesting method, storage and efficient use of stored water, integrated nutrient management in relation to most efficient cropping systems and alternative approaches to mitigate its effects and economics of new dryland technology and constraints in its adoption by the farmers. Therefore, the major objective of the watershed development project is to stabilise the watersheds in a cheap, sustainable and replicable fashion which can be applied to the whole country. (World Bank Report, 1988).

Usilampatty integrated watershed development programme

In Tamilnadu, integrated watershed development programme is being implemented from 1992-93. Great emphasis and importance were given for IWDP in the seventh five-year plan. In Madurai district, Usilampatty IWDP was selected with specific reasons which was considered as an ideal pilot watershed with World Bank assistance under the departmental programme namely Extension programme approach component. The selected IWDP is to be implemented for six years in thirty villages in an area of 3225 ha, with the financial outlay for Rs.23.989 lakhs. (Addl. Director of Agriculture & Watershed Co-ordinator 1992). The total area benefited in ten villages from 1992-93 to 1994-95 was 1125 ha with the financial expenditure of Rs 11.78 lakhs. The total number of farmers benefited were 1370.

Any programme of this magnitude of dry farming technology on watershed basis not only aims at increasing and stabilising yields in the rainfed areas but also involves the behavioral changes of farmers towards new technologies. To bring about behavioral change, we need to provide a constraint free atmosphere. So, study of constraints faced by beneficiaries is of very much useful in identifying prioritised constraints and provide the same to planners. Keeping this in mind, a constraint analysis of beneficiaries of Usilampatty watershed was taken with the following objectives.

- To identify the constraints faced by IWDP beneficiaries.
- To elicit suggestions from beneficiaries for effective functioning of programmes.

METHODOLOGY

Randomly selected as 120 beneficiaries from the watershed area, were the respondents for study. A well structured interview schedule was prepared to collect information on constraints and suggestions. This was pretested on a pilot basis. The finalised interview schedule was administered to the beneficiaries.

FINDINGS AND DISCUSSION

The constraints encountered by the beneficiaries of IWDP are furnished in Table 1. Among the major five categories of constraints, extension constraints was expressed by most of the respondents, followed by programme constraints, technological constraints, subsidy and financial constraints and marketing constraints.

Technological constraints

Majority of the respondents expressed that occurrence of drought was the most serious constraint in watershed programme, followed by seedlings transport mortality, unsuitability of supplied seed drill, non-availability of suitable implements, non availability of required inputs and lack of specific technology. The working of the supplied seed drill was not at the satisfactory level of the respondents as the sand particles block the free flow of seed. There was no provision for distributing the required suitable dryland implements other than the seed drill supplied under this programme. The major input namely fertilizer and improved sorghum seeds were not made available to the farmer at the required time. Lack of specific technology was also expressed as a constraint.

Table 1. Constraints encountered by respondents on IWDP based on mean score

S.No.	Particulars	Mean score	Rank
I	**Technological constraints**		
1	Occurrence of drought	2.98	I
2	Seedling transport mortality	2.73	II
3	Seed drill unsuitable for local condition	2.61	III
4	Lack of suitable implements	2.54	IV
5	Agricultural inputs not available	2.29	V
6	Lack of specific technology	1.92	VI
	Grand mean	2.51	
II	**Programme constraints**		
1	Poor quality in summer ploughing	2.71	I
2	Untimely Government orders	2.51	II
	Grand mean	2.61	
III	**Subsidy and finance constraints**		
1	Lack of credit facility	2.53	I
2	Delayed subsidy allotment	2.33	II
3	Lack of funds for developmental activities	2.30	III
4	Lack of agriculture and allied enterprises	2.27	IV
	Grand mean	2.63	
IV	**Extension constraints**		
1	Lack of qualified and exclusive extension personnel	3.38	I
2	Lack of printed materials like leaflets, folders	3.37	II
3	Lack of training	3.27	III
	Grand mean	3.34	
V	**Marketing constraints**		
1	Low market price for jowar and other produce	2.32	I
2	Lack of storage facility	2.31	II
3	Lack of inclusion of marketing in the programme	2.30	III
	Grand mean	2.31	

Programme constraints

Majority of the respondents expressed the poor quality of work done by the agro-service centres as constraints. The agro-service centres engage tractors on contract basis for summer ploughing. It was reported that the land was not ploughed according to the gradient. The other constraint expressed was the untimely receipt of the government orders for the implementation of the programme.

Subsidy and finance constraints

Among the various components under this category lack of credit facility, delayed subsidy allotment, lack of insufficient funds and absence of technologies and subsidy for agricultural and allied activities were expressed as constraints by respondents.

Extension constraints

The most important constraints in the implementation of the programme was the non availability of printed materials like leaflets, folders, etc about the programme and about advocated specific technologies. There was no training component in this programme for the beneficiaries. The respondents were also of the opinion that the absence of exclusive extension personnel to look after the programme was a constraint since regular extension personnel could not concentrate their work for this programme.

Marketing constraints

Low market price for jowar and other products was the major constraint under this category. Lack of storage facilities in the rural areas was also an equally serious constraint in the study area. Non-inclusion of marketing component in this programme, was also concerned as one of the constraint. Since many constraints were identified for the effective functioning of the IWDP the hypotheses that there exist no constraint for the beneficiaries in getting the project benefit was rejected.

Suggestions for effective functioning of IWDP

The perceived suggestions of the respondents for effective functioning of the programme was also collected. The expressed suggestion of the respondents are presented in Table 2. It could be observed from Table 2 that majority of the respondents (70.83) suggested that special emphasis should be given in the scheme for agriculture and allied enterprises also, like dairying, poultry, sheep rearing etc. which are suited to the area. This would create opportunities for more employment and additional income. Little more than two thirds of the respondents (68.33 per cent) suggested to organise exclusive training for IWDP beneficiaries. This would help the beneficiaries for acquiring better knowledge about the recommended technologies and also to create opportunities for farmer to farmer transfer of technology.

Supply of printed materials like leaflets, folders, etc. about the recommended technologies was suggested by 62.53 per cent of the respondents. The printed materials will be useful to the farmers for future reference. Half of the respondents (51.66 per cent) suggested to distribute useful dryland implements at the subsidised rate under the scheme. They expressed that only one implement (seed drill) was supplied to them under the scheme and that was also not much useful to the beneficiaries. Provision of credit facilities through banks was suggested by 46.66 per cent of respondents. Less than half of the farmers (45.83 per cent) stated that the agricultural produce purchased by the private traders fetches very low price during harvesting season. They also felt that regulated market is situated about 15 km away from the study area. The respondents expressed their opinion that there could be special arrangements under the programme to sell the produce at the village itself during harvesting season. Starting of agro-based industries was suggested by 39.16 per cent of respondents. This could create more employment opportunity as well as get reasonable market price for their farm produces.

Table 2. Suggestion for effective functioning of IWDP

S.No	Particulars	No	Per cent
1.	Training of IWDP beneficiaries	82	68.33
2.	Supply of printed materials like leaflets, folders etc.,	75	62.50
3.	Supply of useful dryland implements	62	51.66
4.	Marketing facilities	55	45.83
5.	Emphasis on agricultural and allied enterprises	85	70.83
6.	Starting of agro based industries	47	39.16
7.	Credit facilities	56	46.66

CONCLUSION

The study thus revealed that the major constraints for the IWDP farmers were lack of qualified extension personnel available to them, lack of printed materials, lack of training in farming, occurrence of drought, seedling mortality during transport, lack of appropriate implements and lack of credit facilities. Assisting in marketing, giving importance to the allied fields of agriculture also, and commencing agro-based industries were the points suggested by respondents for effective functioning of the scheme.

References

Sangawan, P. (1994). Watershed and dryland agriculture, *Journal of Rural Reconstruction*, 3(4): 62-65.
Singh, R. (1993). Role of dryland farming in food production, *Journal of Agricultural Marketing*, 3(2): 22-27.
Venkateshwarlu, R. and M. Fletcher (1985). Watershed based farming, IBH Publishers, New Delhi. 52-58.

Watershed Management: Issues and Policies for 21st Century pp. 128-135

Participatory Watershed Management for Land and Water Care: Planning, Impact Evaluation, Sustainability and Future

Alok K. Sikka

INTRODUCTION

The declining per capita land and fresh water availability coupled with soil erosion and land degradation in India are posing serious threat to environmental, food, social and economic security. About 50 per cent of the country's geographical area is afflicted with different forms of soil erosion and land degradation. This is posing the challenge of maintaining the soil health while obtaining more and more from less and less area and quality of land. Land degradation equally affects the quantity and quality of water availability as it influences hydrologic cycle. Unscientific land use also leads to the unsustainable exploitation of surface and ground waters. Land use decisions, therefore are also water use decisions. For sustainable development, land and water should, therefore, go together as land and water care constitutes the very foundation for building evergreen revolution in the country. It has also been amply demonstrated in India and elsewhere that it is impossible to envisage or implement sustainable solutions for land and water resource development and management without active and full participation of civil society. Development of land and water together with sustainable production system when confined to small natural drainage unit such as watershed leads to sustainable development.

Growth of WSM - As a New Paradigm

Watershed Management (WSM) has emerged as a new paradigm for planning, development and management of land, water and biomass resources with a focus on social and institutional aspects apart from bio-physical aspects following a participatory "bottom-up" approach. Watershed programmes are being implemented under an array of national schemes/programmes. Table 1 presents the growth and emergence of WSM in India. Some of the important ongoing WSM programmes include Integrated Wasteland Development Programme (IWDP); Drought Prone Area Programme (DPAP); Desert Development Programme (DDP); Employment Assurance Scheme (EAS); River Valley Project (RVP) and Flood Prone River (FPR); National Watershed Development Programme for Rainfed Agriculture (NWDPRA); Western Ghats Development Programme (WGDP); Hill Area Development Programme (HADP); International Programmes of DANIDA, GTZ; KfW, DFID (UK), SIDA, Japan, etc; National Bank for Agriculture & Rural Development (NABARD) Watershed Programmes; State funded WSM programmes and various foreign funded Afforestation Projects.

Table 1. Growth of nationally and internationally funded organized watershed management programmes in India

Schemes/Projects	Year of launch	Watershed Nos./area	Sponsoring Agencies
Research Watersheds	1956	42 Nos.	Min. of Agri., GOI/ CSWCRTI, ICAR
Soil Conservation in RVP Catchments	1961-62	29 catchments in 9 States	Min. of Agri., GOI
Operational Research Watersheds	1974	4 Nos.	CSWCRTI, ICAR
Watershed Management in Catchments of Flood Prone Rivers	1980-81	10 catchments in 8 States	Min. of Agri., GOI
Model Watersheds	1983	47 Nos.	CSWCRTI & CRIDA, ICAR
Watershed Development in Rainfed Areas	1984	28 Nos. (3.47 lakh ha)	World Bank (A.P., Karnataka, M.P. & M.S.)
Watershed Development in Ravines Area	1987	0.62 lakh ha	EEC
Drought Prone Area Programme (DPAP)	1987*	91 districts 615 blocks	MRAE, GOI
Desert Development Programme (DDP)	1987*	21 districts 131 blocks	MRAE, GOI
Western Ghats Development Programme (WGDP)	1987* 5 States	158 blocks Commission	Union Planning
Indo-German Watershed Project	1990-91	Maharashtra	Germany
Indo-German Bilateral Project	1990-91	Monitoring	Germany
NWDPRA	1991	2497 Nos.	Min. of Agri., GOI
IWDP (Hills & Plains)	1991	1.12 lakh ha	World Bank
Comprehensive Watershed Development Project	1991	1.13 lakh ha	DANIDA (Karnataka, Tamil Nadu and Orissa)
Rel Majra Watershed Project	1991	1 No.	CSWCRTI, ICAR/ Min. of Env., GOI
Doon Valley Project	1993	1.72 lakh	EEC
Integrated Wasteland Development Project (IWDP)	1994	25 States	MRAE, GOI
Indo-Swiss Participatory Watershed Development	1995	0..35 lakh	Swiss Government
Attapadi Wasteland Comprehensive Environment Conservation Project	1996	507 Km^2	OECF, Japan

MRAE - Ministry of Rural Areas & Employment
*Since Programmes adopted watershed approach
Source: Sikka & Samra (2000)

Government of India has accorded the highest priority to the holistic and sustainable development of rainfed areas, wastelands, drought prone areas, hilly areas and degraded areas through integrated watershed management approach. The current strategy of various ongoing national, bilateral and internationally aided projects of WSM is based on the concept of conservation of rainwater, groundwater recharge, promotion of diversified and integrated farming systems approach, management of common property resources and augmenting family income. Most of the WSM programmes envisage sustainable institutional arrangements at watershed level in the form of local level people institution as an integral component of these projects for promoting participation of civil societies and ensuring sustainability. Though the WSM programmes are being administered by various Ministries such as Agriculture, Rural Development, Environment & Forest and Water Resources, but they all converge to overall objective of sustainable management of land, water and vegetation resources with multiple benefits. Like the large number of WSM programmes, different Government agencies and NGOs are involved in the implementation of these programmes.

PLANNING AND COMMUNITY ORGANIZATION

Bench mark survey

Generally a workable size of 500 to 1000 ha watershed is preferred. Base line/bench mark surveys for physiography, climate, soils, landuse, vegetation, hydrology and socio-economic data are conducted and analysed for preparing watershed plan. Apart from surveys, this information is also supplemented from Participatory Rural Appraisal (PRA) exercise. Multi-data approach involving extraction of data from various conventional sources and new techniques/sources such as remote sensing together with limited ground truth and field survey could be ideally employed in watershed development programmes. The spatial and non-spatial data can be effectively utilized through GIS for watershed planning.

Community organization

Participatory approach is more pertinent in the planning and development of watershed management (WSM) programmes, because it is basically the people's programme and the government agency should participate in that as a facilitator. The concept of participatory process is easy to advocate but difficult in practice. This needs active involvement and dedication. This also demands some psycological and functional adjustments by the government functionaries. Involvement of social organizations, NGOs and VOs is a must to organize and mobilize community support and make this as people's programme and national movement.

This whole process may involve following steps.

Meetings with the Watershed Community: Conduct regular meetings with the farmers in villages to clearly explain the purpose of the programmes, get their feed backs, develop contacts, gather Indigenous Technological Knowledge (ITK), win their confidence and involvement.

Problems Appraisal and Plan Formulation: Conduct Participatory Rural Appraisal (PRA) or Rapid Rural Appraisal (RRA) exercises in the village to gather information, diagnose their problems, needs and priorities to arrive at a common outline of watershed development plan. During the PRA exercise, activities/works to be taken up under Entry Point Activities for rapport and confidence building may also be identified.

External Organisation as a Facilitator: External implementing agency, governmental or NGO, to provide technical support, act as a facilitator to help them in programme implementation and community organization.

Formation of Local Level People Institution: Local level people institutions such as Watershed Association/Water Users Society, Self Help Groups (SHGs), User Groups (UGs) are formed, by laws framed and society is registered for day to day running, management and distribution of benefits and create working capital through revenue generation, people contribution, etc. for repair and maintenance of the works. This will create a self sustaining local level institution to take over the activities after withdrawal of Project.

The process of community organization and plan preparation may proceed simultaneously. The plan should be prepared based on suitability of land for different uses, resource availability, needs of the people incorporating results and perceptions of PRA exercise with community involvement to make it socially acceptable Watershed Plan. However, in practise, this aspect has received less attention and needs to be strengthened.

Components of a typical plan may include the following

Protection and conservation measures: Majority of the project experiences suggest that a blend of structural and vegetative measures is a better option i.e. bio-engineering measures. These include the measures/structures that provide the very base or foundation for production system. This will include all the measures/structures including *in-situ* soil and moisture conservation measures like bunding, leveling, terracing and vegetative barriers, water harvesting structures such as ponds, nalla bunds, small dams, percolation ponds etc., drainage line treatments with engineering structures and vegetation for checking land degradation and conserving water; and repair, restoration/strengthening of existing structures for sustained benefits from previous investments made, if any.

Production measures: These include the activities that are required to make the efficient use of conserved soil and water resources in producing user products such as food, fodder, fuel, fruit, fibre, milk etc. on sustained basis. These may include improved crop cultivation and management practices, afforestation, alternate land use systems like agri-horticulture, horticulture, agro-forestry, pasture development with or without plantation crops, silvipastoral etc. and cultivation/raising of industrial, medicinal and aromatic grasses and plants for providing alternate livelihood support system, development of livestock, dairying, poultry, sericulture, fisheries and other essential income supporting activities.

Convergence approach

Watershed management is being a single window, integrated area development programme. Integrated watershed management cannot perhaps be achieved just by following integration of resources using multidisciplinary approach with the funding or support provided alone under any watershed programme line IWDP/DPAP etc. This may also involve harmonized use of resources available from other on going/existing sectoral and development schemes in the area/district. Such resources can be dovetailed with the watershed programmes that will not only help in useful convergence of various schemes and programmes for overall development of the area programmes but also in effective monitoring. Some of these sectors may include education, health, sanitation, drinking water, roads, etc and most of these can also be dovetailed with the entry point activities.

Monitoring and impact assessment

Watershed monitoring and impact evaluation is essentially a part of the comprehensive process of watershed management. Inadequate monitoring and impact assessment of various watershed programmes is a major concern. Watershed development projects affect social, economic and environmental activities. Traditionally, completion of activities and physical and financial targets rather than results achieved and their impact on biophysical and socio-economic aspects are monitored, while there is a need to monitor and evaluate the impact of these activities on biophysical, socio-economic and environmental aspects (Sikka *et al.*, 2000). Such projects may generate productive (or direct), protective/reclamative, ecological and employment generation benefits. The monitoring and evaluation in a watershed will thus not only help in analysing impacts of current and future activities but also plan corrective measures after mid-term evaluation. This will also help in monitoring and evaluating the performance of various agencies involved at different levels of watershed project.

Impact indicators/indices

The impact indicators should be (a) simple and easy to measure, (b) reliable and replicable, (c) sensitive to changes, (d) simple to compute and (e) easy to understand and interpret.

There are no single definitive indicators or measures and indices for overall impact measurement/assessment in view of complexities and diverse/multiple objectives of such watershed development projects. In view of this, a set of quantitative and qualitative simple surrogate measures, indicators and indices may be evolved and/or used to assess changes as a result of watershed impact. Biophysical/technical, socio-cultural, economic, community participation, institutional and sustainability indicators are developed based on their applicability to objectives of the project. These indicators are developed based on consultations with resource persons, project staff, farmers representative and other stake holders.

Since most of the time bench mark data is not available or inadequate, participatory impact evaluation may be adopted. The evaluation team may largely use PRA techniques for information/data collection during field visits. Sample surveys, field observations, discussion with community members and community organizers and line department staff are used for data/information collection.

Some of the methods that can be used are as follows:

- Transect
- Direct Observation
- Semi-structured interview technique
- Resource mapping
- Scoring, ranking and matrices
- Focus group discussion
- Self assessment through scoring technique
- Sustainability analysis
- Garrett ranking

- SWOT analysis
- Perusal of Records/Registers/Account books maintained by farmers

Sustainability

Programme is sustainable that continues to operate after withdrawal of monetary and/or technical supports of the project. In WSM, an *in-situ* development programme, the participation of local community is "sin qua non", since the different works/activities implemented on private and common lands have to be sustained by the owner of the land in addition to maintaining collectively the works on common lands.

This emphasizes the significance of "participation" for sustainability in WSM programmes.

Collective participation of people imperative due to inter-dependence of beneficiaries in different reaches and spill over of benefits. Transfer of responsibility of running within their communities is a key requisite for ensuring the sustainability.

What is required for sustainability?

- People's participation right from pre-planning stage.
- Community empowerment including integration of women in project activities.
- Local level people's institution.
- Capacity building of the institution and its actors.
- Resource generation or capital for sustenance.
- Belief of deriving tangible benefits from development of PPRs & CPRs.
- Linkages with credit and input institutions and technical/scientific support organizations.
- Suitable withdrawal strategy.
- All these would mean integration of Social Resource Management with Natural Resource Management for achieving sustainable results giving emphasis to protection/ conservation, production measures and social and livelihood support activities.
- Role of project and institutional partners

As a provider of selected services and inputs *e.g.*, capacity building, technical assistance, micro-credit, materials, etc.

As a facilitator and coordinator.

CONCLUSION AND FUTURE CHALLENGES

The encouraging results and the experience of various integrated watershed management programmes in the country strongly suggest their importance and technical, economical, social and environmental viability for sustainable development through people's participation, following a "bottom up" approach with a blend of "top down" input. Participatory WSM helps in local innovations and Indigenous Technological Knowledge (ITK) become part of WSM plans through participatory planning and implementation. WSM needs a low level of technology than major and medium irrigation projects, but a large amount of planning, survey and community participation at different levels to make the programme need based, socially acceptable, eco-friendly and sustainable.

The key features of Integrated Watershed Management should involve equal attention to demand as well as supply management; active and extensive involvement of all stake holders (local community, NGOs and government) in the planning and decision making process and implementation; integration of a broad array of biophysical and social objectives and the factors affecting them; creation of local level People's Institution (Watershed Association/ Water Users Association, Users Group) and their capacity building to undertake; implementation, management and maintenance of common property resources and assets created for sustainable water resource management.

FUTURE CHALLENGES

- Though the growth of watershed programmes indicates divergence in perception, planning and implementation but they all converge to overall objective of sustainable resource management with multiple benefits. However, the programmes still remain piece-meal and lack strict adherence to integrated approach. Our future endeavours have to bring in more coordination and integration by sectoral departments by adopting a participatory approach first at their departmental levels.
- The bottom up approach with a blend of top down approach is being recognized as a participatory watershed management approach through community empowerment. This needs active involvement and dedication and also demands some psychological, attitudinal and functional adjustments by the government functionaries. This participatory watershed management demands openness and transparency.
- Technology for *in-situ* moisture conservation, micro level network of water harvesting systems, combined management of perennial and annual plant cover management through agro-forestry models should be extensively demonstrated and adopted.
- Water resource management rather than development is the major challenge. It therefore calls for a greater attention towards both supply and demand management of water resources through renovation and rejuvenation of existing/abandoned structures/systems, conservation education and training, institutional financing for improved water management methods/practices including micro-irrigation, promotion of viable, cost effective local rainwater harvesting, rainwater conservation and other groundwater augmentation measures.
- Biophysical aspects of watershed research are relatively better attempted and understood than socio-economic, community participation, capacity building, conflict resolution, gender issues, etc.
- Scientific data about the on-site effects is relatively richer as compared to off-site effects of flooding, ground water recharge, environmental externalities and hydro-ecological aspects.
- Comparative performance of grasses and woody species including value added aromatic and medicinal plants/grasses for conservation under varied watershed and socio-economic conditions needs more attention.
- Some other important researchable issues for future may include stress on sub-surface hydrology of watersheds, hydro-ecology, impact of WSM on ground water recharge, use of modern tools like GIS and remote sensing, Decision Support System, social aspects of watershed management and environmental/resource economics.
- Monitoring and evaluation should be made a compulsory component right from beginning and suitable biophysical, socio-economic, participatory and environmental indicators be used for impact assessment.

- Formation of local watershed level people instiiutions to carry out various works and activities and constitution of watershed committees be made essential to develop local institutions. Contributions and revenue generation from community assets can be used to build up working capital. This will create a self sustaining local institution to take over the activities after withdrawal of project.
- Looking at the size and dimensions of different watershed programmes in the country, an increased emphasis should be given to train manpower in watershed management through short term training programmes including sensitization of higher level officers.
- There is a strong need to bring in ethical change in the minds of the people to sow seeds of ethics regarding the use and importance of water so that it will no longer be looked upon as a resource which is in plenty and valueless. This will play a much more important role towards sustained water harvesting and conservation rather than merely following old administrative techniques of laws and regulations alone.
- Social mobilization, more decentralisation and community empowerment needed to be explicitly included in government policies for success of community based watershed programmes.
- Process of social mobilisation may require a re-look at the role of government, institutional and financial mechanism, legal rights and technical needs. NGOs may be involved at the local levels to act as social mobilisers for educating and encouraging communities and actively involving them in WSM programmes.
- It is suggested that basin wise planning and management of water resources should integrate participatory watershed management programmes for sustainable development, and small and micro water harvesting systems made integral part of the water resource planning, development and management at the national/regional levels.
- There is now a challenge to ensure the sustainability of WSM programmes for which appropriate withdrawal and post-project strategies and mechanisms have to be evolved and demonstrated.

References

Sikka, A. K. and J. S. Samra (2000). Participatory Watershed Management—A New Paradigm for Integrated Water Resource Management. Proc. Inter. Conference on Integrated Water Resources Management for Sustainable Development, New Delhi.

Sikka, A. K. Subhash Chand, M. Madhu and J. S. Samra (2000). Report on "Evaluation Study of DPAP Watersheds in Coimbatore District".

Watershed Management: Issues and Policies for 21st Century pp. 136-142

Participatory Watershed Management, Coimbatore: An Experience

M. Jayabalakrishnan

INTRODUCTION

With great noble cause of overcoming drought by tapping all underground resources, balanced eco-friendly atmosphere to control soil erosion and to conserve soil fertility and moisture, the Govt of India strived their best for more than 2 decades by implementing schemes like, Integrated Wasteland Development Programme (IWDP) Drought Prone Area Programme (DPAP) and Desert Development Programme (DDP). But the outcome of result was not satisfactory and not to the expected level. The reasons attributed may be so many, but the striking cause identified was the non-participation of the rural people and non involvement of voluntary organization in implementation of the projects.

Because of these, the assets and construction works built were ruined and the very cause for which the project was undertaken went in vain. Based on the past experiences a historical decision was taken and a turning point has occurred in implementing such projects, say the "Participatory Watershed Approach" from 1995.

Now a days none of the watershed project is carried out without the actual participation of the beneficiaries and hence all the earthen bunds, stone bunds, contour bunds, vegetative bunds and check dams are scrupulously maintained as the assets of their hamlets, villages or panchayats or Blocks as a whole and the benefits derived from them are innumerable.

In Coimbatore District, Annur, Avinashi, Palladam, Sulur and Tiruppur Blocks were identified as Drought Prone Areas. During 1995-96, 15 watersheds were delineated in the above 5 blocks and watershed development projects were carried out by Water Technology Center of Tamilnadu Agricultural University, Coimbatore. During 1998-99, under "Integrated Wasteland Development Programme (IWDP) 9058 hectares of Wastelands were selected in Pongalur Block by District Rural Development Agency, Coimbatore and Community Organisation activities are beginning. In the same financial year under DPAP scheme 10 additional watersheds have been identified and preliminary works have also been taken up. The above said projects are implemented by DRDA (PIA), Coimbatore through the Co-ordination of Voluntary Organization "COODU", a Coimbatore based NGO.

All the Entry Point Activities, Community Organisation and training packages to all watershed Associations and Committee members, Secretary and volunteers, User Groups and Self Help Groups, are being co-ordinated by the NGOs for the past three years. We have the privilege to say a few words on the success stories of people's participation.

ENCOURAGING PEOPLE'S PARTICIPATION

Just drafting proposals, implementing schemes will never be fully successful but for the involvement of realizing the importance of people's participation. The Coimbatore DRDA, along with a few NGOs have taken a leap forward in encouraging people's participation. Our road to this involves a series of non conventional methods (Formal as well as informal discussions), exposure visits and various trainings with watershed development teams, village level institutions etc. We present below a few techniques of our approach.

Community organisation activities

The DRDA Coimbatore is one of the pioneer in the state of Tamilnadu in approaching and co- ordinating all social and Community Organisation in Watershed Development Projects. On identification of watershed area, the Watershed Development, Voluntary Organisation personnel visit the hamlets, revenue villages pertaining to that watershed area and conduct formal and informal meetings with different village people like farmers, artisans, labourers, elders and identify the opinion leaders. Utmost care is taken to involve all the Watershed Community in the project.

Existing Village Level Institutions were identified and the steps, modalities precautions to be followed were discussed with them. Further they are motivated to actively participate in these projects by listing out the beneficiaries that will be derived out of such schemes.

To list out a few methods of approach that are followed by us are as follows :

- Informal discussion with watershed community.
- Formal meetings with existing village level institution.
- Formation and maintenance of Self Help Groups (SHGs) .
- Formation and maintaining of User Groups.
- Awareness campaigns
- Participatory Rural Appraisal Exercise.
- Support to the Existing Village Level Institutions.
- Watershed Institutional arrangement etc.

Convergence approach

Watershed management is a single window, integrated, sustainable area development programme following participatory approach. In Panchayat Raj System, this kind of watershed approach fits well through community empowerment. In a real sense, this involves integrated approach of harmonized use of natural, human and animal resources for sustainable area development. Watershed approach is a very practical unit for integration because it takes the various schemes to the grass root areas. A village Panchayat, which is so far considered a unit for development, has more than two watersheds. Fulfilment of requirements to both the inhabitants of the watersheds as well as for the maintenance of natural resources, the watershed is a practicable approach.

With this in view, the DRDA, Coimbatore as PIA for IWDP and DPAP watersheds has made an earnest attempt in this direction, right from beginning of the project. The steps taken by the DRDA, Coimbatore and progress made so far in harmonizing and integrating various schemes and programmes under IWDP/DPAP watersheds for their convergence towards sustainable area development are presented and discussed. Convergence does not mean only integrating various sectional schemes in one area but also through the effective monitoring of available resources in certain fields like education, health, drinking water and as a whole overseeing the proper implementation of the policy guidelines of the Central and State Governments in the targeted areas.

Eye care programme

All the 18,000 habitants in the 10 DPAP watersheds, old, young and school children etc. were covered under a comprehensive eye programme in which 51 people have been identified for surgery which was conducted free of cost by Sankara Eye Society, an NGO and 19 school children were identified and spectacles provided to them under DPAP project administrative funds. All the school children were given 10 Vitamin 'A' tablets and deworming medicines. This programme has taken the DPAP scheme to the hearts of the inhabitants in the respective watersheds. The field organisation done by COODU-NGO, Coimbatore.

Central rural sanitation programme

The most excluded aspect in any development programme is the provision of a toilet to the households to upgrade the village sanitation. 1000 CRSP latrines were allocated and now these watersheds can proudly be declared as a watershed in which every household possesses a latrine. This major task was achieved through proper Information Education Communication, motivating the Panchayat Presidents, NGOs, block staff, etc.

Supply of improved toolkits to rural artisans

Traditional artisans like barber, blacksmith, cobblers, goldsmiths and carpenters etc. were identified and as a measure of upgrading their technology, 89 artisans were given toolkits in all the 10 DPAP watershed areas.

TAMPCOL as an institution for marketing

Two SHGs consisting of women belonging to SC and backward classes were identified. Each possessing an area of 0.75 acres. A project was submitted to the Chairman-cum-Managing Director, TAMPCOL at Chennai for cultivation of herbal plants, provision of their technical expertise and very importantly a buy-back facility by TAMPCOL. The total project was worked out at Rs. 50,000/-. These 4-month crops utilize less water and give a return of Rs. 12,000/- to the group. This small initiative by DRDA and COODU is spreading like wild fire and many farmers in the neighboring areas have responded positively to take up this activity in their fields in a big way. This particular initiative takes into account community organization as well as marketing tie-up thereby benefiting the rural poor.

Conflict and Compromise: Elavanthi Experience

During 1997-98 in the whole of Tamil Nadu, a secured and largest project under IWDP was sanctioned and 9058 hectares of wasteland were selected in and around Vattamalaikarai river basin of Ponglur block, comprising six Village Panchayats. The whole area was divided into 18 watersheds and preliminary steps were initiated out of 18 watersheds, three were identified in Elavanthivadugapalayam Panchayat. Among the three based on the extent of area, one watershed with 450 hectares in Elavanthi, Karasamadai and Akranam, the second one with 550 hectares in Vadugapalayam, Mosuvampalayam and Kulathupalayam and the third with 440 hectares at Krisipalayam village alone, were divided based on the drainage line. The watershed area and profile is formed based on the extent of area only, not on any other criteria. Based on the area, project fund to the tune of Rs 20 lakhs were sanctioned to Krisipalayam village watershed and Rs. 20 lakhs were sanctioned to Elavanthi and other villages, the second one without knowing the project scopes, some persons belonging to Elavanthi Watershed raised problems and prevented implementation of work throwing false propaganda of funds misappropriation.

Even after compromising meetings for more than 3 months, the people refused to co-operate in the project work. On investigation, it was unfolded that all the problems were due to the Panchayat president's non-co-operation in providing the basic needs of the village. Sincere efforts of WDT members, officials of DRDA and NGOs could not pacify them. It was due to the opposing groups of the president. Then the issue was taken to the Sub-Collector of Tiruppur who in turn deputed the Tahsildar. The revenue officials convened a meeting and warned the problem creating people and explained in detail the future benefits that may be derived to the village and the area. Now, based on the promise and assurance given by the local people, the project is being implemented.

Village folk music teams

Five village folk artisans SHGs consisting of SC beneficiaries have been formed and registered under the Societies Act and a separate account has been opened in the post office. These groups have been given new instruments under the community organization funds for improving their income at a cost of Rs. 12,000/- per group. These groups were able to replace the old dilapidated instruments with new ones.

Red sanders

Dr. Sadhasivam, Scientist and Environmentalist, who is also a consultant at Periyar Maniammal College, Vallam, Tanjore had donated 200 red sander seedlings, which would cost around Rs. 15.00 lakhs after 20 years. These red sanders have been distributed in the watershed areas as a component of homestead plantation to the SCs and other castes.

Conflict and Compromise: Moopripalayam Experience

During 1999-2000 under DPAP 15 watersheds were delineated in Moopripalayam town Panchayat of Sulur Block. Agricultural Engineering Department, Tiruppur is the project implementing agency. One thousand hectares were selected comprising two watersheds and initiation for the formation of community was taken up. Co-ordinating all WDT members, Voluntary orgainsation personnel PIA and DRDA, Coimbatore, experts and the Panchayat members visited all interior hamlets and revenue villages and explained in details the project works to be implemented. The participation of rural mass and their promise for co-ordination and co-operation was nearly appreciable. Awareness campaign and participatory rural appraisal (PRA) were conducted and the involvement of the people was also felt by their discussion and suggestion. We felt the difference of opinion between two groups of people. (*i.e.* two political parties) during the formation of watershed association and committee and final decision could not be arrived at. While all the project works were in progress in other watersheds we could not undertake even initial steps on Moopripalayam watershed during the past four months. It is our duty to share the problems faced by us during the project work.

The local ruling party members lodged a complaint with MLA and in turn, we, all the department people and voluntary organisation personnel explained the happenings to the MLA and he was convinced. The president of the Panchayat, belonged to the opposite party and established his voice and power during every meeting and campaign. We could not come to any conclusion. Then the issue was taken to the District Collector, by some well wishers of that village. It was well explained to the gathering that if such problems continued in implementation of the projects, the scheme will be shifted to trouble free areas where people will cooperate. Thus the issue was amicably settled based on the assurances and promises given by Moopripalayam watershed people and the scheme is in opertion smoothly.

Training

Meticulous training will always complement successful watershed approach. Continuous training was imparted at various levels.

Training to the watershed development teams

WDT members have already been trained in integrated projects by NGO's and have been placed at the DRDA as PIA. They are further tuned through technical training at CSWCRTI, Ooty as well as SIRD. The services of these members are very much essential in watershed development as they endeavour to make the people involve in each and every activity of watershed development.

Training to the office bearers of watershed association

The office bearers of Watershed Association as registered body and they are given training for better implementation of projects in accordance with the socio-economic conditions and nature of the watershed areas. They are also trained to encourage people's participation and to create awareness on the guidelines to local members as well as to the public. They are further trained to connect the beneficiaries as members and to create awareness on the mission and vision of the Association.

Training to watershed secretaries and volunteers

One secretary and one volunteer are appointed for each watershed area. They are given training on the management of watershed associations, group activities. SHGs, User Groups are as well as on maintenance of records. The secretaries and volunteers are further trained on identification and selection of sites in watershed, survey of disabled and affected people, conducting socio-economic surveys, to draft monthly action plans in accordance with the Government of India Guidelines and guidance by the NGO's and Watershed Development Team.

Training to watershed committees

This Committee compresses of 12 to 14 members, the members inclusive of farmers, Panchayat members, SHG, rural artisans, women folk, shepherd etc. One member is inducted into this committee on behalf of the DRDA through NGO. This committee is a very important one and is responsible for the approval of schemes. The committee will also monitor whether the benefits reach the beneficiaries. Training to this committee is given on all the components of watershed managements. This committee collects the community contributions and utilizes the funds for post-project management. This committee is also trained to maintain the financial outlay of scheme works.

Training to user groups

In each watershed area, User Groups are formed comprising of about 20 farmers who are direct group beneficiaries. These groups are trained on watershed concepts and on recent technology in farming systems and involvement in scheme works. Following are the line departments institutions etc. which impart training in Agriculture, Agricultural Engineering, Horticulture, Sericulture, Social Forestry: TNAU, CSWCTRI - Ooty, Animal Husbandry, University Training and Research Centre, Social Welfare Department, District Registrar, NGOs, Women Development Corporation, Health, Pollution, Control Board, AIMS Research, Chennai, Options for Development, Chennai, Gandhi Gram Rural University, Dindigul, etc.

Training to watershed SHGs

About 150 women SHG have been formed by COODU NGO in all watersheds under Tamil Nadu Government Bangaru Ammiyar Ninaivu Mahalir Thittam. These SHGs are given series of training on techniques of watershed approach and participate in implementation of scheme works, also training by woman development corporation through NGO.

Exposure visits to successful watershed areas

In each watershed the farmers and UG, SHG members are given exposure visit on already developed watershed project areas, by taking them and arranging interaction with that watershed community, so that they can develop better ideas in implementing their own watershed project.

Table 1. Details of exposure visit

Sl.No.	Scheme	Place of visit	Number of participants
1.	DPAP	Raligan Sidhi,Pune, Maharastra	50
2.	DPAP/IWDP	Thanjur, Maduranthagam, Salem.	120
3.	DPAP/IWDP	Dindugul, Madurai, Kodaikanal, Trichi.	55
4.	DPAP/IWDP	Sathiyamangalam, Erode Thalavadi	55
5.	DPAP	Dharmapuri, Husur, Salem, Namakkal	60

Self help group activities-village folk music team

The DRDA and COODU identified the scheduled caste people (a particular section) well versed in traditional music. These people were supplied instruments called " Thidumam" through community organisation component. Now these village folks earn their bread by performing in festivals, functions and weddings, not only in Coimbatore but also in nearby districts like Erode, Salem, Namakkal etc. They even get loan from Association for their betterment.

Self help group - women SHG

More than 150 women SHGs were formed and systematically motivated in watershed areas through COODU to our surprise. The total savings of the women SHGs have crossed Rs. 50 lakhs. Thirty SHGs got revolving fund from SGSY and banks for economic activities.

CONCLUSION

One can understand the problems and conflicts that were faced by the implementing agencies like DRDA, NGOs and WDT members. But at the same time, the experience gained out of the different situations, will go a long way in helping the individuals fit to face some other bigger problems, smoothly. Above all, we derive some sort of mental peace having given with golden opportunity to work and mingle with rural mass, villagers, farmers, farming labourers, artisans, leaders and well wishers of watershed areas and to see the smiling faces on completion and at the time of deriving benefits out of the project work.

Watershed Management: Issues and Policies for 21st Century pp. 143-147

Participatory Watershed Management and Development in Tamil Nadu

S. A. Jayakumar

INTRODUCTION

The basic principles of management may be universal like the seven colours of a rainbow. But the blend of colours required for painting different subjects is different. The blend suitable for mountain peaks and pastures will not do for portraits and people. Likewise, the amalgam of the basic principles of management consonant with one setting may not suit a different setting. Further, the selection of right amalgam, though necessary is not enough. The skill of application is equally important. The scientific component of management is indeed universal. But that does not constitute total management, which is only partly science and mainly art. What do we mean by scientific maintenance? It is simply this. Application of the principles of management to the function of maintenance. The fundamental elements of management are plan, organize, motivate and monitor. The importance of effective maintenance is applicable to all the creations of man as well as his interfaces with nature—be it dams, canals, watersheds etc. Today, a holistic "watershed" focus for water management strategies is emerging. The concept is not new, but the key to broadly conceptualizing and implementing integrated water management plans has been elusive. Integrated water management means putting all the pieces together. Social, environmental and technical aspects must be considered. There is a need to guide water management decision-making flexible, holistic and environmentally sound. Water resources professionals must be prepared to offer credible guidance to those who meet it, at the right time and in a comprehensible form. Watershed as a basic unit and systematic planning for land development, soil and water conservation, water harvesting, water management and the like are crucial for successful plant establishment, growth and drinking water problems. The non-governmental organizations (NGOs) Government Departments and other agencies involved more particularly in projects on land development, soil and water conservation, water harvesting and afforestation. The degree of success in participatory programme entirely depends upon the extent of people's involvement and commitment. People's wholehearted participation is one among the crucial inputs for success.

WATERSHED

Every watercourse depends upon a catchment area. This term defines the area from which the water in a particular section of a river is obtained. The demarcation lines between catchments or drainage areas are watersheds. Usually the watershed is a line joining the highest points round a river basin. Thus at every point along this line there is a slope down into the central catchments area. The area has the shape of a basin inclined in the overall direction of flow, into which it gathers water flowing from the watershed.

The lines of highest points that we have defined has its most obvious effect as a watershed in distinguishing surface run-off from precipitation. Potentially, the catchments area and the precipitation area are identical. If, however, rainfall first penetrates the soil before it flows in a stream channel, then catchments area and precipitation area no longer necessarily coincide. Now the catchments area can also include regions from which water flows through underground channels into the individual rivers. It often happens that on the peripheries of their catchments areas, rivers either draw water underground from neighbouring precipitation areas or lose it from their own.

There are different orders of rivers from the many-branched tributaries to the main river, so it is possible to distinguish between watersheds of different order, from those of purely local significance up to continental watersheds.

The watershed for surface run-off is not always distinct. It is most apparent in mountain regions, where mountain crests and ridges separate river catchments. On the other hand, on flat ground there may be no distinct separation of this kind, and a watercourse can join up with more than one river system. This is known as bifurcation. Bifurcations can cause water to flow not only areas of drainage. Bifurcations are, however, exceptional. Where they exist, they have little impact on the discharge of the principal drainage arteries. As a rule, river basins are independent of one another. Even in flat alluvial plains where rivers often divide into several branches, new watersheds develop as a result of obstruction and channel movement and by general raising of the level, particularly between the main drainage arteries and individual, separated river branches. Water bodies and lotic systems in particular are not isolated entities, they impinge on the atmosphere and the land. Across these boundaries there is a constant movement of materials and energy. Across the air-water interface gases, water dissolved nutrients and particles in rain all enter the river. The land-water interface is the main source of material from weathered and eroded rocks, and from dead and decaying plants and animals. Other factors such as transpiration rate of trees, soil and rock type, and on land use all effect water quality. Therefore, it is necessary to study and manage the catchments area or watershed as a whole. This facilitates integrated management of the water cycle within that area.

WATERSHED MANAGEMENT AND DEVELOPMENT

Watershed management is an integration of technology within the natural boundaries of a drainage area for optimum development of land, water and plant resources to meet the basic minimum needs of the people in a sustained manner. The poor in the rural areas who are struggling for survival cannot be expected to pay heed to the conservation strategy unless their daily needs of food, fiber and fuel are met. A still more urgent need is for assured and full employment for all. Integration of many scattered programmes of soil conservation, afforestation, minor irrigation and other development activities into well prepared micro watershed projects based on a study of climate, land, water and plant resources in one hand and man, animal resources on the other, offers hope for brining about sustained natural resources development.

India being a monsoonic country, it rains much and a large quantity of run off takes place removing the top fertile soils and the crops also get damaged. Similarly when there are droughts, crops do not get adequate moisture since there is no proper and well-planned watershed management practices. Floods and droughts are in fact the two sides of the same coin of neglect of land. The basic works needed are : Land shaping to store and use the water; drainage to remove the excess water without affecting the soil. This will help in both extreme situations. Therefore, the priority in watershed management practices are to be given for these two: Water development and water harvesting. If these basic needs in dry lands are met, high crop yields can be effected without many problems.

The area can be divided into two distinct categories to tackle the problems. They are where the rainfall is more than 750 mm and where the rainfall is less than 750 mm. This can be again divided into permeable soil (red, gravely soils) and impermeable soils (clay, black cotton soils). The management practices may be different for different rainfall and soil conditions. The different objectives of watershed management programme are: To control damaging run off; To manage and utilize run off for useful purposes; To control erosion and effect reduction with sediment production; To moderate floods in the downstream area; To enhance groundwater storage wherever applicable and appropriate use of land resources in the watershed and thus developing forest and fodder resources. The problems of each watershed are different and programme for their management could also be different having emphasis on the problems of particular watershed. The important factors which affect the watershed behavior and which need to be studied for developing a management programme are: Size and shape of the Watershed; Topography; Soil and their characteristics; Precipitation; Land use and Vegetation cover.

Apportioning of groundwater assessment from watershed to development unit: As far as groundwater is concerned, the assessment unit is a watershed, there is a need to convert the groundwater assessment in terms of an administrative unit such as block/taluka/mandal. This may be done as follows: A block may comprise of one or more watersheds in part on full. First, the groundwater assessment in the sub areas, non-command and command areas of the watershed may be converted into depth unit (mm) by dividing the annual recharge by the respective area. The contribution of the sub area of the watershed to the block is now calculated by multiplying this depth with the area in the block occupied by this sub area. This procedure must be followed to calculate the contribution from all the sub areas of watersheds occurring in the block, to work out the total groundwater resource of the block. The total groundwater resources of the block should be presented separately for each type of sub area namely non-command areas, command areas and saline/brackish water areas as in the case of the individual watershed. After assessing the watershed it is very useful to classify the area for utilizing with proper planning, cultivation and other drinking water problems.

The concept of treating the land based on watershed basis is a recent one. Many development works like afforestation (tree planting), soil conservation measures, water harvesting techniques and rural water supply programmes were implemented in need based areas in a haphazard and compartmental way. These works were not done in an integrated manner. The result was that whatever works were completed did not last long since, rainfall/run off and erosion were not controlled from the top to bottom in the watershed. Hence in order to get the benefit of various conservation and welfare measures, it is now agreed in principle by all, to take up all the measures (works) in an integrated manner on watershed basis. The result was very spectacular. Government and voluntary agencies are tackling the drought/flood, dry land/wet land and other rural development works on watershed basis. In soil conservation measures the upper reaches of a watershed usually consists of hilly/ mountainous areas. These areas have an undulating topography and are foci for soil erosion. In areas with steep slopes and subject to soil erosion, vegetation does not get established. Due to lack of vegetative cover, soil erosion is accelerated transporting large quantities of silt into the stream below. The uncontrolled run off from the slopping areas cause extensive damage to the adjoining agricultural lands. It is therefore necessary to treat the hill slopes under soil erosion with appropriate soil conservation measures so that they are reclothed with vegetation and the surfaces run off from these areas are controlled. Contour trenching and afforestation, bench terracing and contour stone walls (some terracing) are suitable measures for these areas.

The following are some of the ways and means by which the watershed should be treated in order to conserve soil moisture, store excess rainwater and utilize the excess water for increasing the agricultural production in the selected watersheds. The treatments may be different to red and black soils and the less and more rainfall areas. Based on the soil type and rainfall the watershed plan can be prepared and implemented with the co-operation of the farmers in the watershed (watershed plan). They include:

- *In situ* conservation practices.
- Soil conservation techniques—*viz.* contour bunding/graded bunding.
- Construction of check dams/gully control structures.
- Forming farm ponds in individual farmer's field.
- Construction of percolation ponds.
- Construction of irrigation tanks.
- Utilisation of ground water in the watershed to the maximum extent.
- Application of water by sprinkler and drip irrigation methods.
- Resources management.
- Alternate land use.
- Cultivation of land using agricultural implements and machinery.
- Training of soil conservation officers/agricultural engineers to have the basic skills in watershed development, but they require training in land and water management specifications in the improved technology.

The degrees of people's participation do vary in any developmental programme. To sustain the interest of these committed and to enlarge the scope of awareness to those who reluctantly watch and observe periodical meetings, group discussions field visits are of great value.

CONCLUSION

From foregoing account it can be concluded that the fundamental elements of management are plan, organize, motivate and monitor. The importance of effective maintenance is applicable to all the creations of man as well as his interfaces with nature—be it dams, canals, watersheds etc. Watershed as a basic unit and systematic planning for land development, soil and water conservation, water harvesting, water management and the like are crucial for successful plant establishment, growth and drinking water problems. The degree of people's participation involved in the watershed management and development programme in Tamilnadu.

References

Eberhard Czaya (1983). River of the World, Cambridge.

Gray, N. F. (2000). Water Technology.

Sai Bindu (1999). Scientific Management: Spiritual Way.

Sivanappan R. K. 1989. Soil and Water Conservation and Water Harvesting Manual.

Sivanappan R. K. and P. K. Aiyasamy (1992). Report 2. Models for Greening the Barren and Wastelands Experiences of NGO's, Government Departments and Farmers.

Warren Viessman, Jr. Mark J. Hammer (1999). Water Supply and Pollution Control.

Watershed Management: Issues and Policies for 21st Century pp. 148-157

Socio-Economic Evaluation of Watershed Management Programmes and New Strategies for Rural Development

Y. V. R. Reddy and Shaik Haffis

INTRODUCTION

Watershed is defined as a topographically delineated area draining into a single channel. It is a geo-hydrological unit draining at a common point by a system of streams. Conceptually, watershed development is nothing but a risk management strategy which is meant for protecting the inhabitants of the fragile and deplorable ecosystems of rural India from acute distress caused by recurring droughts and intensity of floods (excess rains). Watershed management is the process of formulating and carrying out a course of action in a right perspective to exploiting full potential of natural, agricultural and human resources of a watershed to provide resources that are desired by and suitable to watershed community.

Watershed development and management, rather a multi-disciplinary activity represents a dynamic strategy, which is much more multifaceted than mere soil and water conservation. Dhruvanarayana *et al.* (1987) rightly endorsed that watershed management (Resources Centred Technology), is a holistic approach aimed at optimizing the use of land, water, vegetation and all associated components in an area which can alleviate drought, moderate floods, prevent soil erosion, improve water availability, increase fuel, fodder and agricultural production on a sustainable basis. Thus, keeping in view of the importance, Government of India initiated watershed programmes during 1983-84. It launched National Watershed Development Project for Rainfed Areas (NWDPRA) during 1986 in 46 model watersheds for rainfed areas in 16 states through Departments of Agriculture to test the strengths and weaknesses of dryland farming technologies through integrated approach, exploit the full potential and sustainability of agricultural productivity and minimizing risks in rainfed agriculture through mitigating drought sensitivity and severity, intensity of resource degradation, etc. Government of India has been spending rupees eleven thousand crores annually since then on watershed programmes.

The basic objectives of watershed development programme are as stated below:

- To improve the productivity of the soil under rainfed condition through improved soil and water management practises;
- To impart stability to crop yields through proper run-off management, restructuring of cropping pattern and land use;
- To restore the ecological balance through resource conservation, afforestation and pasture development; and
- To improve the socio-economic conditions of the inhabitants.

Technical committee (GOI, 1994) reported that among the 46 watersheds, Sukhomajiri, Ralegaon Sidhi, Adgaon, certain tribal areas of Panchamahal, Chevella, Mittemari, Jhabua, etc. were found to be the outstanding successful cases of watershed development. Although, the overall impact of these watersheds was found to be positive and significant, yet, the distinctive features of socio-economic evaluation of watershed management are discussed below for proper understanding and exposition of its impact.

WORKS IN WATERSHEDS IN INDIA

Soil and water conservation structures were taken up in all watersheds ignoring integrated watershed management. The soil and water conservation structures like contour bunding, strengthening of existing bunds, gully plugs, etc., were given more prominence in all watersheds but not developed vegetative cover, dairying, rural skills, cottage industries, etc., due to lack of co-operation among different departments. Most of the structures breached, damaged and vanished due to lack of maintenance leading to inefficient use of public exchequer. If alround development is made in watershed farm, it is easy to execute programmes systematically for rural development. However, there were only a few watersheds which were successful due to co-operation among line departments in the district, peoples' participation, water table development, etc.

BENEFITS

Tangible and intangible benefits are to be identified, studied and evaluated to know the impact of watershed programmes in India. A few sporadic studies were made till now.

Socio-economic factors

Socio-economic factors such as social status, age, education, training, exposure to technologies, land holding sizes, fragmentations, occupation, gender issues, peoples' participation, social system in the village, profitability, feasibility, sustainability, marketability, availability of labour force, resource use efficiency, availability of inputs at reasonable prices and credit, etc. play a role in rural development on watershed basis.

Watershed programme as an instrument for socio-economic change

Earlier programmes such as National Demonstrations, Operational Research Projects, Krishi Vigyan Kendras, Trainers Training Centres, supply of fertilizer/seed kit programmes, lab-to-land programme, etc. had an impact in transfer of technologies to farmers' fields but, it was abysmal. Thus, adaptation of technologies by farmers was not only slow but also low due to many, varied, complex and complicated reasons. Hence, watershed programmes were initiated with a view to change socio-economic conditions of farmers in selected village/s taking into small unit of area varying from 500 ha onwards for easy coverage in transfer of technologies and improving natural resources since 1983-84. Only watersheds managed by ICAR / SAUs covered with all aspects in the watershed but it did not sustain after withdrawal. Watersheds developed by Department of Agriculture, Soil and Water Conservation, Irrigation, etc. of various State Governments mostly concentrated on soil and water conservation structures which were not maintained after withdrawal of the programmme. Though it is a good programme to develop on unit area basis effectively, peoples' participation and withdrawal strategies have been missing due to varied reasons. However, no perceptible changes were noticed, a few benefits occurred to farmers and ground water recharge/ infiltration of rain water into soil increased leading to over exploitation of ground water which will cause severe damage to natural resources in the long run.

EFFECTS OF WATERSHED DEVELOPMENT PROGRAMMES

The effects/impacts of some of the success stories of watersheds on socio-economic conditions of the rural people in general and farmers in particular after the termination of the programme as assessed by 'experts committee' are as follows:

Crop diversification and productivity

A noticeable improvement (2-3 fold increase) in crop yields registered in most of the watersheds after the project than before watershed. Very high crop diversification to the extent of 98 and 75 per cent in case of Chevella and Mittemari watersheds were noticed during 1986-87, respectively (Haffis and Reddy, 1989). In another study, Haffis *et al.* (1990) worked out higher additional benefit-cost ratios (ABCRs) i.e., 10.97 in case of sorghum(L) + pigeonpea(L) and 8.21 in case of sorghum(CSH-5) + pigeonpea(HY-2) in Chevella while 4.40 in case of groundnut(JL-24) + pigeonpea(TTB-7) intercropping system in Mittemari during 1987-88 over the year 1984-85. However, among the different watersheds implemented in our country, many studies showed positive impact of watershed development programmes on crop yields, cropping intensity and cropping pattern changes, which are presented below:

Impact of Watershed on Crop Yields with Respect to Different Components/Activities

Sl. No.	Study made by researchers/Year	Region	Activity	Crops	Impact Yield (Per Cent)	Net Income (Per Cent)	B-C Ratio
1	Reddy and Kanwar (1985)	Andhra Pradesh	Agroforestry		–	–	1.45
			Silvi-agri		–	–	1.30
			Silvi-pastoral		–	–	2.25
2.	Agnihotri (1985)	Shivalik Hills	Vegetative cover	All crops	64q/ha to 85q/ha	–	–
3.	Government of Punjab (1986)	Punjab	Soil conservation	All crops	–	+ 84#	–
				Maize	– 13	–	–
				Paddy	– 30	–	–
				Wheat	+ 6	–	–
				Potato	+ 4	–	–
4.	Itnal and Narayan (1987)	Bijapur (Karnataka)	Farm Pond	Jowar	–	Rs. 1,300/ha	0.95
				Sunflower	–	Rs. 1,800/ha	1.11
5.	Pant (1989)	Madhya Pradesh	Integrated watershed	Sorghum	+127	–	–
				Wheat	+111	–	–
6.	Deshpande and Reddy (1990)	Maharashtra	Integrated watershed	Paddy	+ 18	–	–
				Ragi	+ 34	–	–
				Pulses	+ 7	–	–
				Bajra	+ 46	–	–

(contd. on page 151)

(contd. from p. 150)

			Wheat	+ 11	–	–
			Jowar	– 2	–	–
			Sunflower	+ 64	–	–
7. Rao (1990)	Andhra Pradesh	Integrated watershed	Paddy (K)	+ 54	+ 13#	–
			Paddy (R)	135	–	–
			Sorghum	+ 17	–	–
			Pigeon Pea	+ 7	–	–
			Caster	+ 6	–	–
8. Singh (1991)	Karnataka	Integrated watershed	All crops	–	Rs 1,970/ha	–
9. Singh *et al.* (1991)	Punjab	Integrated watershed	Forestry	–	–	2.74
			Animal Husbandry	–	–	1.14
			Soil conservation	–	Gross margin 0.90	–
			Horticulture	–	Rs 825-2,780	7.05
			All components	–	–	0.95
10. Mahnot *et al.* (1992)	Rajasthan	Integrated watershed	Maize	+ 21	–	1.03
			Wheat	+ 20	–	–
			Oilseeds	+ 52	–	8.37
11. Singh *et al.* (1993)	Kandi (Punjab)	Integrated watershed	Maize	+ 5 –14 q/ha	–	1.72
			Paddy	+5-13 q/ha	-	(avg)
			Blackgram	+4-6 q/ha	-	
			Wheat	+8-22 q/ha	-	
			Gram	+4-9 q/ha	-	
			Mustard	0-8 q/ha	-	
12. Singh *et al.* (1995)	Udaipur (Rajasthan)	Integrated watershed	Maize	+2 q/ha	+Rs 1,100	-
			Paddy	+2 q/ha	Per year	1.76
			Jowar	+2 q/ha		(avg)
			Blackgram	+4 q/ha		
			Wheat	+2 q/ha		
13. Nalatawadmath *et al.* (1997)	Bellary (Karnataka)	Integrated watershed	Bajra	+16 q/ha	+50	1.75
			Jowar	+13 q/ha		(avg)
			Sunflower	+22 q/ha		
14. Joshi and Bantilan (1997)	ICRISAT Asian Centre	Integrated watershed	Chickpea	-	+40	-

Note: # Gross income.
Source: V. Ratna Reddy (2000). "Sustainable Watershed Management: Institutional Approach".

All these review of studies revealed variation in magnitude of the impact across regions and crops depending on the nature of various components adopted as can be seen in the table.

Employment

As revealed by review studies, the significant impact of watersheds on cropping pattern followed by the farmers resulted in higher cropping intensity leading to rise in employment and reduction in seasonal migration of labour anything from 20 per cent to one per cent. Krishna Rao *et al.* (1990) observed that the own farm employment of male workers varied directly as the size of the farm increased. Similarly, the employment pattern followed by female workers was at par with male workers.

Income

More employment opportunities were created in the watershed areas due to very high level of crop diversification, substantial increase in cropping intensity and crop yields leading to rise in income generation of the rural people to a considerable extent. Walker *et al.* (1990) found that the incremental net income ranged between 49 and 203 per cent of the base level. Krishna Rao *et al.* (1990) found that per household income derived by the farmers in Maheswaram Watershed in Ranga Reddy district of Andhra Pradesh was directly proportional to farm size. All the studies showed higher BCRs (See Table).

Standard of living

Standard of living of the people is normally income sensitive, which is a real indicator to know the economic condition of the people. After the watershed programme, an improvement in standard of living (quality of life) of the people was noticed in terms of change in tastes, fashions/ styles of living, food habits, dressing, and other social etiquettes in case of farmers in particular and rural households in general which was attributed mainly due to rise in income of the people.

Social improvement

Watershed development programme had made a mark in social improvement of the people in the rural areas by improving food habits, level of education, awareness, health care, participation of women in decision-making, organizational building-capacity of women, etc. The rural inhabitants in watershed areas derived a greater extent of socio-economic benefits than the costs incurred/investment made by the Government in different activities/ components of the programme. Ratna Reddy (2000) observed that the costs of obtaining the information to explain institutional change in the context of watershed management where collective action is pre-requisite for institutional arrangements are not large compared to collective economic benefits, let alone social benefits. Thus, watershed development programme has brought fortunes for the rural development in India by improving the socio-economic status of the rural people. However, by adopting the watershed as a unit, different measures are to be adopted and executed carefully in different topo-sequences of terrain according to their capability without causing any hindrance or loss to ecological balance leading to undue environmental stress. Otherwise fortunes will turn into misfortunes. As reported by Anil (1995), Aral Sea in Russia was gradually dried up when its water was utilized to the watershed area (farm fields), which showed negative effect on ecological balance of the entire area. Ratna Reddy (2000) opined that the complex nature of watershed development and management calls for a comprehensive understanding of the situations under which watershed technology becomes economically viable, socially acceptable and ecologically sustainable in the long run.

To achieve the basic objectives underlined in the watershed development plan, such as equality, gender issues, etc. an effective implementation mechanism through integrated approach is indispensable by proper planning, execution, administration and monitoring.

Strategies

Integrated Watershed Management with development of agriculture, horticulture, agro-forestry, silvipasture, silvi-culture, dairy, fisheries, sericulture and non-farm sector activities like improvement of skills of artisans, making pappad, juices, soaps, detergents, food plates by leaves, etc., will give fruitful results, but marketing poses a major problem for these products. Hence, Government may initiate steps to develop and establish marketing network so as to encourage and develop cottage industries in rural areas for prosperity and economic stability. Rural technology and local wisdom can easily and selectively be assimilated. Empowerment of women and weaker sections, equity and sustainability issues are to be addressed at all steps of planning, execution and monitoring. The holistic and multi-dimensional aspects of watershed management was achieved in most of the watersheds by integrating multifarious government schemes like IRDP, TRYSEM, DWCRA, PMRY, Special Employment Programmes, Literacy Programmes, DPEP, ICDS, Biogas Schemes, etc. In addition to these schemes, some other activities like sanitation, control of diarrhoel diseases, control of iodine-deficiency diseases, use of advanced technology, ensuring food security, universalisation of primary education and development of rural industries were developed and assimilated within the ambit of Rajiv Gandhi Mission on Watershed Development. All these activities under watershed management resulted into remarkable achievements and sustainable developments leading to expansion of watershed development programmes to many more parts/areas of the country for overall development of rural India.

Peoples' participation

It is heartening to note that the role of farmers' contributing (donor) in public investment is highly appreciable as reported by Assistant Director of Agriculture (ADA), Jamnagar and in Gollahalli (Karnataka) where rich farmers really felt the importance of watershed development programme that made them ease to freely contribute their resources in the form of supplying cement bags at different levels of magnitude for the construction of check dams, nala bands, etc., thus, resulting in reduction of burden on public expenditure/ exchequer which is really a welcome step on the part of the farmers. Thus, peoples' participation may be encouraged and involved in all activities through different modes/ methods of continuity in development on watershed basis.

Suitable water conservation structures

Construction of percolation tanks, vegetative checks, brushwood checks, loose bolder structures, percolation wells, farm ponds, etc., may be given a preference over cement structured check dams causing stagnation of water and controlling the percolation of water. To prevent soil erosion, bunding and vegetation is 'a must' from upper reaches to lower reaches following which top-soil loss can be controlled. Hence agro-forestry systems suiting to different regions based on land use and classification may be developed for soil and water conservation in addition to developing plantations, silvipastural systems and thereby livestock production for better socio-economic upliftment of rural people.

Ground water balancing

Though ground water table has increased due to watershed programmes in India, ground water exploitation has also been accelerating due to lack of knowledge among farmers and Government policies. Hence, Government may come out with a feasible and viable policy in the use of ground water which is considered as more or less common property failing which desertification may start in the long run in semi-arid and arid regions due to over exploitation of the ground water. Hence, ground water policy is required to be framed and implemented to avoid over exploitation of ground water by farmers without knowing implication. Ground water table balancing is also to be studied to avoid exploitation.

Agricultural technologies: international standards

Embodied and non-embodied technologies of international standards are to be developed and transferred so as to sustain and compete in the international market, as it is not possible to consume within the country.

Women empowerment

Women empowerment through different activities and policies is a must for development of society with 'equity', 'equality', etc., as Self Help Groups became common feature in most of the watersheds but the development of cottage industries will utilize resources more effectively and efficiently.

Regional balance and economic principles

While planning watershed programmes for construction of check dams and such other water conservation structures, it is also necessary to look into entire region rather than only watershed area so as to see Benefit-Cost Ratios of lower reaches vis-à-vis proposed programmes in upper reaches. Economic viability and opportunity cost and/ or Comparative Advantage Principle should be applied in construction of water conservation structures on watershed basis. Hence, Government of India may provide guidelines to all Governments of all States for managing natural and common resources to the best advantage of the region/ state/nation so as to avoid uneconomical proposition for prosperity of the people in rural areas.

Withdrawal strategies

As there is no withdrawal strategy and subsequent maintenance of soil and water conservation structures, etc., in India, watershed effects vanished within five years after withdrawal. Therefore, there is a need to develop withdrawal strategy and enforce through legislation so as to decide the agencies like Watershed Associations/ Gram Panchayats/ Departments of Agriculture/Soil & Water Conservation and Irrigation, etc., to look after and maintain with corpus funds, etc.

Income generation

Income can be generated through corpus funds and land taxation to the tune of Rs. 5 to Rs. 10 per hectare in addition to collecting the amount from real beneficiaries in watershed to sustain and maintain after withdrawal by any agency.

GOs - NGOs

Government Organizations and Non-Government Organizations may work together for effective, efficient and economic way of utilizing resources for rural upliftment. Though, there is no doubt that the agriculture development in rainfed areas is possible on watershed basis only in India if agriculture development is clubbed with cottage industries in rural areas, watershed programme will become an instrument for socio-economic change in India. It will revolutionarise for growth, development and prosperity of people in rural areas of India.

Focus on rainfed/less favoured areas

During green revolution era, greater investment was made on irrigation and irrigated areas which paid off well and rapid advances were made in food production. However, additional investments in these areas have faced diminishing returns and social and environmental problems. Hence, rainfed and less favoured areas offer opportunity for socially and profitable investments. The success of investments depends on comparative advantage across the diverse areas. Therefore, strategies for developing and disseminating technologies must take into account so as to make strategy participations, demand drivers, indigenous knowledge, risk reducing and natural resources. No single strategy will cover in all such areas and effective local planning based on the assets, liabilities comparative advantages of different options coupled with decentralization and good governance has to be evolved to improve the socio-economic condition of rural people on watershed basis.

References

Anil Agarwal (1995). Will the Aral Sea ever come back ? *Down to Earth*, **4**(12): 27-38.

Dhruvanarayana, V. V., R. P. Singh, S. P. Bharadwaj, Srinivas Sharma, A. K. Sikka, K. P. R. Vittal and S. K. Das (1987). Watershed Management for drought mitigation, ICAR Publication, P.I.1987.

Government of India, Ministry of Rural Development (1994). Report on the Technical Committee on Drought Prone Areas Programme, New Delhi.

Krishna Rao, G. V., P. B. Parthasarathy and Y. V. R. Reddy (1990). Impact of dryland farming technology on production, income and employment in Ranga Reddy district of Andhra Pradesh. Paper presented at IUFRO Working Groups on Integrated Land Use and Forest Policies (S4. 08- 04) and Economics of Integrated Watershed Management (S4. 07-03) held at IUFRO World Congress in Montreal.

Rajesh Rajora (1998). "Integrated Watershed Management", Text Book, Rawat Publications, New Delhi.

Ratna Reddy, V. (2000). Sustainable Watershed Management : Institutional Approach, *Economic and Political Weekly*, **35**(38) 3375-3444.

Shaik Haffis and Y. V. R. Reddy (1989). An index to study the impact of watershed programme on crop diversification in drylands. *Indian Journal of Dryland Agriculture and Research and Development*, **4**(1): 79-86.

Shaik Haffis, Y. V. R. Reddy, S. C. Mathur and P. B. Parthasarathy (1990). Studies on diversification of cropping systems and its economics in watershed programmes. *Agricultural Situation in India*, **45**(5): 313-318.

ANNEXURE - I

Evaluation Methods

1. Ex-anti and post project evaluation with and without Project Area Evaluation.
2. Rapid rural surveys.
3. Concurrent surveys.
4. Sampling techniques in selection of farmers.
5. Analysis: Tabular Analysis, BCR, A.V., IRR, BEP, PBP.
6. Multivariant Analysis, Regression Analysis and Econometric models.

ANNEXURE - II

Chart 1: Hydrological chart of water resources in India (units in cubic kms).

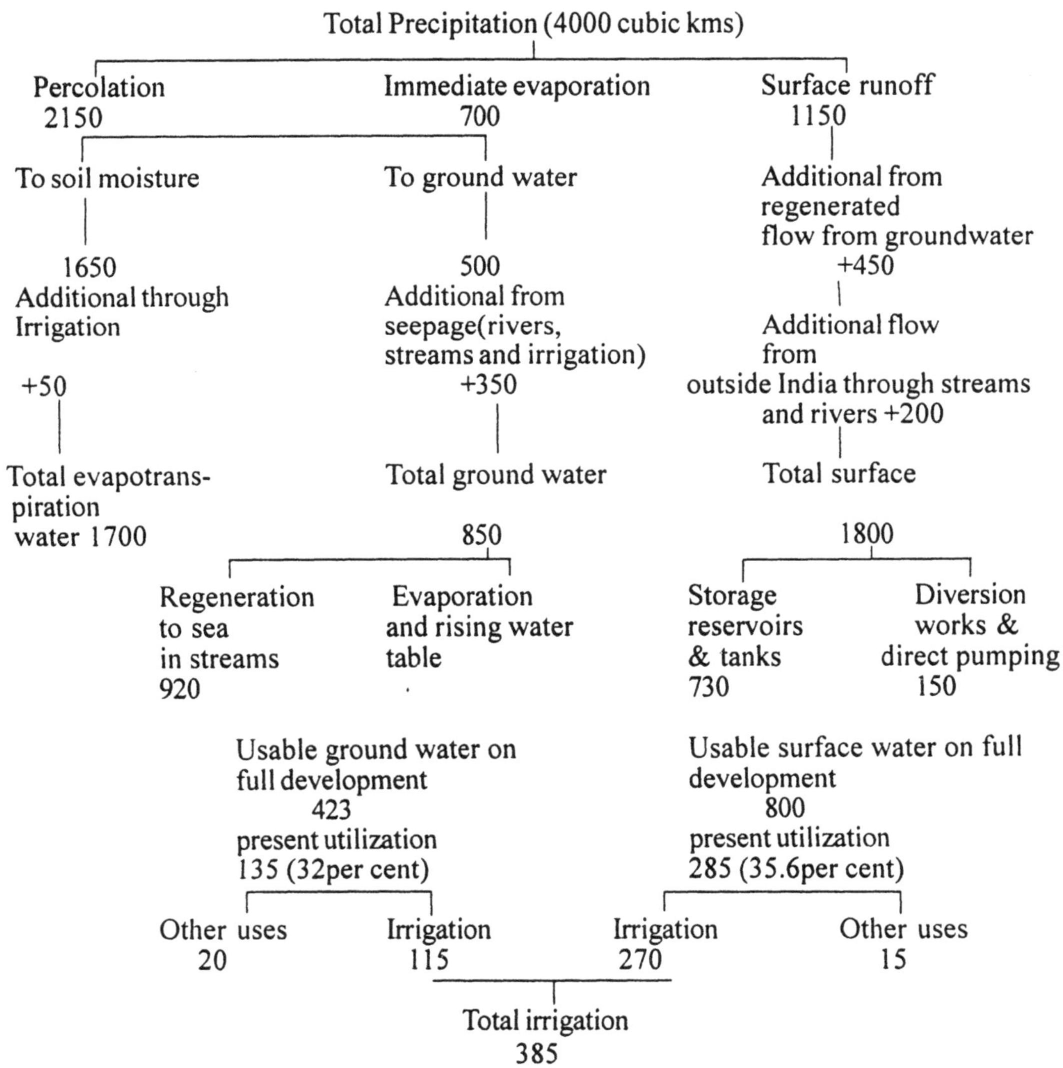

ANNEXURE - III

Chart 2: Broad Approach of Water Management.

- Water Management
  - Ground water withdrawal consistent with aquifer charge
    - -Conjunctive use of surface and ground water
    - -Artificial ground water recharge
    - Reduced conveyance losses
  - Water harvesting
  - Water conservation
    - Irrigated Agriculture
      - Canal water management
    - Rainwater conservation
      - *In situ*
      - *Ex situ*
      - On-farm water management
        - - reduce application losses
        - · scientific irrigation scheduling
        - - precise land leveling
        - - excess water drainage
  - Conjunctive use use of saline, alkali and sewage water with good quality water

ANNEXURE - IV

Chart 3: Integrated approach to arrest water table decline.

- Integrated approach for control of declining water table
  - Reduce ground water draft water
    - - Conjunctive use of surface and ground water
    - - Crop diversification
    - - Minimize demand and supply gap
    - - On-farm water management
    - - Policy for ground water exploitation
    - - Feasibility of exploiting deep aquifers
  - Develop new projects
  - Interzonal water transfer
  - Surplus water storage
  - Water recycling
  - Increase ground recharge
    - Network of surface drainage systems
    - Percolation tanks, water harvesting structures and watershed management in upper catchments
    - Diverting surplus water from rivers, streams to declining water table areas,
    - Spreading basins, wells, shafts, etc.

Watershed Management: Issues and Policies for 21st Century pp. 158-181

Achieving Food Security through Watershed Development

C. Ramasamy

INTRODUCTION

India with her population with one billion plus has the monumental task of providing food to all. But ensuring food security to the vulnerable sections of about one-third of total population who live below poverty line is a much greater and formidable task. Government of India has implemented major agricultural and rural development programmes to increase food production to match with the population growth since 1950s. India has been successful in meeting the growing food demand. But lack of purchasing power on the part of the low income groups is still widespread despite the implementation of a number pro-poor food security programmes.

Programmes such as Public Distribution System, School Free Meals Scheme, Food for Work and Nutrition Schemes (targeting women and children) have strengthened food security in the country. Non-governmental sector has also been playing an important role to provide food security either directly or indirectly to the poor population. However, people living in rainfed / dry farming areas are still prone to food insecurity. Food insecurity in rainfed and drought prone areas persists mainly because of low productivity of rainfed land, small size of holdings, and low levels of farm and off-farm employment opportunities. To tackle poverty and food security issues, in recent years, watershed development has become the focal point of agricultural and rural development efforts in rainfed areas of India. Development projects managed under various government ministries as well as by the non-government sector increasingly fall under the label of watershed projects which aim not only to improve the food security of people in these areas but also help to achieve conservation of natural resources such as water, soil, flora and fauna.

EVOLUTION OF THE CRS AGRICULTURAL PROGRAMME

Catholic Relief Services (CRS), Baltimore, USA, a Non Governmental Organisation largely supported by USAID provides assistance to India under the Indo-US Agreement since 1951. Initially, Title II food resources component of CRS programme were used to support the family feeding programme, evolved to include maternal and child health, education, agricultural and emergency and non-emergency general relief activities. CRS/India has used food aid as an incentive to motivate and mobilise communities to undertake relief and development activities.

The CRS agriculture programme began in Bihar during the devastating drought of 1966. Food was then used to partially compensate people for their unskilled labour in single activity and single-farmer-oriented activities in a number of scattered villages. Since the beginning of 1997, CRS/India concentrated its efforts on community based approach, aiming at holistic development of all people by resorting to watershed development as a strategy.

Strategies

CRS considered the following strategies and implemented them to achieve food security to the vulnerable sections of population in rainfed areas:

Cluster Area Approach / Watershed development approach

Before actually undertaking watershed development, CRS followed cluster area approach wherein cluster area was a well defined geographical area which is contiguous in nature, identified for programme implementation. Cluster is only a transition towards watershed. In the second stage, cluster area approach was gradually replaced by watershed approach, involving the communities in selected catchment areas.

Community Organisation

CRS promoted and strengthened of local community organisations to increase their capacity to take control over the project planning, implementation and management which is key to project success and sustainability.

Technical Feasibility Studies

Technical feasibility of individual projects were studied to ensure sound and better quality agricultural projects.

CURRENT STATUS OF PROGRAMME

CRS agricultural programme is implemented in 44 under developed districts spread over nine states of India. The total numbers of agricultural projects which are under implementation are 401. Agricultural programme is in progress in seven states viz., Andhra Pradesh, Bihar, Madhya Pradesh, Orissa, Rajasthan, West Bengal and Uttar Pradesh. In accordance with the strategy of shifting from individual oriented to community oriented watershed programme, more and more resources are allocated to watershed projects. By the end of 2001, about 63 per cent of resources are expected to be channeled to watershed projects. Presently, CRS agricultural projects are implemented in about 664 villages by 180 Operating Partners (Local NGOs) supervised by 25 Counter Partners, (District level NGOs - mostly (hurch based). In 2000, about 275 staff were working in these projects. Some details relating CRS programme are shown in Annexures I and II.

GOAL AND OBJECTIVE OF THE STUDY

Overall goal of this evaluative study is to assess the impact of CRS agricultural programme strategies and interventions, and the progress made in achieving the intended results which are measured through both impact and outcome indicators set for the programme process indicators to track the progress of the programme were also collected. The list of indicators for the evaluation are attached in Annexure-II. The results were compared with the baseline estimates derived at the beginning of the programme in 1997. The project period was five years (1997-2001) specific objectives were:

- To assess the impact of watershed development in the project regions on food security of the vulnerable sections.
- To identify the factors that positively affected the impact or restricted reaching the expected impact and to make recommendations for improving the future CRS programmes and to draw lessons in order to make use of them in other watershed development programmes.

DESIGN OF THE STUDY

Sampling

Collection and analysis of relevant information by scientific methods are important to realise the above objectives of the study. Both primary and secondary data were collected by sample surveys of different stake holders, field visits by evaluative team and by accessing information in secondary sources.

As CRS/India targets major per cent of its resources to five states (Bihar, Madhya Pradesh, Orissa, Rajasthan and Uttar Pradesh) the sample selection for primary data collection has been limited to these states. Multi-stage random sampling technique was followed for sample categories of respondents. The details of sampled units are shown in Table 1.

Table 1. Details relating to sampling

Respondent Category	Sample Size
A. Study sites	
States	5
Districts	15
Villages	60
B. Stake holders	
Cooperating Partners (CPs)	10
Operating Partners (OPs)	37
Village Extension Workers (VEWs)	37
Focus Groups	20
Watershed Committees	20
Forest Protection Committees	6
Women's Groups	23
Farm Households	1200

The analysis of data were done for four CRS administrative zones namely, Mumbai, Kolkata, Hyderabad and Lucknow. Pre-tested questionnaire was used to conduct personal interview and elicit information from these sample respondent units. The design of the study was developed in consultation with External Expert and Technical Adviser (Agriculture) of CRS Delhi. Specific Study Questions Matrix for Agriculture was developed by them and has been used by Field Survey Agency to develop questionnaire for field data collection.

In order to ensure the projects to cover vulnerable sections, the project was implemented in states with larger poor population (Bihar, Madhya Pradesh, Orissa, Rajasthan and Uttar Pradesh). Then districts in these states were selected by constructing Relative India of Backwardness (RIB). The criteria considered for backwardness are given elsewhere in the paper. Further, in the selection of watersheds for project implementation, micro-level criteria such as concentration of SC/ST population and other related factors. Again details are given in a later section in this paper.

Evaluative framework

The evaluation team suggested an evaluation design and related methods to achieve the goal and objectives of the evaluation. Broadly, following three types of assessment were envisioned.

Quantitative assessment

Final evaluation survey to assess the change over the programme period, a comparative analysis between baseline and final survey carried out. Specific indicators for measuring the economic, environmental, social and process impact were developed and used.

Qualitative assessment

Focus groups, key informants, managers and stake holders interviews were carried out to identify the factors that influence the impact of the programme, to understand the effectiveness of the programme strategies, structures, systems and interventions, the ambiguity or clarification of quantitative data, and review of progress reports, monitoring data and other documents of the programme. These informations formed the critical input in understanding the processes of implementation.

Watershed development activities

As watershed projects require three to five years to implement in full, multi-year and multi-plans were prepared and implemented. This ensures sustained improvements in the agricultural production of target communities. The multi-activities encompassed: (*i*) designing and construction of water and soil conservation techniques, (*ii*) improving vegetative cover (*iii*) organising economic enterprises such as crop production, horticulture, agro-forestry, pasture development, pisciculture, livestock production, nursery development, etc., (*iv*) supporting formation of Self Help Groups for credit needs and creation of mechanisms for input supply and maintenance of structures and (*v*) enhancing the technical capabilities at CRS and CP levels for improved watershed planning, implementation and management.

MAJOR FINDINGS

Economic impact

The key result expected of the watershed intervention is to increase agricultural production in the defined geographical area. The incremental production will improve food security to the vulnerable sections who are the major beneficiaries of watershed development. The

CRS assisted watershed development projects have resulted in increased quantity of food over 1997 baselines. This occurred through yield increases based on watershed activities to such as water harvesting, expansion of arable and irrigated land, project-induced rises in water tables as well as increased cropping intensity, increased livestock ownership, crop diversification and increased income.

The direct impact of watershed intervention can be assessed using several indicators which are in order.

Yield

The key objective which is achieved through watershed approach is, basically, to conserve soil and moisture which add to soil health. Improved soil fertility and adequate moisture improve the yields significantly. The available yield data sets collected during the course of 1997-2001 by 'Annual Results Surveys' aimed to measure annual progress in the project and the data gathered for Final Evaluation Survey provide evidence for significant increase in crop yields which are shown in Table 2.

Table 2. Trends in crop yields in watershed areas

Kgs. / ha.*

Crop	1997 (Baseline)	1998	1999	2001
Pulses	600	265	638	889
Paddy	824	1193	1133	1641
Wheat	1085	1674	1440	2032

*Yield estimates are averages of project sites

There is a clear positive yield trend. Paddy and wheat yields have doubled by 2001 as compared to base yield due to watershed intervention. The yield has risen by almost half of base yield in the case of pulses which are totally rainfed and often raised in unfavorable lands in comparison with paddy and wheat. Thus the investment on watershed has paid sizeable dividends. The 1997, 1998 and 1999 Results Reports showed variations in productivity for a given crop. The variances are principally due to external conditions such as weather and soil properties in different project sites and within project site, for different crop seasons. In some cases, yield measurement is made complicated due to the mixed cropping systems followed in different zones.

The farm level data on yields collected through final evaluation field survey reinforce the fact that crop yields have increased significantly consequent to watershed development as can be seen from the Table 3.

The yield of most important crop, paddy, has increased by more than one-third of original yield in the post project period.

The key achievements of watershed approach are improvement in irrigation and land quality. Once these two factors are adequately established, it is important that the inputs such as better quality seed, manures and plant protection measures must be applied supported by better agronomic skills of the farmers to realise the potential yields. The figures in the

Table 3. Average yield of different crops after and before implementation of CRS (Based on total sample)

Crop	Current Avg. Yield (kg/ha)		Avg. Yield before the project (kg/ha.)		Total No of cultivators
	Kharif	Rabi	Kharif	Rabi	
Amar	933		595		155
Bajra	883		704		72
Paddy	1641	2173	1152	1869	897
Jowar	929		806		41
Kulthi	706		479		37
Maize	1190		1180		36
Mustard		726		500	45
Wheat		2032		1896	94

Table 4 indicate that still large percentage of farmers have not realised the importance of inputs. In the future watershed activities, more attention must be paid to transfer of modern crop technologies to watershed areas for adoption of agricultural production technologies so as to maximise agricultural production in watershed project areas.

Table 4. Farmers perceptions on factors contributing to improvement in yield

Factors	Percent of farmers
Irrigation	70.5
Chemical fertilizers	18.2
Manures	78.5
Timeliness of sowing	42.6
HYV Seed	18.3
Plant protection measures	31.3
Land quality	68.7
Technical skills	59.0
Institutional credit	14.2

*(Percent of Sample Farmers Reported for All Four Zones Pooled)

Arable land

The lands treatments under CRS watershed programme have made the fallow lands arable again. The arable land due to CRS aided watershed programmes in the project areas put together has grown significantly over the period 1997-2001 and can be found in Table 5.

Table 5. Expansion in arable land* (For all Four Zones)

Year	Area (ha)
1997	5879
1998	3978
1999	3862
2000	5750
Total	19469

*Based on Yearwise Results Reports

Cumulatively, 19469 ha. have become arable. This has expanded the agricultural production potential in the watershed. The land treatments such as bunding, trenching and levelling have helped to achieve the objective of sustainability in the use of land and water resources.

Irrigation Potential

One of the results of the programme is the growth in irrigation potential meaning that the area that could be brought under irrigation if proper irrigation structures are created, due to water conservation, both surface and groundwater. The year wise irrigation potentials created due to the programme are shown in Table 6.

Table 6. Irrigation potential created and area irrigated (For all Four Zones)

Year	Irrigation Potential (ha.)	Area Irrigated (ha.)
1997	5879	100
1998	3978	651
1999	3862	2016
2000	4922	3116
Total	18641	5883

*Based on Yearwise Results Reports

Irrigation is fundamental to raise land productivity. And the full potentials of associated inputs such as manures, seeds etc., can be fully exploited only in the presence of irrigation. Watershed development is the most appropriate strategy to achieve this objective. The watersheds which have been intervened with CRS assisted programme have added substantial irrigation potential, a prerequisite resource for realising food security through expanded production and yield improvements.

To bring area under irrigation, further investment is needed to make use of the potential created. The investment may be in the form of storage structures for surface irrigation and /or groundwater extraction by digging new wells. Hence only 32 per cent of the potential created has been brought under irrigation in the projects. Table 6 shows the trends. The gap between irrigation potential and actual area irrigated points out the need for more investment on creating irrigation structures.

Cropping Intensity and Diversification

The water conservation activities in the project areas have not only enhanced the irrigation potential but expanded the irrigated area providing scope to raise the second crop (winter season) in treated areas and this has raised the crop area sown more than once. This has also given opportunity to diversify the crops grown. After watershed development, wheat is grown in the winter season and farmers have taken up vegetable production in both monsoon and winter seasons; which is expected to add to their food and nutritional kitty. The perceptions of farmers regarding change in cropping pattern can be found in Table 7.

An analysis of perceptions of farmers indicate that 64.6 per cent and 75.3 per cent of farmers in Kolkata and Mumbai zones consider that increased irrigation facility as the major factor which has caused diversification of cropping pattern. According to Functional Groups at Watershed level in Kolkata, Hyderabad and Lucknow zones, vegetables are widely grown in post-project period. Other respondents viz., Operating Partners (Local NGOs involved in the project) and Village Extension Workers (VEW) brought to the notice the emergence of other enterprises such as fruits and fodder production, agroforestry, fisheries and livestock. It may be seen that double cropped area has increased from 30 per cent to little over 50 per cent in various zones.

Table 7. Change in cropping intensity and cropping pattern (Diversification)

Zone	Percentage of command area double cropped		Percentage of households reporting change in cropping pattern
	Before	After	
Kolkata	23.5	36.1	64.6
Hyderabad	23.5	56.8	52.8
Lucknow	25.1	46.0	58.3
Mumbai	19.0	27.7	75.3
All India	21.3	34.7	68.9

*Based on Yearwise Results Reports

Other Economic Impacts

Other impacts observed during the field visits by the Final Evaluation are:

- As a result of greater access to water, land value has increased in the post-project period.
- Farmers own more micro-livestock (more chicken, pigs and sheep)
- Farmers are producing and storing more fodder for livestock (more haystacks observed in fields)
- Increased savings have been achieved. Saving habits have improved, thanks to activities of Self-Help Groups.

Impact on environment

One of the objectives of CRS programme was to ensure that environmental impact of watershed interventions are positive. Watershed Development Project of 1997-2001 aims to strengthen the institutional capacity to carry out environmental review before the project is launched. It also emphasises on diagnosis of environmental problems and development of programmes that enhance natural resources. An environmental review was carried out for each of the watershed project supported by CRS and strict compliance by village level implementing agency was monitored.

Reversing Acquifer Depletion

There is a growing concern about the water scarcity and aquifer depletion in the country. One of the expected direct impact of the watershed programmes at national level are rise in groundwater tables and water conservation at land surfaces significantly adding to environmental benefits. Moisture conservation and acquifer recharge are vital for growth and survival of flora and fauna in the watershed geography. Almost entire groups of respondents *viz*., farmers, functional groups, Village Extension Workers (VEWs), Operating Partners, (OPs), District Level Implementing Agency and Coordinating Partners (CPs). District Level Implementing Agency indicated that water conservation has brought in qualitative improvements in environment and also more awareness among local population.

One of the arguments put forth was that water conservation may lead to breeding of more mosquitoes due to water stagnation and drainage problem. This negative consequence was not, however, reported by the beneficiaries. It may be because that possible water collection sites are widey dispersed.

Watershed approach to water management that include levelling and bunding, live barriers on bunds for stabilization, and different methods of small-scale irrigation are resulting in higher groundwater levels during non-drought years and stabilization of groundwater levels. Data in Table 8 below reflects the decline in depth to groundwater table from the surface indicating the improvements in acquifer recharge in watershed treated areas. Results Reports (1997-1999) and Monthly Reports for 2000 have been the sources of data for Table 8.

Table 8. Trends in aquifer recharge

Year	Depth to groundwater table from surface (cm)
1997	554
1998	385
1999	375
2000	425

The improvement in water table has been achieved despite the fact that many parts of CRS working area were affected due to drought for the last 2-3 years.

Vegetative Cover

A key resource conservation/protection measure has been to increase surface vegetation through introduction of agro-forestry and horticultural trees, fallow-pasture land conversion and regeneration of flora. This has been achieved through implementation of CRS assisted watershed programme. Table 9 provides the improvements in vegetative cover in project areas.

Table 9. Vegetative cover

Year	*Area improved with vegetation (ha.)
1997	514
1998	471
1999	665
2000	474
Total	2124

The vegetative cover has also added to ecology and environment of the project villages. This is one of the positive impacts of watershed development. The soil and water conservation measures have resulted in regeneration of flora in the watershed areas. The watershed communities were observed to be happy about the regeneration of vegetation in their habitats. This achievement was possible because of aquifer recharge, agro-forestry activities and community-regulated grazing of livestock. The conservation of moisture led to more area under pulses and the stalk of the pulses partly compensated the fuel demand on household resulting in lesser demand for fire wood thus reducing the pressure on forests and vegetation in the village. About 62 per cent of VEWs, 60 per cent of Operating Partners (OPs,) 100 per cent Forest Protection Committee and 82 per cent of farmers confirmed the regeneration of vegetation due to watershed intervention. For example in Simla watershed (Jamshedpur district), forest based communities have stopped felling trees as they realised the long-term negative impact of deforestation.

Improved Soil Quality

The projects have improved the soil quality by arresting erosion and conserving soil moisture that have increased vegetative growth and added to organic matter to the degraded soils. In the pre-watershed / cluster projects, degraded soils are a common scene. Watershed projects have changed the whole scenario and the positive impact of soil conservation activities are clearly visible and perceptions of grass root functionaries on the impact of watershed on soil quality are shown in Table 10.

As per table, more than 50 percent of VEWs and OPs, and cent per cent of functional groups confirm the improvement in soil health due to project activities. The soil conservation measures such as trenching, bunding, levelling, live fence establishment, gully plucking, zigzag water flow technique etc. have reduced soil erosion. It was noted during field visits by evaluation team that massive gully plucking activities are responsible for decreased depth and size of gullies. In the longterm, this will reduce the negative impact of floods and downstream runoff and regimentation. One of the positive externalities is due to change in cropping pattern with more area sown to pulses which are nitrogen fixing in nature. This means that cultivation of pulses enrich the soil with nitrogen. Pigeonpea and Kulthi also add lot of organic matter to soil.

Table 10. Perceptions on soil quality

Respondent category	Percentage of respondents indicating improvement in soil quality
Functional Groups*	100
VEWs	54
OPs	51

*Watershed Committees (WC), Water Users Group (WUG), Forest Protection Committees (FPC), and Women's Group (WG)

Thus, there are clear evidences of positive environmental impact after the watershed intervention. Further, there is no evidence of environmental threats due to CRS supported watershed development activities. And the potential for health hazards was negligible due to small scale operation. CRS, CPs, OP staff and community members are quite aware of the environmental implications of project because of their total participation in watershed planning and programme implementation. The strategy of Initial Environment Examination (IEE) innovatively introduced and followed in CRS assisted watershed projects has created awareness on environment among community members. In the course of Final Field Evaluation visits, it was observed that, surprisingly, almost all members of community had good knowledge on environmental implications and they are concerned about improving their physical environment through these interventions. There were also instances wherein individual farmers have offered their land area to locate irrigation tanks which adds enormous benefits to the society. As regards training, though organised for CPs and OPs, communities have not yet received formal training on environmental compliance.

Social impact

Improved income and employment at household level are expected to change the social behaviour of local community. Increases in foodgrain production and additional employment may result in food security, lesser seasonal migration, and enhanced allocation of monetary income for better health and vertical movement in education.

Reduction in Seasonal Migration

During the pre-project period, beneficiary household members were migrating outside the village seeking employment to tide over the lean season. This results in holding children at the house to look after the younger siblings and also to take care of animals. Employment benefits generated out of the projects keep household members get occupied in the lean season, in their own land, thus considerably reducing the migration. This can be considered as one of the major impacts of watershed projects.

Majority of functional groups at community level confirmed reduction in seasonal out-migration. The pooled information for all four zones relating to seasonal out-migration are shown in Table 11.

Table 11. Perceptions of Community Functional Committees on seasonal out-migration (Pooled Analysis for All-Zones)

Functional Committee	Per cent of Functional Committees indicating reduction in migration
Watershed Committee	86.50
Water Users Group	100.00
Forest Protection Committee	100.00
Women's Group	86.50

According to larger proportions of farmer beneficiaries and VEWs, increased agricultural output and additional employment achieved owing to land and water conservation are primarily attributed for reduction in seasonal out-migration.

Impact on Education: School Enrolment

In post watershed situation, a considerable improvement was seen in school enrolment. Analysis of survey data collected for final evaluation from farmers in the watershed villages led to the finding that 75 per cent, 100 per cent, 98 per cent and 89 per cent of farmers in

Kolkata, Hyderabad, Lucknow and Mumbai zones respectively reported the increase in school enrolment.

Partly, this is due to withdrawal of children from labour force of the family. Almost 85 to 100 per cent farm households, functional groups, VEW, and OPs viewed that watershed development has been catalytic for improved school enrolment in all four zones. Earlier when adult members migrate seeking employment elsewhere, children were taking care of small children and animals. Reduction in seasonal out-migration of adult members has helped to release the children from the tasks resulting in enrolment in schools. In addition, higher household income has encouraged children to pursue education. About 70 to 100 per cent of Farm Households, Functional Groups, VEWs, and OPs opined that withdrawal of child labour and improvement in household income have been the primary factors for increased enrolment through reduction in time needed to collect fuel, wood, fodder and water because local availability of these items due to watershed development has improved which are now done by adult members replacing children.

Equity in Distribution of Benefits

The key objective of the CRS assisted programme is to extend the benefits to disadvantaged sections such as small farmers, landless labourers, women, children and SC/ST population in the project areas. The objective has been achieved by resorting to two significant new dimensions to the targeting strategy. At the macro/district level, there is a commitment to consolidate the geographical spread of the programme and ensure that resources are targeted to the neediest areas. Criteria for district level targeting were finalized with the development of a Relative Index of Backwardness (RIB) in which five poorest states were targeted with respect to the following indicators: (*a*) rural female literacy, (*b*) per capita value of output of major crops, (*c*) infant mortality rate and (*d*) proportion of rural SC/ST in the population. During 1997-2001, 80 per cent of watershed development resources were programmed for districts meeting these criteria.

In selection of watersheds, micro level criteria adopted were: (*a*) low percentage of households access to winter crop irrigation, (*b*) high SC/ST/OBC percentage in watershed / cluster, (*c*) low percentage of households with access to tractor and power tillers, (*d*) high percentage of households having agriculture as primary occupation and (*e*) majority of the cultivable lands owned by farmers with less than five acres.

Improved Health and Nutrition

There is strong perception among communities of improved nutrition due to food aid given as a part of watershed development activities and due to increased food production (higher yields, increased crop and livestock production). Mothers report that their children eating bulgur (wheat flour) which is given as kind wages by the NGO, supplied by CRS from USA under food aid are healthier than neighbours whose children are eating rice. Bulger is a whole grain, which when combined with legumes, nuts or certain vegetables such as cabbage, greens and potatoes provide complete protein while rice has had the protein layer removed and provides only carbohydrates in diet. Besides, communities report availability of potable water which has increased due to well construction as part of the project.

Now more income has made households in the watershed area to go for better health services. Though there is not much change in incidence of diseases, people are aware of and seek decent health services. About 87 to 91 per cent of farmers, 87-100 per cent of functional committees and 75 - 100 per cent of VEWs and 100 per cent of CPs asserted the improvement in health and nutrition of the population living in project areas. The two dominant factors identified by different respondent categories, for improvement in health and nutrition are more consumption of fruits, vegetables, milk and meat. In Kolkata region, 40 per cent of the OPs consider increased availability of these horticultural products as the major factor. But unambiguous conclusion is that watershed development activities have made positive impact on health and nutrition of poor people. Women community in the course of field visits interviews by Evaluation Team reported reduction in alcoholism among men folk of their respective villagers.

Benefits to Women

In backward and underdeveloped regions where CRS partners are working, the status of women were low. They are not involved in public discussions and decision making in community development issues before watershed interruption. The positive consequences of the watershed strategy leading to empowerment of women was seen in agricultural sector. In post implementation situation, women are able to come out of their houses, socialize and discuss common problems and issues. Women's involvement also raised their status and respect within family and community. Their attitudes have become out-ward looking.

Step-up in agricultural production in post-project period has ensured more food availability to vulnerable groups namely, women, children and SC/ ST population. There is also improved awareness among the vulnerable groups on the watershed activities, increased food availability and benefits of community action.

In almost all cases, project sites have been so chosen that SCs, STs and OBCs will benefit. Besides, at all watershed sites, different committees have been formed which involve varius socio-economic groups including landless, marginal farmers and women. In many cases, training and exposure visits were organised for community members, including the vulnerable groups in order to enable them to participate in the planning and implementation of watershed projects. About 57 per cent of watershed committees reported that landless labourers are benefited and 95 per cent of watershed committees confirmed the benefits flowing to small and marginal farmers. Watershed projects have benefited the women to a greater extent. The perceptions of respondents categories on benefits accrued to women are presented in Table 12. First, improvement in food availability is conspicuous, adding to better health of women. Second, their employment status has improved within the village helping them to avoid seasonal migration. This has led to better care of their children.

In most of the watersheds, efforts have been taken to form Self Help Groups/Mahila Mandals which have encouraged group savings, thus catering to the small scale financial needs of poor households. In sum, the CRS supported watershed programmes have significantly contributed for empowerment of women.

Table 12. Perceptions of stakeholders on benefits accrued to Women*

(Per cent)

Respondent	Improved food availability	Improved income	Better access to fuel, fodder and water locality	Community mobilization	Improved awareness
Farm Household	33	77	15	–	58
Functional Committees	52	58	61	58	–
VEWs	64	82	20	–	54
OPs	50	84	34	–	84

* The figures were compiled from two sets of tables generated by Survey Agency and data collected by Evaluation Team during field visits.

Acquirement of Technical and Management Skills

The watershed approach has provided an opportunity for community members to acquire agronomic and other technical skills. In post-project situation, they are confident in gully plugging, contour bunding, and construction of micro-tanks, spillways, wells, infiltration and water conservation trenches. Acquirement of technical and management skills have led to the following positive developments.

- Community members showed evidence of greater understanding of agro-forestry with potential economic benefits: *eg.* planting of amla, anona and mango trees in upper watershed areas; planting of papaya and banana in interspecies, planting of live barriers with agave. OPs have advised on inclusion of early bearing drumstic and tamarind in watershed projects.
- Cropping pattern changes were observed which improve sustainability aspects through incorporating annuals and perennials, crop diversification for stability of production, soil fertility enhancement through cultivation of nitrogen-fixing pulses (legumes), etc.
- The regular zigzag placement of surface water collection methods ensured even greater infiltration to aquifers and avoided formation of underground stream flows that would carry water out of the watershed.
- Channel bunds are being planted with different crops/grasses to stabilize bund structures. One of the more successful short-term crops planted on the bunds has been finger millet because it also fills the food-kitty of farm households.
- Watershed interventions are done from ridge to valley for various sound environmental reasons.
- Village watershed committees were allocating separate grazing areas to prevent livestock damage to newly formed watershed structures and they are guiding the maintenance of bunds to protect their investment and maintain yield and income benefits.
- As an example of reforestation for environmental protection and profit, in Upper Debnala watershed (Jamshedpur CP) both *acacia* and cashew plantations are being developed.

At CRS, CP and OP level, training courses were organised for staff on management, environment and agriculture as programme transitions do require adequate institutional capacity to implement watershed projects effectively. The staff were also given exposure visits to other watersheds.

Impact on Civil Society and Local Organisation

People participatory watershed approach to development has achieved a subtle but clear attitudinal change from individualistic to group-oriented priorities among rural poor people. Community mobilisation/organisation is a key to the participatory but sustainable watershed development. It helps in effective implementation of the project and also enables the community members to share the benefits. Besides, it facilitates upward mobility of community members.

In all watersheds, now, participatory approach to development has been adopted by involving the communities/people in planning and implementation of the project. Functional committees have been formed to facilitate their full participation in the programme though in some watersheds the watershed committees have not grown into full form. The existing Village Committees and Self-Help Women Groups do the functions of Watershed Committees. Now, there is a greater harmony, a focussed common purpose, a greater social awareness, and an acquisition of higher technical knowledge and more participation in collective action among the watershed community members in the project villages. After watershed approach, many watershed communities have acquired group problem-solving capabilities. According to 68 per cent of women groups there is greater harmony within the community. And 74 per cent reported greater social awareness in the community. More than 85 per cent of all functional committees reported improvements in managerial capacity among community members.

One hundred per cent of CPs and OPs, 97.3 per cent of VEWs, 95 per cent focus groups, 95 per cent of watershed committees, 83.3 per cent of forest protection committees and 100 per cent of water users groups and women groups reported that CRS programme has improved their managerial capability at the community level. Formation of functional committees has enhanced the organisational and managerial capacity. Some have organised groups to protect natural resources and prevent destructive practices like tree felling.

Despite the greater desirable impact of the watershed approach on civil society and organisational and managerial capabilities of local communities, the watershed committees have not taken full shape and involvement of women in the committees are still low.

While formation of Self-Help Groups among women members is promoted in watershed projects, some SHGs have excess funds and are unsure of how to spend them. There lies tremendous scope for imparting skills to the women's groups to run micro enterprises using the surplus funds. The community approach has helped to generate a certain amount of employment for village youths as VEWs are appointed from the programme villages, which also lend them credibility.

Process of project implementation

The process in implementation of any development project is critical for it improves the quantity of and quality of benefits which are accruing to the communities in addition to imparting sustainability to the project beyond the project period.

Participatory Project Design

Involving all stakeholders in project design will ensure more practicability, better understanding of local realities, and make them owners of the project. For example, the local communities may identify a right species of agro-forestry trees for their location, and watershed committee can decide where to place the bunds, trenches, water ways etc. in consultation with project engineers, VEWs and OPs so that maximum physical efficiency is achieved. The beneficiaries of watershed having land at different points in the watershed will come to know what benefits they will get out of the project. For example, the nature and quantum of benefits for farmers at upland and lowland in a watershed will be different. And this reality will be acceptable to them in the initial stage of the project itself. In all CRS assisted watersheds, all the stakeholders were involved in project design and the approach made them participatory in the whole process.

Functional committees were part of the project design. Table 13 adequately reflects their involvement.

Table 13. Participation of functional committees in project design*

Functional Groups	Participation in project	Involvement in effective discussion	Difficulties in functioning
Watershed Committee	86.50	86.50	21.60
Watershed Users Group	100.00	100.00	0.0
Forest Protection Committee	100.00	100.00	42.3
Women's Groups	86.5	86.5	21.6

*Pooled for all four zones

Despite achievements made in design of the projects, some errors were made in some of the watersheds. The errors are improper selection of tank sites, starting soil conservation from valley to the ridge, failure to provide spillways, inappropriate crop and other enterprises, poor planning for marketing of products and missing the delineation of common grazing lands. More than 86 per cent of all functional committees indicated that they effectively participated in discussion and finalisation of project designs. Proportions of Watershed Committees, Forest Protection Committees and Women's Groups who pointed out some difficulties in functioning of committees were 21.6 per cent, 42.3 per cent and 21.6 per cent respectively. These quantitative information reinforced the qualitative observations made during fields visits of final evaluation team and MTR.

Functional Committees

Formation of committees such as Watershed Committee (WC), Water Users' Group (WUG), Forest Protection Committee (FPC) and Women's Group (WG) are emphasised and followed up in watershed projects. The members are expected to involve in planning and implementation of the projects. Their role become more critical in day-to-day activities in terms of mid-course correction in watershed design and distribution of benefits and community mobilization for various activities such as irrigation, savings and land development.

Various committees are functioning in all watersheds. Often, a single watershed includes more than one village. Currently, in many watersheds villages, committees play the role of watershed committees. When more than one village is involved, the village committees have not formed any federation or constituted an apex committee. It is expected that the village committees will graduate into watershed committees in the course of time. It is important that the Watershed Committees are full-fledged and play their expected roles. Similarly, Water Users Group, Forest Protection Group, etc., are not in existence in all watersheds. According to household survey, 65 per cent of farmers indicate that WCs meet once in a fortnight. But other committees such as WUG, FPG and WGs meet monthly. Overall, they function satisfactorily but there is scope to improve their performance.

In most of the watershed committees, WGs in the form of Self Help Groups (SHGs) have been formed and their activities are so far confined to microfinance activities. They have not forayed into organising small-scale enterprises. In all watersheds, functional committees play a larger role in planning, development and management of watershed projects. The farmers survey results suggest that 75 per cent of farmers feel that the recommendations by functional committees are accepted.

Monitoring, Evaluation and MIS Systems

MIS are expected to provide relevant and accurate but timely information both for decision making and accountability purposes. Under CRS programme, MIS provided information to several types of programme stakeholders: the primary donor, USAID, CRS, CPs, and OPs and the communities where the decision regarding development activities are carried. The system was designed to supply information to the various user-levels to assist in programme management capturing both impact and process indicators. By middle of the project period, all Zonal CRS Offices had hired an MIS Coordinator to support zonal sectoral teams in the management and compilation of information received through periodic reporting. And, focus was on implementation of the system through partner training. In addition, there were supervisory/support visits for CRS and partner staff. Each M & E system consists of the following components. (*i*) baseline survey (*ii*) on-going data collection and periodic (monthly, quarterly, annual) reporting from partners, (*iii*) supervisory visits by CRS and intermediary partners and (*iv*) evaluation.

The Management Information System (MIS) is considered an integral part of an effective project implementation to follow use of input, to assess the output and impact and trace implementation process. It is designed to help monitor the programme by way of getting adequate feedback, making mid-term corrections, solving emerging problems and getting to know the direction and likely impact of the project. Because of the high variability of agro-environmental conditions in different areas, a longitudinal (change over time) rather than a cross-sectional (control / group) system is used for gathering data. The MIS contemplated in CRS assisted watershed projects compares impact indicators over a long-term cycle and tracks both impact and output indicators over annual intervals. MIS system consists of Project Register (PR), Sample Field Record (SFR) and Food For Work Progress/Completion Reports (FFWPCR). PR provides data on project level progress of activities and is maintained by the community level workers. This register is being maintained in all watershed projects. The

SFR system provides information on selected indicators to measure impact of the programme as well as external factors such as rainfall, pests and crop cycle practices of farmers. These information are reported by OPs in their monthly reports to CPs and in turn to CRS.

Information on external factors are used to explain variations in impact and to improve the current and future projects by offering training or other inputs. FFWPCR is a tool to monitor progress on activities (process /output), supported by Project Register, in all projects. FFWPCR also helps to measure impact of the programme through selected indicators. In addition, an annual survey is conducted by CRS and partner field staff to gather information relating to impact indicators. Baseline information gathered at the beginning of the watershed project provide baseline data which is the bench mark to measure the progress of the project. Since the agricultural programme is being implemented involving various agencies namely, CPs, OPs and the community, programme performance depends to a large extent on mutual support and better interaction.

Human Resource Management

Successful implementation of a project requires appropriate technologies, management mechanisms and human skills. A best combination of these requirements will maximize the benefits which will flow out of the project. In a project, human resources (managers and project staff) must have adequate knowledge, skills and managerial capability to carry out activities efficiently. CRS assisted watershed programme has conceived a proper model of human resource deployment. Partnership is the core of all CRS sector programme. Implementation requires a process of negotiation based mutual understanding. However, in all levels of multi-tier structure there lies more scope to improve the programme in terms of adequate staff strength, quality of personnel, allocation of works and overall capacity building. When CRS has opted for a shift in strategy to watershed approach, there is a need for specialized skill areas, development of MIS, and new outreach methods to communities. Wherever possible, CRS has attempted to integrate gender equity strategies into the design of the new community-based approach.

Food for Work (FFW) as a facilitating factor

The CRS programme has a very high rate of food utilization. Food resources are highly valued by partners and participants. FFW is used as a facilitating factor for implementing the agricultural programme. The proper management of it is critical to achieve the objectives such as food security to the vulnerable groups, crash employment to prevent community migration and capacity building of the participants. Observations at the field level indicated that FFW is not only used to ensure food security but helps to create physical structures at watersheds (dams, wells, bunding, channel etc.) ensuring water and soil conservation, contributing to improving land productivity and overall economic improvement. FFW as a strategy has created new awareness and organisation of people to plan and participate in various developmental activities and fulfilled achieving food security, more income and employment generation and reduced seasonal migration.

Networking/Collaboration

Under watershed projects, it is encouraged that CPs and OPs may develop linkages with government and other development agencies to mobilize support for their agricultural projects. Some of the agencies may complement each other which may improve the efficiency in implementation in terms of cost-saving through avoiding of replication of activities or may enhance the quantum of benefits through interaction between the agencies. But the achievement in this respect is rather poor. Few OPs are however keen to establish linkages with government departments. In watersheds where public forestlands are part of it, OPs have already initiated dialogue with Forest Department to get their cooperation. In most places, OPs are engaged not only in agriculture but undertake activities relating to health and education. Alongside, where developmental agencies, government and non-governmental organisations are working in the same watershed area, exploring the scope of inter-sectoral and inter-agency linkages will maximise the impact of CRS - supported agricultural programme. Many partners believe that agricultural development in a given area will, subsequently, promote health and education as the income of the households rise. Generally, though projects show weak sectoral integration largely due to failure of human relations, farmers observe that in watersheds, where agriculture, health and education activities are carried out simultaneously, benefits increased significantly.

Exit strategy and sustainability of watershed activities

In most development projects, the concern of donors, policy makers and the public has been the life of the project beyond the project period, until then the project gets technical, management and financial support from project staff, institutions and donors. Long term benefits of the project will not be a reality unless the project activities or its functions are sustained. In CRS assisted watersheds, a closer examination confirmed that mechanisms have been built into the project/system that there exists scope for sustainability of the projects once the external support is withdrawn.

Functional groups, which are either an outcome of watershed project or existing prior to the project, have emerged as a powerful mechanism to run the show in the future. Functional groups and the community members appear to be confident about the viability of the project. They are categorical that since they have already tasted or understood the real benefits of the community based watershed approach, they will not allow any meltdown in sustaining watershed project. It was observed that the beneficiaries have already put in efforts in the annual maintenance of structures and to keep alive the regeneration of flora and fauna in the geographical unit. They have acquired skills and wisdom in solving technical and management problems.

There is fairness in distribution of benefits among the community members. Even if the flow of benefits may result in some inequalities which are inevitable (farmers in lower slopes of watershed will benefit more than the farmers in the upper levels), the consequences are informed at planning stage itself, so that all members are aware of the benefits they will get. Equity considerations thus built into the system added strength to the sustainability of project. Community members and functional committees are participating in design, planning and implementation of the project. They are made the principal partners and involved in day-today decision making. Training offered to them and the acquisition of technical knowledge and managerial skills acquired by them have resulted in internal capacity building to manage the project and solve problems.

Women SHGs either formed as part of the project or in existence before is a real strength to project management. Women's groups provide a strong support to project activities. Though there is inadequate representation to women in the watershed committees, women's groups have potentials to utilize the savings from their micro finance activity to establish nursery and simple processing and marketing enterprises to strengthen production activities in the project area. Probably, many projects are in middle of the way and women and SC/ST members will have their full share in the functional committees when the projects mature. At present, there is no evidence of creation of a community fund in most of the watersheds. This issue must be critically looked into by CRS, and a strategy / mechanism must be designed to ensure the creation of a community fund which will provide lot of strength to the community to sustain the activities into the future. It was expected that, in the project areas, collaborative linkages with other agencies will facilitate sustainability of activities. So far, nothing solid has happened. A move by CPs to deal with district level authorities of government may help to make use of resources supplied through government departments to sustain watershed programme.

CONCLUSION AND FUTURE OF CRS AGRICULTURAL PROGRAMMES

Are Goals and Objectives Achieved?

The findings of final evaluation study provided substantial evidence that CRS agricultural programme has achieved its goal of improved food security to the vulnerable groups through participatory community based approach in backward regions. The achievements during (1997-2001) matched with the intended project objective of promoting sustainable improvements in households agricultural production of resource poor farmers through natural resource development and management. The improved conservation of soil and water and resultant higher crop yields and food paid as wages are key factors which brought in benefits to the intended beneficiaries. No major constraints worth mentioning has restricted reaching the expected impact. The programme components of strategies, structures, systems and interventions have worked largely as expected, and no major changes in these components are required except fine tuning and minor modifications. As such no external factors have affected the relationships between outputs and impacts of the programme. The recommendations of Mid Term Review Report have been given serious consideration and efforts made to implement them though not to the intended level.

The monitoring system is working to a satisfactory level. The system of monitoring as visualized at the planning stage are being implemented. It must however be emphasised that the system offers scope for improvement in terms of better interfacing between partners, reducing communication gap, simplifying the reporting system and allowing operational flexibility at CP and OP level. The data generated to be reliable, valid and used for decision making purposes.

In the course of implementation of the project (1997-2001) CRS, CPs, OPs and committee members at community level have acquired sufficient technical, administrative capabilities that have, by and large, had a positive impact on the society. Additional training, brainstorming workshops among stakeholders and expositor visits to successful projects are prerequisites to fine tune and improve their capabilities on all aspects. Adequate mechanisms are in place achieving the sustainability of the projects. Though community is reasonably equipped to manage the watershed affairs, the weak spot is mobilisation of financial requirements for maintenance and further improvement of watersheds. Innovative and community friendly methods of accumulating enough community funds is need of the hour.

Most of the planned activities are completed in mejority of the projects well within the budget. Overall, these projects are implemented in a cost-effective manner. Nowhere inefficiency in resource use was observed but there are delays in approval and budget sanction, which require correction.

Impact of the CRS Agricultural Programme

The agricultural programme has made a major impact on economic, environmental and social fronts. Due to watershed activities, there is perceptible positive change in water and soil conservation leading to yield increase of crops, expansion in arable and irrigation potential. The end result was increase in agricultural and food production contributing to food security of population in project areas.

Realisations on environmental front were reversal of declining ground water table and flourishing of vegetation, and initiation of agro-forestry and horticultural activities.

There was reduction in seasonal out-migration, increased school enrolment, and better health and nutritional status due to watershed interventions.

The approach has ensured equity in distribution of benefits as evidenced by the larger flow of benefits of agricultural programme to less advantaged sections of the population and women. Another silver lining is empowerment of community members with new technical and managerial skills required for agricultural production and watershed management. Participatory approach through involvement of functional committees at watershed level in planning and implementation of agricultural developmental activities has inculcated harmony and spirit of accommodation among community members. They have realised the merits of collective action and group approach.

Project Implementation

Community participatory approach in watershed projects had assured incorporation of more practical aspects and local realities in design of the projects. Formations of committees such as WC, WUG, FPC and WG have improved the technical and management efficiency in watershed activities. Involvement of committees have created among the community members a sense of ownership of the projects.

Monitoring, evaluation and MIS systems which are in place in CRS agricultural programme have contributed significantly to improve efficiency in programme implementation. There are still some gaps, which are to be addressed for improving the mechanisms of monitoring, evaluation and MIS. The achievement of the programme on human resource development and management has been impressive. The trend must be continued in the future. FFW has emerged to be a useful facilitating factor in achieving the objectives of CRS agricultural programme.

Linkages of CRS partner with governmental and non-governmental organisations have been weak. Innovative and bold initiatives are required to strengthen inter-agency and inter-sectoral (within CRS) linkages.

CRS's approach in partner network has withstood the test of time and has emerged successful. CRS respects and recognizes the role and contribution of its partners, which has cemented the network. However, it is important that more meaningful interactions are required to improve the efficiency in project implementation and make a more desirable impact on targeted population.

Exit Strategy and Sustainability

In CRS assisted watersheds, a closer examination confirms that mechanisms have been built into the project/system to achieve sustainability of the projects beyond the project period. However, though community members have acquired technical and organisational capabilities to manage the watersheds, there are no appropriate mechanisms for building up financial resources to maintain the watersheds into the future. Innovative and community-friendly measures are required to build adequate financial resources to sustain activities and management of watershed projects. Majority of respondents in various groups reported that CRS agriculture programme created a sustainable impact because of adequate community awareness and social mobilisation, formation of self-help groups of women and men, diagnostic and indicative design features. Some of the threats that may arise are influence of larger farmers in watershed management, occurrence of drought in certain regions and lack of effective communication between partners. These issues must be addressed then and there.

Future of CRS Programme

CRS agricultural programme has been planned and implemented scientifically. Consistent efforts were made to qualitatively improve the programme with new concepts and strategies. What is accomplished during (1997-2001) has provided a strong base for better planning of future activities. Rich experiences gained in the process of implementation of innovative strategies such as community-based watershed approach, initial environmental examination and food for work have instilled enormous confidence to all stakeholders of the programme.

CRS agriculture programme in its form with more new features has emerged as a most successful model providing multiple benefits. Hence the programme should be extended to larger geographical areas. Solving minor gaps in the existing programme will add to further perfection in the programme. CRS approach has shown new innovative methods for successful implementation of watershed projects. Other implementing agencies must pluck a leaf or two to implement their programmes successfully.

ANNEXURE I

Statewise Beneficiaries, Tonnage and Percentage of Total Resources Utilized

State	Beneficiaries	Food Metric tonnes	Per cent of total resources
Bihar/Jharkhand	11026	2239	14
Orissa	8788	1785	11
West Bengal	4266	866	5
Madhya Pradesh/Chhattisgarh	30064	6221	36
Himachal Pradesh	882	179	1
Uttar Pradesh	2717	552	3
Dadra & Nagar Haveli	1857	377	2
Gujarat	11151	2264	13
Maharashtra	355	72	5
Rajasthan	11540	2343	14.5

ANNEXURE II

Geographic Spread of CRS Agriculture Sector Programme (after bifurcation of Bihar and MP)

State	Number of					
	CPs	OPs	Districts	Blocks	Projects	Watersheds/ Prewatersheds
Bihar & Jharkhand	3	7	5	9	9	9
Himachal Pradesh	1	9	4	4	12	0
Gujarat	2	19	9	17	44	30
Madhya Pradesh & Chhattisgarh	7	77	8	40	242	42
Maharashtra	2	2	2	2	4	1
Orissa	3	8	5	8	8	8
Rajasthan	2	9	3	7	16	12
Uttar Pradesh	3	13	6	10	15	2
West Bengal	1	7	2	4	8	8
Dadra & Nagar Haveli	1	6	1	1	8	5
Total	25	157	45	102	365	117

ANNEXURE III

Impact Indicators

A.	**Economic Impact**
1.	Yield per ha (Karif/Rabi/seasons for major crops - 8 crops (Baseline vs. Final Evaluation Survey)
2.	Expansion in arable land (ha)
3.	Irrigation potential created
4.	Area brought under irrigation
5.	Cropping intensity and crop diversification
6.	Other indicators: Land value, addition to livestock, fodder production, household saving
B.	**Impact on Environment**
7.	Rise in groundwater table
8.	Area expansion under vegetative cover
9.	Improvement in soil quality (Perceptions of respondents)
C.	**Social Impact**
10.	Reduction in seasonal migration
11.	Increase in school enrolment
12.	Equity in distribution of benefits
13.	Improved health and nutritional status
14.	Benefits to women
15.	Improvement in technical and managerial skills of community members
16.	Organisational capabilities
D.	**Process Indicators**
17.	Participatory watershed project design
18.	Formation of functional committees
19.	Improvements in Monitoring, Evaluation and MIS Systems
20.	Human resource management
21.	Improvements in logistics
22.	FFW as a facilitating factor
23.	Linkages inter-sectoral and inter-agency
24.	Measures for sustainability
25.	Shift in programme strategy

Watershed Management: Issues and Policies for 21st Century pp. 182-191

Some Impact Indicators and Experiences of Watershed Development in Drought Prone Areas of Tamil Nadu

K. Ramaswamy and K. Palanisami

INTRODUCTION

Soil and water conservation measures play a vital role in resources management and improving socio-economic condition of the rural folk through watershed development programmes. Government of India (GOI) and Government of Tamil Nadu funded Drought Prone Area Programme (DPAP) with the Water Technology Centre (WTC) of TNAU as the Project Implementing Agency (PIA) and DRDA, Coimbatore as the nodal agency. Also, on-farm research in 84 watershed areas has been taken up by the WTC as a part of research component in National Watershed Development Programme for Rainfed Areas (NWDPRA). The main focus of this paper is to describe the key indicators used to evaluate the impact of watershed development programmes in drought prone areas of fifteen micro watersheds (8463 ha) of Coimbatore district. The experiences gained so far in execution of all the watershed measures are grouped as under:

1. Issues on people participation.
2. Transfer problem of agricultural technologies.
3. Issues on engineering measures.

The speciality of the scheme is to involve users and beneficiaries in planning process itself besides developmental activities which will likely to yield better results than target oriented Governmental programmes. Hence, watershed development associations and committees involving village Panchayat President/Members have been formed by the Society Registration Act for each watershed. Initially Entry Point Activities (EPA) were done to build up the rapport between PIA (TNAU) and village people. The EPA works include: school building repairs, cleaning of roads, drinking water supply, drainage facilities, desilting of tanks, etc. Also, the awareness camps and training of village youths, women, self-help and user groups were periodically done with reference to watershed technologies and related skills. Each watershed was allotted a sum of Rs.20 lakhs at the rate of Rs.4000 per hectare as per GOI guidelines for regular development works, amounting to Rs.300 lakhs for 15 watersheds.

BACKGROUND OF SELECTED WATERSHEDS

Coimbatore district falling in Western Agro-climatic Zone of Tamil Nadu comprises of low rainfall areas accounting for 600-650 mm mean annual rainfall with 20-30 rainy days. The zone has undulating topography sloping towards West to East with small hillocks at a few places having an altitude ranging from 171 to 1525 m above Mean Sea Level (MSL).

The Western and Northern parts of the zone are bounded by the Western ghats bordering Kerala and Karnataka, states with peaks ranging from 1000 to 2700 m above MSL. The geological formations of this region consist of hard rocks of charnokites, pennisular geniss and the youngest intrinisives.

In majority of blocks like Sulur, Palladam, Avinashi, Annure and Tiruppur, industrial economy complemented the agricultural economy after 1970's. Agricultural technology has played an effective role in crop production from 1940's to 1970's due to the availability of ground water to the crop needs. There was considerable shift from agriculture to other industrial activities at present. This is mainly because of depletion of ground water in this region. There is no sources of surface water supplies like canals to augment the ground water recharge in these particular blocks where irrigated area is less than 20 to 30 per cent of the cultivable area. The tanks comparatively low in number harbour *Acacia* species; encroached and silted up with disappearance of many of them. Ultimately, the areas are becoming drought prone with natural resources deterioration and poor socio-economic conditions of rural-folk.

Three micro watersheds in each block of Annur, Avinashi, Palladam, Tiruppur and Sulur are the areas coming under the programme. The area of each watershed ranges from 500-750 ha. Most soils vary from sandy loam to sandy clay loam in texture except in parts of two watersheds (Arasur) of Sulur block where black cotton soils exist. More than 50 per cent area in Palladam and 30 per cent in Avinashi watersheds constitute shallow soils of 45-75 cm depth which is underlain by stony strata followed by hard rocks. More generally, the ground water is extracted through tube wells and all the existing open wells are dried (except a very few in each watershed). The average water table during the most part of the year is 75-100 m. The dryland crops of sorghum and pulses (Blackgram/green gram/cowpea) are common during North-East monsoon period. Cotton crop in Avinashi, Bengal gram in Sulur block, Groundnut in parts of Tirupur and Avinashi are some other common crops. Coconut, banana and vegetable crops are grown in batches of irrigated tracts in all watersheds.

TREATMENT MEASURES

The nature of developmental works vary from summer ploughing and contour bunding in upper reaches to major check dams farm ponds and percolation ponds in lower most valley portions. The proposed measures and the achievements in 15 DPAP watersheds of five blocks are given in Table 1.

MAJOR ISSUES IN IMPLEMENTATION

Issues on human behaviour in watershed development

- The inherent socio-economic differences between existing communities/castes coupled with political rivalry make many watershed associations function inefficient or ineffective. Under these circumstances, speed and quality of construction is impaired due to emotional feelings.
- The people in upper reaches of watershed are normally dryland farmers with poor seasonal agricultural returns. Hence, the possibility of using them as supervisors or in some assignment of execution of works shall be explored. Also, it should be ensured that atleast 50 per cent farmers from upper slope should get gainful

employment in other related agro-based industrial/watershed association establishments.

- The frequent festivals traditionally followed from ancient times also interfere in peak season achievements, as the people give more importance to the celebrations of clan deities rather than work. The time frame may be extended for such happenings as they are to be recognized.
- When socially and economically weaker sections are becoming watershed Association Presidents/Leaders due to reservation policies of Government, they have to face hurdles in executing the works effectively by the affluent sections in the same watershed. This is not only due to the long-term benefits enjoyed by the so-called intermediate and immediate castes of the watershed but also not to digest the equality among the castes.
- Also the good quality works initially done by socially-economically weaker section leaders are tempted to gain more economic profits in doing the works due to many compulsions. This led to medium to poor quality works in later part. In one case construction was destroyed by the other rival groups (example: Destruction of check dams constructed in Vallipuram and Kuppepalayam watersheds).
- In some cases, good quality works were being ensured and checks were being done among different groups of people mutually about the quality of work, as time progressed and more awareness was created. This happened many times only after initial clashes among the different groups.

Table 1. Accomplished measures in watershed development

S. No.	Name of the Measure	Achievement
I	Land Resources Development	
	1. All kinds of bunding (ha)	874
	2. Land levelling & shaping (ha)	150
	3. Summer ploughing (ha)	2300
	4. Retaining well (Running metre)	4386
II	Water Resources Development	
	1. Percolation and farm ponds	70
	2. Drainage line treatment (Major & Minor check dams)	395
	3. Gully control (Loose rock check dams)	228
	4. Desilting and renovation of existing tanks	18
III	Afforestation and Pasture Development	
	1. Agro forestry / Social forestry (ha)	195
	2. Road side plantation (ha)	30
	3. Fruit tree / Horticulture plantation (ha)	120
	4. Fodder demonstration plots (No.)	420
	5. Nursery raising / Supply of seedlings (A central nursery was established)	
IV	Other Activities	
	1. Vetiver/Grass plantation	750
	2. Sericulture (ha)	30
	3. Crop demonstration (ha)	1500
	4. Entry point activities (Number)	98

Major issues on agriculture related technologies

- The plantation measures need mass adoption approach (every family to be involved) for further maintenance after the project period.
- Model farms with proven watershed technologies should be established as demonstration units in each micro watershed.
- Adoption of dryland technologies like improved seeds/seed treatment, *in situ* moisture conservation, good package of practices have a lag time of 3 to 4 years after construction of structures to yield full results in upper reaches of watershed. Horticulture-silviculture based systems to middle reaches and low water consuming crops with efficient water management practices without any further increase in number of bore wells for exploitation is the immediate step to be taken up for preventing further deterioration of agriculture in these areas.

Issues on engineering measures

- Accurate prediction of water yields and peak rates of runoff could not be done due to poor data base with the time-frame scheduled (already some basic studies have not been carried out) which often results in under estimation or over estimation of the cost component in the order of 20-30 per cent.
- More often designs of water impounding structures based on rigorous theoretical criteria should be modified as per the convenience of adjoining patta lands (due to narrow space/leakage/prevention of flows/spoiling of adjoining field crops).
- Stability of bunds is rarely attained by earth movers like JCB machines compared to chain dozers and also poor establishment of grasses due to barren and compacted soils on side slopes and top, forming rill erosion.
- More number of small water storage structures/check dams are preferable than big structures.
- The traditional technologies coupled with refined techniques should be encouraged based upon local resources available.
- The prefabricated structures for different components could be developed like head wall, side wall, apron, wing walls, etc. for check dams so that curing problems due to water scarcity resulting in creacks development will be avoided.
- As water itself is a costly input which should be transported three / four kilometers by vehicles, the water costs need to be included in estimates for DPAP watersheds.

DEVELOPED INDICATORS FOR IMPACT EVALUATION

Watershed management projects have been using many principal monitoring and evaluation indicators. The overall impact on ecological or social or economic parameter is difficult to be captured by a single master indicator. The objective could however be achieved by using composite indicators as listed in Table 2.

Watershed eco-index (WEI)

This index is used to represent the fraction of green area created in the watershed.

$$\text{Induced WEI} = \frac{\text{New green area created}}{\text{Watershed Area}}$$

The values of induced WEI ranges from 4 to 43 per cent depending upon the establishment for all the 15 micro watershed during the initial project period (Table 2). However, it has fallen to 1.5 to 21per cent within three years of establishment which include one year of post - project period also, depending upon the maintenance of the plantation by the people both during project and post-project. The index varies from watershed to watershed.

Table 2. Watershed eco-system index in Coimbatore DPAP Watersheds

S.No.	Name of the watershed	Total watershed Area (ha)	New green area (ha)	Induced watershed Eco-index (WEI) (during the project)	Actual WEI (Post-project)
1.	Kattampatti-I	460.00	120	0.26	0.18
2.	Kattampatti-II	467.50	124	0.26	0.16
3.	Kuppepalayam	672.50	118	0.18	0.12
4.	Karumapalayam	752.50	91	0.12	0.10
5.	Cinneripalayam	542.85	120	0.22	0.17
6.	Naduvacheri	767.50	25	0.04	0.015
7.	Arasur-I	605.00	138	0.23	0.22
8.	Arasur-II	590.00	110	0.18	0.17
9.	Rasipalayam	560.00	120	0.21	0.13
10.	Kodangipalayam-I	455.00	129	0.28	0.26
11.	Kodangipalayam-II	465.00	20	0.05	0.04
12.	Ichipatti	405.00	175	0.43	0.21
13.	Vallipuram	390.00	124	0.31	0.19
14.	Merkupathy	708.00	109	0.15	0.10
15.	Sokkanur	640.00	115	0.18	0.15
	Total	8463	1637	0.20	0.20

The efficiency of eco-system improvement of each watershed may be represented as the ratio of actual WEI to the induced WEI. This will reflect mainly the involvement of the community for the project maintenance with a little influence to the climatic externalities. As there is no appreciable variability in the variations of rainfall/temperature occurrence during the study in the same district, this efficiency of Watershed Eco-system Improvement is considered to be an important indicator of participation by the community. The efficiency is relatively high to Arasur and Kodangipalayam I watersheds due to the reasons mentioned above (0.94 and 0.93 respectively).

Induced surface water storage indicator

Construction of chokedams, farm ponds, percolation ponds and deepening/renovation of existing water storage structures enhanced the available storage capacity in the watersheds to store runoff water for surface water use and groundwater recharge. Watershed wise surface storage capacity multiplied by the number of fillings in a year will give the total surface water storage.

$$\text{Indicated Surface Water Storage Indicator} = \frac{\text{Total surface water storage (in a given year)}}{\text{Total Watershed Area}}$$

$$= \frac{\text{Total surface water capacity} \times \text{No. of fillings}}{\text{Total watershed area}}$$

The values of the indicator ranges from 0.08 cm to 0.29 cm (Table 3) for the watersheds under study.

Table 3. Induced surface water storage indicator for DPAP Watersheds (1999)

Name of the watershed	Additional storage created (ha/cm) with one filling	Area of watershed (ha)	Induced surface water storage indicator (cm)
Kuppepalayam	119.57	672.50	0.18
Kattampatti-I	92.99	460.00	0.20
Kattampatti-II	136.29	467.50	0.29
Nadurachery	82.95	767.50	0.11
Cinneripalayam	82.76	524.85	0.16
Karumapalayam	63.28	752.85	0.08
Rasipalayam	43.21	560.00	0.08
Arasur-I	70.55	605.00	0.12
Arasur-II	106.31	596.00	0.18
Ichipatti	46.13	405.00	0.12
Kodangipalayam-I	129.43	455.00	0.28
Kodangipalayam-II	48.77	465.00	0.11
Sokkanur	125.60	640.00	0.20
Vallipuram	97.07	390.00	0.25
Merkupathy	136.69	708.00	0.19
Total	1381.54	8462.85	0.16

Channel water flowing time factor

Owing to the treatment measures and structural/biological interactions, the visible time of water flow in major and medium gullies, channels leading to the common drainage point is substantially increased in many treated watersheds. Two rainfall events before and after treatment were used to measure the visible time of surface flow in channels and the mean value of all major gullies for each watershed is used to arrive the channel water flowing time factor as given below.

$$\text{Channel Water Flowing Time Factor} = \frac{\text{Mean of all times of flow in major gullies/ channels after treatment}}{\text{Mean of all times of flow in the same gullies before treatment for the same rainfall}}$$

Three watersheds subjected to treatment were used for the computation of the above parameter (Table 4). The higher values indicate the good retention of water or indirectly a measure to know the increased time of concentration of flows.

Increase in ground water table

On the basis of data collected from sample survey of dug walls after interviewing with farmers, it was found that the water level in the wells has risen in the range of 1.0 to 2.8 m with a mean rise of 1.8 m. Range of increase in water table for different watersheds is given below (Table 5). However, these data are to be again confirmed with regular monitoring of data with different commonly occurring seasonal rainfall. The ratio of the averages before and after treatment should be used as indicator for measuring the effectiveness.

Table 4. Channel water flowing time in selected DPAP Watersheds

S.No	Name of the watershed	Channel mean water flowing time (Hours) Before treatment	After treatment	Channel Water Flowing Time factor
1.	Karumapalayam	2.7 (6.1)	46.5 (6.0)	17.22
2.	Cinneripalayam	2.9 (6.5)	49.2 (6.0)	16.97
3.	Sokkanur	1.8 (6.7)	39.3 (6.6)	21.83

Note: Figures in parenthesis indicate the rainfall in cm.

Table 5. Water table increase in selected watersheds of Coimbatore

S.No.	Name of the watershed	Range of water table rise (m)	Mean water table rise (m)
1.	Kuppepalayam	1.5 - 2.0	1.75
2.	Kodangipalayam-I	2.0 - 3.0	1.50
3.	Arasur-II	1.8 - 2.8	2.30
4.	Karumapalayam	1.5 - 2.5	2.00
5.	Sokkanur	1.0 - 1.5	1.25
6.	Ichipatti	1.5 - 2.5	2.00

Duration of water availability in the wells as influenced by percolation ponds

Duration of water availability over the year has been increased in number of wells as revealed by the sample survey conducted from 27 wells of two watersheds (Table 6) due to the DPAP developmental works.

Table 6. Duration of water availability as influenced by the percolation ponds

Duration of well alive for	Kattampatti		Arasur-II		Total	
	Before (Nos.)	After (Nos.)	Before (Nos.)	After (Nos.)	Before (Nos.)	After (Nos.)
12 months	2	3	2	6	4	9
9 months	8	11	6	6	14	17
6 months	5	1	4	–	9	1
Total	15	15	12	12	27	27

Silt deposition

Checkdams have arrested silt in their upstream in the gullies to stabilise the gully beds during the one monsoon season under study after the treatment. Some of the small check dams were found to have been silted nearer to the crest level in Arasur, Rasipalayam etc., Silt was also arrested and deposited in the pondage area of percolation ponds. After heavy rains in the monsoon season of Oct-December, 1999, a substantial amount of silt was arrested by these structures. On an average, 104.7, 304.11 and 456.59 m^3 of silt was deposited behind loose rock boulder checks, minor check dams and major check dams respectively in the chosen 8 watersheds for the study (Table 7).

Table 7. Rates of silt deposited in check dams in selected watersheds

Watershed	Gully control measures (LBCD)		Minor Check Dams		Major check dams	
	Nos.	Silt deposited (m^3)	Nos.	Silt deposited (m^3)	Nos.	Silt deposited (m^3)
Kattampatti - I	17	6.95	10	18.35	7	56.62
Kattampatti - II	12	23.54	6	34.13	12	12.48
Kuppepalayam	15	3.13	25	37.44	8	173.16
Karumpalayam	20	41.52	24	49.96	7	20.63
Cinneripalayam	25	15.81	27	54.61	–	–
Arasur - I	6	13.75	8	45.89	7	38.36
Arasur - II	0	0	4	22.56	6	91.32
Rasipalayam	0	0	24	41.17	8	64.02
Total	95	104.70	128	304.11	55	456.59
Average		1.10		2.38		8.30

Diversified land use potential index and cropping intensity

The impact on land use change could be calculated by the fraction of land use under horticulture, afforestation etc. from the traditional agriculture or Government fallow lands. For the fifteen watersheds, 100 ha of patta (private) land has been diversified into horticulture plantation and 24 ha into Mulberry plantation for sericulture. The overall cropping intensity was also found to have increased from 57 to 75 per cent after the project.

Additional irrigated area

As a result of enhanced water availability the additional area ranging from 7 to 21 per cent in 15 DPAP watersheds was brought under irrigation.

Employment generation potential

Two kinds of employment opportunities i.e. causal and regular were generated through implementation of various soil conservation and other works. Casual labour employment was created during the implementation of the works such as bunding, leveling, check dams, ponds/tanks etc. apart from skilled workers like watershed secretaries/volunteers owing to the diversified land use system, regular employment opportunities, crops etc were also generated. On an average, about 12000 man days of casual employment per watershed and about 4000 man days of regular employment can be expected in each watershed as calculated from work-expenditure details and actual surveys.

Financial analysis

The benefit - cost ratio (BCR) and the internal rate of return (IRR) of all the 15 watersheds worked out to 159:1 and 38 per cent.. The financial indicators of different blocks (three watershed pooled per block) are indicated (Table 8). The analysis showed that the project is economically viable.

Table 8. Financial indicators of DPAP watersheds, Coimbatore

Financial indicator	Name of the block					Overall
	Annur	Avinashi	Sulur	Pollashi	Tiruppur	
BCR	1.62	1.57	1.68	1.61	1.47	1.59
IRR	43	34	48	38	29	38

Crop productivity index (CPI)

Crop yields were improved depending upon the crop and watershed conditions. This index was calculated by the ratio of the actual farmers yield to the normal yield of a crop as per the package of practices. Overall CPI was increased from 0.62 for the pre-project period to 0.70 after the project, registering an increase of 13 per cent in productivity level of crops. Per cent increase in crop productivity level varies from 3 per cent in Bengal gram to a maximum of 55 per cent in fodder sorghum due to increased level of input usage and better soil moisture regimes.(Table 9)

Table 9. Crop Productivity Index (CPI) In DPAP Watersheds, Coimbatore District

Name of Crop	Before			After			Change (per cent)
	Average yield kg/ha	Potential yield kg/ha	CPI	Average yield kg/ha	Potential yield kg/ha	CPI	
Jowar	12000	30000	0.40	18500	30000	0.62	55
Maize	2800	50000	0.56	3200	50000	0.64	14
Paddy	3050	4750	0.64	3400	4750	0.72	13
Coconut	85	97	0.88	92	97	0.95	8
Banana	25000	45000	0.56	32000	45000	0.71	27
Brinjal	16000	27500	0.58	18250	27500	0.66	14
Chilly	6250	12500	0.50	6800	12500	0.54	8
Tomato	18850	35000	0.54	20250	35000	0.58	7
Bhendi	6800	13500	0.50	7400	13500	0.55	10
Sugarcane	113000	12000	0.94	118500	12000	0.98	4
Ragi	2100	4500	0.47	2250	4500	0.50	6
Cotton	1050	1800	0.58	1100	1800	0.61	5
Turmeric	22500	27500	0.82	26000	27500	0.95	16
Cow pea	3250	5000	0.65	3850	5000	0.77	18
Green gram	550	750	0.73	625	750	0.83	14
Bengal gram	625	1000	0.63	650	1000	0.65	3
Overall			0.62			0.7	13

Crop diversification index (CDI)

Season-wise crop diversification index was worked out based upon the area under each crop in different seasons for the period before and after the project. CDI values nearer to 1 shows the complete diversification.

Particlulars	Before	After	Per cent
CDI (I season) (January - July)	0.83	0.98	18
CDI (II season) (August - November)	0.50	0.64	28

The higher percentage change in the second season is due to better land management and improved water availability.

Fertilization index of crops (FIC)

This index denotes the extend of crop nutrients (NPK) applied to the crop in relation to the normal recommended level of nutrients to that crop. Pre-project FIC varied between 0.4 and 0.90 while this was observed to have increased in the range of 0.66 to 1.16 after the project. Overall FIC increased from 0.67 to 0.86, indicating an increase of 28 per cent in fertilizer application rate (Table 10).

Table 10. Fertilization index of crops (FIC) under DPAP Watersheds in Coimbatore District

Name of crops	FIC		
	Before	After	Change (per cent)
Fodder sorghum	0.43	0.66	53
Maize	0.54	0.82	52
Paddy	0.59	0.72	22
Coconut	0.54	0.76	41
Banana	0.91	1.16	27
Brinjal	0.60	0.95	58
Chilly	0.52	0.66	27
Tomato	0.63	0.87	38
Bhendi	0.84	1.02	21
Sugarcane	0.86	0.99	15
Mulberry	0.49	0.60	22
Cotton	0.79	0.93	18
Turmeric	0.79	0.95	20
Cow pea	0.84	0.92	10
Overall	0.67	0.86	28

CONCLUSION

There is a positive impact on many issues of watershed developmental works carried under DPAP in Coimbatore district. Parameters developed in this paper may be used to evaluate the impact of various watershed development measures including people participation. These are to be calculated periodically with a systematic effort over years and future suggestions can be incorporated for implementation. However, these indicators are only suggestive but not exhaustive. Some more indicators need to be necessarily added and meaningful composite and participatory evaluation should be done for studying the watersheds responses under different treatments.

References

Ramaswamy, K. P., Gnanamurty and A. Mhamed Ali (1997). Watershed development - A soil and water conservation perspective. Bulletin published by the Water Technology Centre, Tamil Nadu Agrl. University, Coimbatore.

Sikka, A.K. (2000). Evaluation Report of DPAP Watersheds in Coimbatore district. CSWCRTI, Ooty.

Watershed Management: Issues and Policies for 21st Century pp. 192-210

Impact of Watershed Development Programme in Augmenting Groundwater Resource in Drought Situation

Bisrat Alemu, M. G. Chandrakanth and Mahadev G. Bhat

INTRODUCTION

Earth's total volume of water is 1400 million cubic kilometers of which 97.3 per cent is salt water and 2.7 per cent is fresh water. Of this fresh water 75.2 per cent is frozen in polar regions, 22.6 per cent is groundwater and 2.2 per cent is surface water. Accordingly there is a wider scope for groundwater extraction than surface water all over the world. In India, groundwater contributes to 56 per cent of the area irrigated while surface water contributes to 44 per cent (Directorate of Economics and Statistics, 2000). In a few pockets in the dry agroclimatic zones, groundwater irrigation is growing by 10 per cent. This is due to increasing demand for fruits and vegetables from the ever growing urban population in proximity.

Hard rock areas are geological formations of hard rocks (inter alia, granite, gneiss, schist) which have very low groundwater recharge. Such areas constitute 60 per cent of the geological formation in India (Nagaraj and Chandrakanth, 1997). Though groundwater use in the hard rock areas is a nature's benediction, due to difficulty in defining property rights, the recharge, extraction and use pose efficiency and equity problems as the areas are fraught by poor or low recharge, over exploitation and uncontrolled extraction and use. With weakening institutions, over-exploitation due to the 'tragedy of the commons' is resulting in cumulative well interference reflected through initial and premature failure of wells. The government policy to offer soft loans to construct/drill irrigation wells coupled with a policy of flat electricity tariff have triggered the over exploitation of groundwater in several pockets. The tragedy of the common phenomenon due to non exclusive ownership of groundwater is leading to excessive depletion relative to the social optimum. The cost of extraction of groundwater is also increasing with the current rate of extraction and the cumulative extraction to date.

Policy support for the implementation of Watershed Development Programme (WDP) has dampened the tragedy of the commons by augmenting groundwater recharge, thereby improving the efficiency and equity. This paper highlights the economic contribution of WDP to address the commons tragedy in groundwater extraction.

BACKGROUND OF IRRIGATION

The earliest irrigation in India was in 2500 BC. The history of water use for irrigation is as old as the history of irrigation tanks in southern India. Karnataka state has the largest number of irrigation tanks. These tanks were built by village communities, kings and local leaders among others, with altruistic and social motives as an act of piety conferring religious merit.

Among the public works, the king or local leader was advised to offer the first priority to construction of irrigation tank, then to construction of wells and construction of temple[1]. In Karnataka more than fifty per cent of the irrigation tank structures were built between eleventh and thirteenth centuries. With 40,000 irrigation tanks spread over 32,000 villages of Karnataka in itself is a reflection of the ingenuity of farmers to have chosen watershed approach to recharge the hard rock aquifers. Numerous stone and metal inscriptions in Karnataka highlight the construction of irrigation tanks (Appadorai, 1936) with altruistic motives to deliver 'use', 'non consumptive use' and 'option' values. After creating the irrigation tanks as stock resources to conserve rain water, maintenance of irrigation tanks by desiltation and repair of tanks were considered vital. The institution of 'charity' enabled such public works. In addition, farmers desilted the irrigation tanks and applied the silt as manure to their farm lands fields periodically. These practices were largely responsible for recharge of groundwater and this enhanced the longevity of irrigation wells along with sustenance of yield of water (Chandrakanth and Jeff Romm, 1990).

At present, construction of irrigation tanks and maintenance by desiltation are nostalgic memories, with the advent of chemical fertilizers and high yielding crop varieties responsive to agrochemicals. It is very rare that farmers apply tank silt as manure to fields.

WATERSHED DEVELOPMENT PROGRAMME

'Watershed' is a delineated geographical unit with aerable and non aerable lands from ridge to valley allowing the rainwater to flow through a common drain. Watershed treatment involves inter alia afforestation, bunding, construction of percolation tanks, check dams, ravine reclamation structures, boulder checks, rubble checks, vegetative checks for *in situ* moisture conservation. In this process the running water is made to walk, walking water to crawl, and crawling water to infiltrate so that soil moisture and groundwater regime improve. As an organic structure, watershed treatments are offered from ridge to valley irrespective of the ownership of land considering the technical requirements, suitability and people's acceptance. WDP is people centered, where people's participation is crucial in the location, management, cost and benefit sharing of the WDP.

Karnataka is the premier State to have successfully initiated watershed development winning nation's acclaim through the National Productivity Council's awards for the best managed WDPs[2]. The State created "Dryland Development Board" during 1985 with plans to implement a WDP in every district. The programme executing officials were drawn mainly from Departments of Forest, Agriculture, Horticulture and Revenue. Considering the success achieved in implementation and the impact, the World Bank funded the watershed development project in Kabbalnala in Kanakapura taluk. University of Agricultural Sciences, Bangalore took an active part in initiating "Operational Research Project" with agronomists, soil scientists, agricultural engineers and agricultural economists working together in a group

---

[1]As advocated by mother of Laxmidhara Amathya during his infant hood: "KEREYAM KATTISU BHAVIYAM SAVESU DEVAGARAMAM MADISU" implying the importance of tanks, wells and temples in that order for the benefit of the people.

[2] For instance the Mittemari Watershed Programme in Kolar District, won the National Productivity Council award for the best managed watershed programme during 1987. Similarly the State has won such awards for other watersheds too.

to transfer the technology from 'lab to land'[3]. The State Government created an exclusive "Watershed Development" Department in 2000 with a separate budget commitment. In addition to the State funding, this Department also receives funding from World Bank and other donors for executing WDPs in different parts of the State. The Non Governmental Organizations have also contributed towards watershed development[4] in the State.

The WDP is implemented to improve the farm incomes from rainfed lands in scanty rainfall areas. The rainfed areas of India support 40 per cent of population, contribute 36 per cent of agricultural exports, 44 per cent of the food production, 91 per cent of coarse cereals, 80 per cent of oil seeds and 65 per cent of cotton (Arunkumar, 1998). Since 70 percent of India's agriculture is based on rainfall and seventy per cent are small and marginal farmers (holding below 5 acres) located in these areas, WDP are designed to improve equity in income and wealth among rural people. The watershed project costs are a modest $ 130 per acre while the surface irrigation projects implemented by Irrigation Department on the other hand cost ten times this size.

Research studies dealing with externalities in groundwater have largely dealt with the application of 'optimal control' theory in determining the future path of extraction under various assumptions (Negri, 1990). Other studies have proven the technological (Rama Mohana Rao *et al.* 1996) and financial (Katar Singh, 1991 and Arunkumar, 1998) benefits from WDP in India. The aim of WDP is to improve and conserve *in situ* moisture conservation. But the potential of the WDP to dampen the negative externalities in groundwater extraction, thereby reduce the cost of extraction and enhance the net returns is not addressed by any study. Further, earlier studies have not included the ammortized cost of WDP as a component of cost of water for irrigation. Even after considering the WDP costs and internalizing them along with the negative externality, is the WDP worthwhile? This study answers this question empirically considering the Basavapura WDP in Gowribidanur taluk, Karnataka, southern India implemented in 1994 (Bisrat, 2000). The potential role of WDP in reducing the externalities and thereby reducing the effect of drought and the tragedy of the commons is thus the focus of this study. The benefits due to Watershed Development Programme on irrigation wells located in area with high interference among wells are delineated. In this study, the physical impacts are translated in economic terms (Table 1).

Significance

Groundwater irrigation is *sine-quo-non* for hard rock areas starved of other irrigation facility. Groundwater has been over-exploited and these areas are fraught with negative externalities in groundwater extraction due to cumulative interference. In such areas endurance of losses by small and marginal farmers due to initial and premature failure of irrigation wells is an equity issue. The prime motive of this study is to analyze the role of WDP in reducing the degree of effect of drought and tragedy of the commons and in the process improve efficiency and equity.

---

[3]Professors Pandurangaish, Ramana Gowda, Ranganatha Sastry, BR Hedge, AM Krishnappa and YS Arunkumar of the University of Agricultural Sciences, Bangalore have been contributing to watershed development in Karnataka. Kudos for the efforts of Mr GS Kedar and Dr Subramanya, from the Indian Administrative Services, and the project directors and implementing officials of several watershed development programmes all over the State for their interest and commitment in demonstrating the ability of the watershed development programme in bringing efficiency and equity.

[4]MYRADA among the NGOs has provided able leadership for such programmes.

Table 1. Physical and economic benefits due to watershed development programme on irrigation wells located in area with high interference among wells

Physical benefit	Economic benefit
Enhanced life of irrigation well	Reduced amortization cost as the cost is spread over larger number of years.
Higher yield of irrigation well	Reduced cost of extraction as the cost is spread over larger volume of water extracted. Increased net returns due to increased gross irrigated area devoted to high value water intensive crops.

Hypotheses

With the assumption that in the context of WDP the aquifer being benefited is a 'common property', the following hypotheses are tested:

- Reduction in extraction costs 'with' over 'without' WDP reflects the magnitude of negative externality internalized by the watershed development.
- Increase in net returns 'with' over 'without' WDP reflects the magnitude of negative externality internalized by the watershed development.

Description of area

Basavapura watershed, Gowribidanur taluk, Kolar district, Karnataka located in the eastern dry agro climatic zone is selected for an empirical assessment of the watershed development impact. Area of this watershed is 948 acres with 692 acres of arable and 257 acres of non-arable land. The annual rainfall is 679 mm. Relative humidity is moderate in March (55per cent) and high in July (98per cent).The watershed has 420 (16 per cent) acres of irrigated land and 272 (39 per cent) acres under rain fed agriculture. The annual utilization of groundwater for irrigation in the taluk (the smallest administrative unit) is 87 million cubic meters (MCM) with 77 wells per MCM of groundwater. Central Ground Water Board (CGWB) initiated watershed development work in Basavapura during 1993-94[5]. Civil works were completed by soil conservation wing of the Department of Agriculture during 1994. A sum of Rs. 844, 600 ($ 18,360) was invested in the programme since 1994. Major portion (81 per cent) of the expenditure was spent on the construction of checkdams and percolation tanks which mainly influenced groundwater recharge.

Sampling

Participatory Rural Appraisal (PRA) mapping of watershed was undertaken to locate water harvest structures and different types of irrigation wells. The PRA mapping covered inter alia location of different types of wells in the up stream and down stream (both functional and non-functional); well depth; location of water harvesting structures; distance between wells; distance between wells and water harvesting structures; farm size; and farmer's name. Consulting the PRA map, a sample of 40 farmers who had irrigation well/s densely

[5]The initiative taken by Dr VS Prakash, the then Hydrogeologist of CGWB, in this regard is commendable.

placed was drawn from the watershed. Another sample of 20 farmers was drawn outside the watershed who owned irrigation wells densely placed for comparison.

Measures of externality

Externality in groundwater irrigation (Chandrakanth, Shivakumaraswamy and Ananda, 1998; Chandrakanth, Shyamasundar and KK Ananda, 1998; Chandrakanth, Sathisha and Ananda, 1998; Chandrakanth, Sushma Adya and Ananda, 1998; Chandrakanth, Basavaraj and Ananda, 1998; Chandrakanth and Arun, 1997; Shivakumaraswamy and Chandrakanth, 1997) is due to forced investment on additional well(s) by the farmer due to initial and/or premature well failure resulting from the commons tragedy. Externality leads to well failure which refers to a:

- Well that dries up because of new well(s) coming in (but not because of decline in rainfall) or
- Well that loses a large degree of its yield because of new well(s) coming in (but not because of decline in rainfall) and/or
- Well that is deepened because of new well(s) coming in, in this study.

The role of WDP is to reduce the effect of drought and degree of tragedy of the commons and in the process reach efficiency and equity. In the first stage, this needs estimation of the negative externalities in groundwater extraction due to tragedy of the commons and in the second stage estimation of how WDP has been able to dampen the externality. In the third stage the influence of the watershed in achieving efficiency and equity in farm incomes is demonstrated.

Estimation of negative externality

Negative externality of the commons tragedy is the difference between 'with and without' WDP and between 'before and after' WDP in the following economic variables: cost of groundwater extraction: a reduced cost implies impact of WDP; net returns: an increased net return implies impact of WDP

Cost of Groundwater Extraction (Irrigation Cost)

Farmers extracted groundwater from different types of wells viz., dug well, dug cum bore well, bore well, filter point bore well[6]. At the time of field data collection (Mar-Apr 2000), farmers were not paying for electricty used to pump groundwater. Thus, there are no explicit costs of irrigation as there were no operational costs of running irrigation pumpsets[7] except maintenance costs. Thus, the major component of (explicit) variable cost of pumping is avoided. However, due to cumulative interference leading to the tragedy of the commons,

[6]Open well, can be round or square shaped, with an average depth of 30 feet, and a radius of 10 to 15 feet, where water is lifted with contrifugal pump); dug-cum-bare well is bore hole of 4 to 6 inches diameter drilled inside adugwell to a depth of 10 to 50 feet, where water is lifted with a centrifugal pump; (surface) bore well is bore well of 6 inches diameter drilled from surface to a depth of 200 to 300 feet, where water is lifted using submersible pump, filter point well is bore well drilled along the river water course to a shallow depth of 30 to 50 feet).

[7]In another study it has has been estimated that it cost one dollar to extract and lift one arce inch of groundwater on a conservative basis.

farmers are incurring external costs of groundwater irrigation imposed by each other. The cost of cultivation is the sum of cost of human labour, bullock labour, machine hours, seeds, fertilizers, manure, plant protection measures, bagging, transporting and packing, the amortized cost of irrigation considering the groundwater used on the farm, and the opportunity cost of working capital at 10 per cent.

Net Returns

Gross return for each crop is the value of the output at prices realized by farmers. Net returns from well irrigation are the gross return from gross irrigated area minus the cost of production of all crops. The difference between the net returns 'with and without' WDP, as well as 'before and after' WDP gives the measure of positive externality or the avoided negative externality.

Cost of Groundwater Extraction

In the absence of electricity meter and water flow meter, estimation of water yield from borewell was difficult. About 87 per cent of the functioning wells in the sample (within and outside watershed) were borewells. Field measurements of water yield from 36 borewells (12 borewells in each village within and outside the watershed) were taken for different delivery pipe sizes and water pressures as expressed by farmers. After 10 minutes, the pump is put on, if the water fell between 0 to 0.5 meter from the source, the pressure was considered as 'low pressure' (LP), between 0.5 meter to 1 meter as 'medium pressure (MP)' and beyond 1 meter, as 'high pressure'(HP) (Fig. 1). The resultant water yields were averaged to obtain a representative figure.

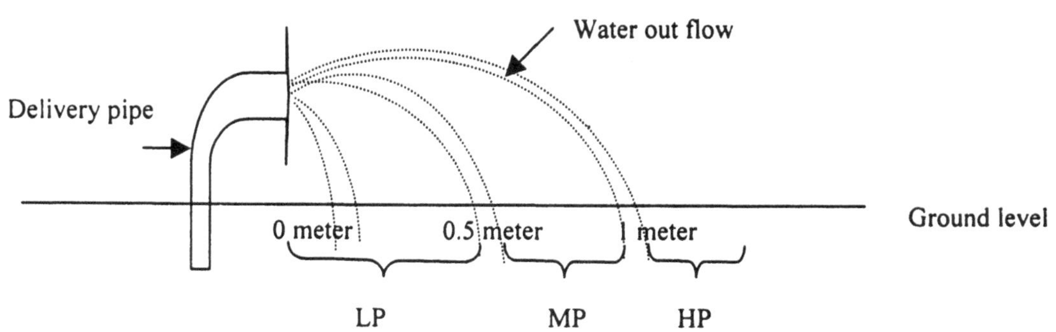

Fig. 1. Pressure of groundwater flow from irrigation bore wells indicated by farmers for estimating the groundwater yield in Basavapura Watershed, Karnataka

Note: Pressure refers to horizontal distance (in meters) between pumped groundwater from the outlet and the ground when pump is on. LP = Low pressure, MP = Medium pressure and HP = High pressure

The water yield of the bore wells was measured by recording the number of seconds taken to fill a bucket of water of say ten litres for each of the wells. This was extrapolated to obtain the groundwater yield in gallons per hour (GPH). The yield of water from other types of wells was also estimated[8]. The total groundwater extracted on a farm is the water applied for each crop over all seasons (summer, *kharif, rabi*) = [(area irrigated in each crop) × (frequency of irrigation per hour) × (number of months of crop) × (number of hours to irrigate) × (Average yield of well in Gallons Per Hour] ÷ 22611[9].

Cost of Irrigation

The cost of irrigation is the amortized cost of (both functioning and non functioning) irrigation well(s), conveyance, storage structure, and annual repairs and maintenance costs. The cost of irrigation depends on the type of well (dug well, dug-cum-bore well, borewell, filter point well), current status of well (functioning or non-functioning), year of construction, average age/life of well, and the chosen discount rate.

Well age and well life

'Age' of irrigation well refers to wells 'functioning' at the time of collection of field data (March-April 2000). The age of the well is the year 2000 minus the year of well construction or sinking or drilling. The average 'age' of wells included 'age' of those wells, which are still functioning given by $\Sigma(f_i X_i) \div \Sigma(f_i)$,

where,

f = Number (frequency) of functioning wells yielding irrigation water in each age group;
X = Age group of wells (0, 1, 2, 3, ..., *n* in years)
i = ranges from zero to n, where n refers to the longest age of well in the group

Such of those wells constructed/drilled during 2000 and still functioning at the time of field data, were assumed to have zero age, as the effect of interference is to increase both the initial and current failures. However, few wells were drilled during the year 2000. As opposed to 'Age', 'Life' of irrigation well refers to the number of years a well already functioned and is no longer functioning at present (time of field data collection). Well life is applicable to totally 'failed'/or 'abandoned' wells after they functioned for some time, including initial failures. Well life was estimated using the same formula as for 'Age', considering only wells, which functioned and failed prematurely, or which never functioned due to initial failure. In this study, 'non-functioning' wells and 'failed' wells are used synonymously as there was no instance of a well which failed initially or prematurely, but became functional later on.

Amortized cost of irrigation

Amortized cost of irrigation is computed as the sum of amortized cost of well, amortized cost of pumpset and accessories, amortized cost of conveyance, amortized cost of over ground storage structure and annual repairs and maintenance cost.

Amortized cost of well = [(Compounded cost of well) × $(1 + i)^{AL} \times i] \div [(1 + i)^{AL} - 1]$

[8]For the estimation of yield of dug well, farmers were asked to indicate the water column height, the well would regain after 24 hours of pumping. For round shaped dugwells the expression ($\pi r^2 h \times 6.2288$) ÷ 24 gives the yield of water in gallons per hour where r = radius of dugwell (in feet), h = height of water column the dugwell regains after 24 hours of pumping (in feet). The volume of water impounded in the well after 24 hours of pumping varies from season to season depending on rainfall, recharge etc.

[9]One acre inch has 22,611 gallons of water.

Where

AL = Average life of wells

Compounded cost of well = (Initial investment on well) × $(1 + i)^{(2000\text{-year of construction})}$

The amortized cost of dug cum bore well is computed as:

Amortized cost of DCBW = {[Estimated cost of DCBW × $[1 + i]^{AL}$ × i] ÷ {$[1 + i]^{AL} - 1$}

Estimated cost of DCBW is the cost of dugwell portion of the investment in DCBW given by:

$$DCBW_{cost} = \{AC_{DW} \times [AL - YEAR_{imp} - Year_{cons})] + IMP_{cost}\} \times (1 + i)^{(2000 - \text{year of imp})}$$

Where,

$DCBW_{cost}$ = Estimated cost of DCBW at current prices (2000)

AC_{DW} = Amortized cost of DW

$YEAR_{imp}$ = Year of improvement of dugwell

$YEAR_{cons}$ = Year of construction of dugwell

IMP_{cost} = Historical improvement cost (such as cost of in bores in the dug well)

The discount rate of two per cent is used in amortization reflecting long term sustainable rate. Similarly investment on conveyance, pumpset, electrical installation, groundwater storage structures were amortized. Labour cost of irrigation was merged with the cost of other cultural operations. The annual cost of irrigation is from all wells on the farm. Investment on WDP is amortized considering the life of WDP as 15 years and distributed over the estimate of total groundwater extracted by all farms in the watershed[10]. The total cost of irrigation is then distributed over the total groundwater extracted on the farm, to obtain the cost per acre inch of water extracted = [Total amortized cost of irrigation] ÷ {[Total acre-inches of water used] + [Amortized cost of WDP per acre inch of water]}.

Access to groundwater

The WDP is expected to improve the accessibility to groundwater resource for farmers. Two types of accessibility measures were used in this study: (1) physical accessibility and (2) economic accessibility. In this study, the physical access to groundwater resource is defined as the volume of groundwater used per acre of gross irrigated area (GIA). Economic access to groundwater refers to the volume of groundwater extracted per rupee of amortized cost of irrigation.

Physical access is hypothesized to increase with the yield of irrigation well and to decrease with cost per acre inch of water. The relationship in log linear form is estimated as:

$\text{Log } Pa = \text{Log } \alpha + B_1 \text{Log } Wy + B_2 \text{Log } Cw$, where Pa = Water used per acre of GIA; Wy = Well yield in GPH and Cw = Cost of water per acre inch.

Factors influencing economic access to groundwater *inter alia* are water used on the farm and distance of irrigation well from water harvest structures. Economic access is hypothesized to increase with water used on the farm and decrease with the distance of

[10]The average water extracted per acre of gross irrigated area is 9 acre inches. The total gross irrigated area of the entire watershed was 863 acres. Thus, the total groundwater use is 7798 acre inches and the cost of watershed project per acre inch of groundwater extracted is 20 cents (Rs. 9.53). An exchange rate of 1 Rupee = US $ 0.02 is taken.

irrigation well from water harvesting structures. The relationship is assumed to follow a log linear form:

$\text{Log Ea} = \text{Log}\,\alpha + B_1\text{Log Wuf} + B_2\text{Log Wd}$, where Ea = Economic access (groundwater extracted in acre inches per rupee of amortized cost of irrigation), Wuf = water used on the farm in acre-inches, Wd = distance of well from water harvesting structures in meters.

RESULTS AND DISCUSSION

Before the commencement of the WDP, the sample of 40 farmers were having 52 irrigation wells, of which 16 (31per cent) were dug wells, 31 (59 per cent) were bore wells, two (4 per cent) were dug cum bore wells and the rest three (6 per cent) were filter point wells. Among the 52 irrigation wells, 15 (29 per cent) were non functional and 37 (71 per cent) were functional earlier due to the implementation of WDP (Table 2). With the introduction of WDP, the number of wells increased to 72 (at a growth rate of 6.4 per cent per year). Among them, 16 (22 per cent) were dug wells, 51 (71 per cent) were bore wells, two (3 per cent) were dug-cum-bore wells and the rest 3 (4 per cent) were filter point wells. During 2000, five (31 per cent) dug wells, five (10 per cent) bore wells and one (50 per cent) dug-cum-bore well was totally abandoned due to well failure.

Among the 16 dug wells, the number of non-functioning wells were nine (56 per cent) before implementation of watershed programme. After programme implementation four dug wells got recharged yielding groundwater and the number of non-functioning dug wells decreased to five (31 per cent). Outside the watershed, before the WDP, farmers had a total of 22 irrigation wells (all of them functional) of which 10 (45 per cent) were dug wells and 12 (55 per cent) were bore wells. After the WDP, number of irrigation wells increased to 34 of which 13 (38 per cent) were non-functional, wherein on an average, two wells non-functioning per year.

Distribution of different types of wells

The proportion of functioning wells increased from 71 per cent to 85 per cent due to the watershed programme. The proportion of non-functioning wells declined from 29 per cent to 15 per cent (Table 2). Outside WDP area, all irrigation wells were functional before 1994. Here, the proportion of non-functioning wells is 38 per cent currently due to the scarcity induced by cumulative well interference effect. All dug wells outside WDP area completely dried up after 1994. Inside WDP four dug wells and one dug cum bore well got recharged.

Age and life of irrigation wells

Bore well is the popular mode of water extraction after the 1980's both within and outside WDP area. Due to the influence of water harvesting structures constructed for groundwater recharge, the life of dugwells, bore wells, dug cum bore wells and filter point wells increased by 4, 5, 3 and 6 years in WDP (Table 3). The dug wells outside WDP have a mere 40 per cent of the life of dug wells in WDP area. The life of bore wells outside WDP was seven years; while the life of bore wells within WDP area was nine years. Since1994, 20 new bore wells were drilled in the WDP area. Outside WDP area, 12 new bore wells drilled . The age and life of bore wells within WDP area was 4.45 and 4.3 years which is 12 and 24 per cent

Table 2. Number of functional and non-functional wells in Basavapura WDP, Karnataka, 2000

(in numbers)

Type of well	Within WDP Area						Outside WDP Area					
	Before WDP (in 1994)			After WDP (in 2000)			Before WDP (in 1994)			After WDP (in 2000)		
	F	NF	TOTAL	F	NF	TOTAL	F	NF	TOTAL	F	NF	TOTAL
Dug well	7 (18.9)	9 (60)	16 (30.8)	11 (18)	5 (45.5)	16 (22.2)	10 (45.5)	–	10 (45.5)	–	10 (76.9)	10 (29.4)
Bore well	27 (73)	4 (26.7)	31 (59.6)	46 (75.4)	5 (45.5)	51 (70.8)	12 (54.5)	–	12 (54.5)	21 (100)	3 (23.1)	24 (70.6)
Dug-cum Bore well	–	2 (13.3)	2 (3.8)	1 (1.6)	1 (9)	2 (2.8)	–	–	–	–	–	–
Filter point well	3 (8.1)	–	3 (5.8)	3 (5)	–	3 (4.2)	–	–	–	–	–	–
Total	37 (71)	15 (29)	52 (100)	61 (85)	11 (15)	72 (100)	22 (100)	0 (0)	22 (100)	21 (62)	13 (38)	34 (100)

Note: Figures in parentheses represent the percentages to the respective total.
F —(Functional wells), NF — (Non functional wells)

higher than the age and life of bore wells outside WDP area, respectively. Farmers have taken advantage of the increased life and age of wells in WDP and have extracted higher volume of water (105 acre inches) compared with before programme implementation (70 acre inches) and outside WDP area (75 acre inches). This further influences the economic access of farmers having irrigation well within WDP, the frequency of investment to drill additional wells and the associated investment in irrigation.

Table 3. Average age and life of wells in Basavapura Watershed, Karnataka, 2000

(In years)

Type of Well	Within WDP Area				Outside WDP Area	
	Before WDP (in1994)		After WDP (in 2000)		After WDP (in 2000)	
	Average Age	Average Life	Average Age	Average Life	Average Age	Average Life
Dug well	48	41	49	45	38	18
Bore well	4.4	4	10	9	8	7
Dug cum bore well	13	10	19	13	–	–
Filter-point well	14	14	20	20	–	–

Note: Age of wells (year 2000 – year of construction)
Life of wells (year of failure – year of construction)
Year 1994 was the year of watershed development programme

Yield of wells

The yield of bore wells was 1150 gallons per hour (GPH) before implementation of WDP and increased to 1426 GPH by 24 per cent (during May 2000) (Table 4). The yield of dugwells before WDP was 264 GPH and increased to 446 GPH by 69 per cent after the WDP. Outside WDP the yield of bore wells was 1470 GPH before WDP (1994) and decreased to 1242 GPH (by 15 per cent) later on by 2000. The yield of dug wells before WDP was 1008 GPH and all the dugwells non-functioning due to groundwater depletion by 2000. In the WDP yield of dug wells in Kharif increased (by 70 per cent) from 264 gallons per hour (GPH) to 446 GPH. Among the nine dug wells, which completely dried up before WDP, four wells got recharged. In the WDP, the yield of bore wells in Kharif increased from 1150 GPH to 1426 GPH an increase of 24 per cent. Yield of filter point wells in Kharif increased marginally by five per cent and was not influenced by water harvesting structures, as they were located on the river path

Cropping pattern within and outside WDP

Shift of cropping pattern towards light or relatively low water intensive crops is the basic coping mechanism available to farmers as the first best solution to cope with scarce water resource for irrigation. However, in the WDP, gross irrigated area (GIA) devoted to water intensive crops like paddy and sugarcane increased from 33 acres to 114 acres (by 245 per cent) as a result of the implementation of WDP (Table 5). The reverse happened outside WDP, i.e., the GIA under water intensive crops like paddy and sugarcane decreased marginally from 30 acres to 26 acres (by 13 per cent) due to groundwater depletion. Thus, the presence of water harvesting structures enabled the beneficiary farmers to adopt water intensive cropping pattern in consonance with the substantially higher volumes of groundwater resource available due to groundwater recharge

Table 4. Groundwater yield of wells in Basavapura Watershed, Karnataka, 2000

(In gallons per hour)

Well type	Within WDP						Outside WDP					
	Before WDP (in 1994)			After WDP (in 2000)			Before WDP (in 1994)			After WDP (in 1994)		
	SF	LF	Over all	SF	LF	Over all	SF	LF	Over all	SF	LF	Over all
Dug well	371	157	264	401	491	446	1477	538	1008	0	0	0
Bore well	1106	1194	1150	1421	1431	1426	1314	1625	1470	940	1544	1242
Dug-cum-bore well	..	626	626	..	626	626	..	..	..	..	..	..
Filter point well	1698	1867	1783	1867	1867	1867	..	..	..	..	..	..

Note: SF = Small Farmers; LF = Large Farmers
The WDP in Basavapura was initiated during 1994.

In the WDP, farmers devoted 48 per cent of their irrigated area to water intensive crops like paddy, sugarcane and mulberry, which fetched them the highest net return. This is due to the larger volumes of groundwater they are reaping, reinforced by the degree of groundwater recharge from the surrounding water harvesting structures. Here, the water intensive crops like sugarcane and paddy share 24 per cent of the total gross irrigated area and contribute to 34 per cent of the net return generated.

Table 5. Cropping pattern in Basavapura WDP, Karnataka, 2000

(In acres)

Crops	Within WDP Area				Outside WDP Area			
	Before 1994		After 1994		Before 1994		After 1994	
	Area	Proportion	Area	Proportion	Area	Proportion	Area	Proportion
Maize	108.00	30.51	117.75	25.23	79.50	35.89	78.50	40.89
Mulberry	66.00	18.64	109.00	23.35	38.00	17.16	30.50	15.89
Paddy	24.00	6.78	88.00	18.85	27.50	12.42	24.00	12.50
Ragi	27.00	7.63	54.75	11.73	9.00	4.06	8.00	4.17
Sugarcane	9.00	2.54	26.00	5.57	2.50	1.13	2.50	1.30
Sunflower	15.00	4.24	24.00	5.14	30.50	13.77	17.50	9.11
Grass (fodder)	14.50	4.10	3.00	0.65	11.00	4.97	1.50	0.78
Flower	3.00	0.85	10.50	2.25	2.50	1.13	6.50	3.39
Chilli	4.25	1.20	4.75	1.02	7.00	3.14	5.00	2.60
Tomato	3.50	0.99	4.00	0.86	4.00	1.81	4.00	2.08
Onion	58.50	16.53	3.50	0.75	9.50	4.29	10.50	5.47
Groundnut	5.50	1.54	18.50	3.96	0.50	0.23	0.50	0.26
Brinjal	–	–	2.25	0.48	–	–	–	–
Potato	15.75	4.45	0.75	0.16	–	–	3.00	1.56
Total	354.00	100.00	467.00	100.00	222.00	100.00	192.00	100.00

Note: The WDP in Basavapura was undertaken in 1994

Investment on water harvesting structures through WDP

The amortized cost of watershed development structures (Table 6) is the annual fixed cost component of the investment in the WDP and is included in the annual cost of irrigation to estimate the economics of irrigation. The amortized cost of WDP per acre-inch of water extracted by farmers was Rs 9.53.

Table 6. Investment on Watershed Development Program in Basavapura Karnataka, 2000

Activities	Physical Unit	Coverage	Total Expenditure
Check dams	No	18	623289
Percolation tank	No	1	63259
Earthen bund	No	1	1367
Boulder checks	No	93	51070
Rubble checks	No	35	47160
Vegetative checks	Sq.m.	460	13930
Gully revetment	Sq.m.	790	44495
Total investment	Rs	–	844600
Total amortized investment per year*	Rs	–	74024
Total well farmers within watershed	No	74	
Total gross irrigated area in the watershed	Acre	863	
Average water used per acre of gross irrigated area (For sample farmers)	Acre inch	–	9
Total water used on the total gross irrigated area in the watershed	Acre inch	–	7767.5
Amortized cost of watershed treatment per acre inch*	Rs	–	9.53

Note: Discount rate is 2 per cecnt; life of watershed structures considered is 15 years.
Amortized cost of WDP per acre-inch = [Rs. 74,024 ÷ 7767.5 acre inch] = Rs. 9.53

Groundwater used, irrigation cost and net returns

Before WDP in Basavapura groundwater used per acre of GIA for small and large farmers was 5.95 and 5.96 acre-inches, respectively. After WDP water use per acre of GIA increased to 8.3 acre inches (by 39 per cent) for small farmers and to 9.21 acre inches (by 55 per cent) for large farmers (Table 7). Cost of irrigation per acre-inch of groundwater used, before WDP implementation in Basavapura micro watershed was Rs.180 and Rs. 164 for small and large farmers, respectively. Cost of irrigation per acre-inch of groundwater used decreased due to the recharge effect of WDP to Rs. 105 (by 42 per cent) for small farmers and to Rs. 81 (by 51per cent) for large farmers.The net returns per acre-inch of groundwater used for small and large farmers in Basavapura was Rs. 1,226 and Rs. 1,037 before WDP. After WDP the net return per acre inch of groundwater used increased marginally to Rs. 1,260 (by 3 per cent) for small farmers and to Rs. 1,485 (by 43 per cent) for large farmers. The net returns per acre of GIA increased from Rs. 7,298 to Rs. 10,505 (by 44 per cent) for small farmers and from Rs. 6,181 to Rs. 13,678 (by 121 per cent) for large farmers after WDP.

The groundwater extraction per acre of GIA was 5.96 acre-inches before WDP, and increased (by 50 per cent) to 8.96 acre-inches due to the recharge after WDP. The groundwater extraction by small farmers was 5.95 acre inches per acre of GIA before WDP and increased (by 39 per cent) to 8.3 acre-inches. The groundwater extraction by large farmers was 5.96 acre-inches per acre of GIA and increased by 55 per cent to 9.21 acre-inches. The groundwater extraction per acre of GIA by small and large farmers is higher than the groundwater extraction by their peers outside WDP area whose groundwater extraction per acre of GIA was 5.6 and 8.8 acre inches, respectively (Table 7).

Table 7. Impact of watershed development programme in Basavapura, Karnataka, 2000

Particulars	Within WDP Area							Outside WDP Area (in 2000)		
	Before WDP (in 1994)			After WDP (in 2000)			per cent change			
	Small farmers	Larger farmers	Overall	Small farmers	Larger farmers	Overall		Small farmers	Larger farmers	Overall
Gross irrigated area (acre)	5.8	11.4	8.8	7.5	15.1	11.7	0	5.7	13.6	9.6
Cropping intensity (per cent)	192	146	156	216	164	175	19	181	159	165
Groundwater used per acre of gross irrigated area (acre inch)	5.95	5.96	5.96	8.3	9.21	8.96	50	5.6	8.8	7.86
Irrigation cost per acre inch of groundwater used (Rs)	180	164	168	105	81	87	-48	257	74	112
Net returns per acre of gross irrigated area (Rs)	7298	6181	6508	10505	13678	12758	96	6051	8757	7961
Average distance of wells from water harvesting structures (meter)	NA	NA	NA	269	188	225	NA	NA	NA	NA
No. of farmers	18	22	40	18	22	40		10	10	20

Note: Irrigation cost and net returns before WDP was calculated based on the current prices. Irrigation cost is amortized cost per acre inch which includes amortized cost of irrigation wells plus the amortized cost of watershed treatment.

NA: Not available

Irrigation cost per acre inch of groundwater in Basavapura watershed was Rs.168 before the WDP, and decreased (by 48 per cent) to Rs. 87 due to the increased availability of groundwater after WDP. The net return per acre inch of groundwater used was Rs. 1097 in Basavapura watershed before WDP, and increased to Rs. 1424 (by 30 per cent) due to the WDP and efficient utilization of groundwater resource. The cost of irrigation per acre-inch of water decreased by 42 and 51 per cent for small and large farmers, respectively after WDP in Basavapura. The cost incurred by farmers outside the watershed programme is 22 per cent higher than farmers in the Basavapura WDP (Table 7). The net returns per acre of

GIA increased in Basavapura watershed after WDP from Rs. 6,505 to Rs 12,758 by 96 per cent. Due to increase in water yield of irrigation wells, farmers in Basavapura watershed began cultivating water intensive crops to reap higher net returns per acre of GIA. The net return per acre-inch of groundwater used increased by 3 and 43 per cent for small and large farmers, respectively after WDP implementation. The net return per farm gained by small and large farmers in Basavapura is 57 and 42 per cent higher than small and large farmers outside WDP respectively. Water harvesting structures have improved economic access to water resource for irrigation by increasing groundwater recharge, reducing the cost of irrigation and increasing the net returns per acre of gross irrigated area (Table 7).

Upstream-downstream benefits

Irrigation cost per acre inch of groundwater used for farmers in the up stream and down stream was Rs 292 and Rs 139, respectively before WDP. After the WDP due to the recharge effect of water harvesting structures, this decreased by 67 per cent to Rs 97 for farmers in the up stream and by 40 per cent to Rs 83 for farmers in the down stream. The irrigation cost per acre inch decreased by 51 and 29 per cent for small farmers in the up stream and down stream, respectively, the irrigation cost decreased by 83 and 42 per cent for large farmers in the up stream and down stream, respectively, after the WDP (Table 8).

The net returns per acre of gross irrigated area for farmers in the up stream and down stream was Rs. 6,508. After WDP, the net returns per acre of GIA for farmers in the up stream increased to Rs. 10,810 (by 66 per cent) and to Rs. 13,672 (by 110 per cent) for farmers in the down stream. The irrigation intensity for farmers in the up stream and down stream was 198 per cent and 174 per cent, respectively. After WDP implementation the irrigation intensity increased to 219 per cent (by 21per cent) for farmers in the up stream and this increased marginally to 175 per cent (by 1per cent) for farmers in the down stream (Table 8). Net returns per rupee of irrigation cost for farmers in the up stream and down stream was 4.4 and 7.52 before WDP implementation. After the WDP, this increased to 13.26 (by 201 per cent) for farmers in the up stream and to 17.9 (by 138 per cent) for farmers in the downstream. The higher rate of increase in the upstream is an apparent pointer to the contribution of WDP, since the downstream benefits are obvious (Table 8).

Physical access to groundwater

Physical access to groundwater measured by groundwater used per acre of GIA is found to be directly proportional to water yield of irrigation well and inversely proportional to amortized cost of water per acre inch. For a per cent increase in the cost of water per acre inch, physical access fell by 0.19 per cent (Table 9). More precisely, if the cost of water per acre inch increases by a rupee (from the geometric mean of Rs. 87 per acre inch) the water used per acre of GIA falls (significantly) by 0.01926 acre inch (= 435 gallons). This indicates that farmers do respond to negative externalities in groundwater extraction despite the WDP. Analogously, if the electricity provided for pumping groundwater is priced, the farmers will make adjustments in their crop pattern to maximize net returns to scare groundwater. In the absence of WDP, this response would have been elastic. This also demonstrates the indispensability of groundwater for irrigation in the watershed. For a per cent increase in

Table 8. Impact of watershed development programme for different classes of farmers in the up stream and down stream in Basavapura, Karnataka, 2000.

Particulars	Before WDP (in 1994)						After WDP (in 2000)					
	Up Stream Farmers			Down Stream Farmers			Up Stream Farmers			Down Stream Farmers		
	SF	LF	overall	SF	LF	overall	SF	LF	overall	SF	LF	overall
Gross Irrigated area (acre)	5	4.8	4.98	7.2	12.8	11.4	7.3	15.4	9.4	7.9	15.0	13.2
Cropping intensity (per cent)	185	118	153	202	151	157	219	194	207	210	158	164
Groundwater used per acre of gross irrigated area (acre inch)	4.99	5.21	5.0	7.3	6.0	6.2	7.4	9.88	8.4	10.14	9.06	9.2
Irrigation cost per acre inch of groundwater used (Rs.)	221	508	292	141	139	139	109	85	97	100	80	83
Net returns per acre of gross irrigated area (Rs.)	7240	4207	6508	7378	6346	6508	10458	11313	10810	10592	14219	13672
Number of farmers	12	4	16	6	18	24	12	4	16	6	18	24

Note: Irrigation cost is amortized cost per acre inch which includes amortized cost of irrigation wells plus the amortized cost of watershed treatment. Net returns per rupee of irrigation cost was derived to compare the net returns per acre inch of groundwater with the irrigation cost per acre inch of groundwater, and calculated as: [Net returns per acre inch of groundwater used ÷ irrigation cost per acre inch of groundwater].

well yield, the physical access rose by 0.77 per cent (Table 9). If the well yield increases by one more gallon per hour (from the geometric mean of 1502 GPH), the water used per acre of GIA increases significantly by 0.0046053 acre inch (= 104 gallons).

Table 9. Analysis of physical access to groundwater in Basavapura watershed, Karnataka, 2000

Dependent variable: Natural Log of groundwater used per acre of gross irrigated area (Geometric mean groundwater use per acre = 8.96 acre inches)

	Coefficient	t-value	R^2	Geometric mean
Natural Logarithm of Intercept	–2.673*	–3.392	0.62*	
Independent variables				
1. Natural Logarithm of well yield (Gallons per hour)	0.772*	7.247		1502
2. Natural Logarithm of cost of water per acre inch (Rs.)	–0.187*	–3.418		87

*Significant at 1 per cent

In 2000 there were 61 functioning wells irrigating a gross area of 467 acres, extracting 105 acre inches per farm. Before the WDP, 37 functioning wells irrigated a gross area of 354 acres extracting 70 acre inches per farm. At the same time, there were 21 functional wells irrigating a gross area of 192 acres extracting 75 acre inches per farm outside WDP areas. This shows the influence on the physical access to water resource through watershed programme. The cumulative interference effect on productivity of functioning wells is dampened by the groundwater recharge in the surrounding water harvesting structures. The extraction of groundwater in the WDP area is higher than outside WDP area. This is due to the cultivation of water intensive crops like paddy and sugarcane. This shows that the physical access to groundwater resource has improved.

Economic access to groundwater

Economic access to groundwater increased with water used on the farm. For a per cent increase in water used on the farm, economic access increased by 0.57 per cent. The distance of irrigation well from water harvest structures did not significantly influence the economic access in the watershed. This apparently is a positive indication of the distribution of benefits across farms irrespective of the distance of the irrigation wells from the water harvest structures within the watershed where WDP is undertaken. This is a positive development from the equity view point.

The results of the study amply prove both the hypotheses of the study confirming the role of watershed development programme. Using 'Before-After' figures, after the WDP, the cost per acre inch of groundwater reduced by (Rs. 168 minus Rs. 87 =) Rs. 81, i.e. a reduction by 48 per cent, when compared with the cost before WDP. Using 'With - Without' figures, with the WDP, the cost per acre inch of groundwater reduced by (Rs. 112 minus Rs. 87 =) Rs. 25, i.e. a reduction by 22 per cent. Using 'Before-After' figures, in the upstream, the cost per acre inch of groundwater reduced by (Rs. 292 minus Rs. 97 =) Rs. 195, i.e. a reduction by 67 per cent. In the downstream, the cost per acre inch of groundwater reduced by (Rs. 139 minus Rs. 83 =) Rs. 56, i.e. a reduction by 40 per cent. These are indications that WDP is responsible for reducing the cost of groundwater, even after including the negative externality cost due to the 'tragedy of commons'.Using 'Before-After' figures, after the WDP, the net return per acre of gross irrigated area (GIA) increased by (Rs. 12758 minus Rs. 6508 =) Rs. 6250, i.e. an increase by 96 per cent, when compared with the net returns before WDP. Using 'With - Without' figures, with the WDP, net return per acre of GIA increased by (Rs. 12758 minus Rs.7961 =) Rs. 4797, i.e. an increase by 60 per cent. Using 'Before-After' figures, after the WDP, in the upstream, the net return per acre of GIA increased by (Rs. 10,810 minus Rs. 6508 =) Rs. 4320, i.e. an increase by 66 per cent. In the downstream, the net return per acre of GIA increased by (Rs. 13,672 minus Rs. 6508 =) Rs. 7164, i.e. an increase by 110 per cent. Thus, on a conservative basis, considering the least savings in the cost of groundwater of 22 percent or the least of enhanced net returns of 60 per cent, there is a clear indication of the WDP impact in reducing the tragedy of the commons which lead to negative externality due to well interference.

Table 10. Regression analysis of economic access to groundwater

Dependent variable: Natural Logarithm of (Groundwater extracted per rupee of amortized cost of irrigation)
(Geometric mean = 0.011 acre inch per rupee of amortized cost of irrigation)

	Coefficient	t-value	R^2	Geometric mean
Natural Log of Intercept	–5.983*	–6.373	0.359*	
Independent variables				
1. Natural Log of water used on the farm (acre inch)	0.57*	4.332		105
2. Natural Log of well distance from water harvesting structures (meters)	–0.203	–1.433		225

*Significant at 1 per cent

CONCLUSION

Considering the WDP experience of six years, and conservative estimates in the reduction in the cost of groundwater by 22 per cent and enhanced net returns to the tune of 60 per cent, the Basavapura Watershed Development Programme has amply proved its contribution towards reducing the effect of drought and reducing the commons tragedy. The negative externality due to partial and complete failure of irrigation wells has been reduced due to watershed development programme. Thus, construction of water harvesting structures through watershed development approach enhances the groundwater recharge in hydro-geological situations even if there is cumulative interference effect among irrigation wells.

As long as there is a WDP, proximity of irrigation wells to water harvesting structures is not a requirement for deriving the benefit from recharge as seen in this study. The watershed development programme contributed richly to physical and economic access to groundwater resource for irrigation. The watershed development programme has helped to reduce the gap between the small and large farmers in respect to physical access to groundwater resource. The small farmers in fact have been able to reap higher net returns per acre of GIA.

References

Appadorai, A. (1936). Economic conditions in Southern India I. (1000-1500 AD). p. 295.

Arun V. and M. G. Chandrakanth (1993). Valuation of externalities in borewell irrigation in hard rock areas of Karnataka - An empirical study, water management: India's groundwater challenge, VIKSAT/Pacific Institute Collaborative Groundwater Project, Ahmedabad, Dec. 14-16.

Arunkumar, Y. S. (1998). Economic valuation of watershed development, Unpublished Ph.D thesis, Department of Agricultural Economics, University of Agricultural Sciences, Bangalore.

Bisrat, A. M. and M.G. Chandrakanth (2000). Economic access to groundwater irrigation in watershed development in Karnataka., Research report submitted to Ford Foundation, Department of Agricultural Economics, University of Agricultural Sciences, Bangalore.

Chandrakanth, M. G., G. Basavaraj and K. K. Ananda (1998). Groundwater scarcity and coping strategies in Channagiri taluk of Shimoga district - An economic analysis. Report submitted to The Ford Foundation, New Delhi, Dept. of Agri Economics UAS, Bangalore, pp. 1-42.

Chandrakanth, M. G. and B. Shivakumaraswamy (1997). Well interference and aftermath: An economic analysis of well irrigation in hard rock areas of Karnataka, *Artha Vijnana*, **39**(3): 341-359.

Chandrakanth, M. G. and V. Arun (1997). Externalities in groundwater irrigation in hard rock areas, *Indian Journal of Agricultural Economics*, **52**(4): 761-771.

Chandrakanth, M. G. and Jeff Romm (1990). Groundwater depletion in India-institutional management regimes, *Natural Resources Journal*, **30:** 485-501.

Chandrakanth, M. G., K. M. Sathisha and K. K. Ananda (1998). Resource eeonomics stUdy of valuation of well interference externalities in central dry zone of Karnataka. Report submitted to The Ford Foundation, New Delhi, Dept. of Agri. Economics, UAS, Bangalore, 1-30.

Chandrakanth, M. G., B. Shivakumaraswamy and K. K. Ananda (1998). Economic implications of unsustainable use of groundwater in hard rock areas of Karnataka, Report submitted to The Ford Foundation, New Delhi, Dept. of Agri. Economics, UAS, Bangalore, 1-30.

Chandrakanth, M. G., M. S. Shyamasundar and K. K. Ananda (1998). Interplay of markets, externalities, institutions and equity in groundwater development - An economic study in the hard rock areas of Karnataka. Report submitted to The Ford Foundation, New Delhi. Dept. of Agri. Economics UAS, Bangalore, 1-43.

Chandrakanth, M. G. Sushma Adya and K. K. Ananda (1998). Scarcity of groundwater for irrigation: Economics of coping mechanisms in hard rock areas. Report submitted to The Ford Foundation, New Delhi, Dept. of Agri. Economics UAS, Bangalore, 1-40.

Directorate of Economics and Statistics (2000). Agricultural Statistics at a glance 2000. Agricultural Statistics Division, Directorate of Economics and Statistics, Ministry of Agriculture, Government of India, Krishi Bhawan, New Delhi 110001.

Katar Singh (1991). Dryland watershed development and management: A case study in Karnataka, *Indian Jr. of Agric. Econ.*, **46**(2): 121-131.

Kerr John M., M. G. Chandrakanth and R. S. Deshpande (1997). Managing Groundwater Development in Karnataka, in J. M. Kerr, D. K. Marothia, Katar Singh, C. Ramasamy and W. R. Bentley (Eds.) Natural Resource Economics Theory and Application in India, Oxford & IBH Publishing Co. Pvt. Ltd., New Delhi, 421-443.

Nagaraj N. and M. G. Chandrakanth (1997). Intra and Inter-Generational equity effects of irrigation well failures: Farmers in hard rock areas of India, *Economic and Political Weekly*, **32**(13)A: 41-44.

Nagaraj N., M. G. Chandrakanth and Gurumurthy (1994). Borewell failure in drought prone areas of Southern India: A case study. *Indian Journal of Agricultural Economics*, 49(1): 102-106.

Nagaraj, N. and M. G. Chandrakanth (1995). Low yielding irrigation wells in peninsular India - An economic analysis. *Indian Journal of Agricultural Economics*, **50**(1): 48-58. (Won the Best Full Length Paper award for 1995, from Indian Society of Agricultural Economics, awarded at the Annual Conference of the Society Peechi, Jan. 1997)

Negri, D. H. (1990). "*Stragedy*" of the Commons, *Natural Resource Modelling.*, **4**(4): 521 - 537.

Rama Mohana Rao, M. S., R. N. Adhikari, S. Chittaranjan and M. Chandrappa (1996). Influence of conservation measures on groundwater regime in a semi arid tract of South India. *Agri. Water Management.*, **30**: 302-312.

Watershed Management: Issues and Policies for 21st Century pp. 211-227

Priorities for Future Directions in Watershed Development

John Kerr, Shashi Kolavalli, Ganesh Pangare and Vasudha Lokur Pangare

INTRODUCTION

This paper is based on a pair of recent studies of watershed development projects. Kerr *et al.* (2002) conducted a detailed survey of mature projects in Maharashtra and Andhra Pradesh; a second survey in Karnataka, Rajasthan and Orissa examined the approaches to watershed development that the newest generation of projects takes, focusing on the nature of participation (Kolavalli and Kerr 2002).[1] While the first survey focused on outcomes, the second one focused on processes, as many of the projects were too new to assess their impacts on conservation and productivity. This chapter presents lessons learned from these two studies. It describes the best practices found in the most successful projects and discusses the opportunities and constraints facing government efforts to implement participatory watershed development. Some specific steps that all projects can take immediately are presented, along with suggestions for bureaucratic reforms that will be needed as government projects aim to become increasingly participatory in the future.

The study of Maharashtra and Andhra Pradesh was based on a survey in 1997 of 86 villages, covering a range of watershed projects as well as control villages with no project. The focus was on projects that preceded the Ministry of Rural Development's 1994 participatory guidelines, because the research aimed to examine the results of project efforts after several years. Projects covered under the study included those from the Ministry of Agriculture's National Watershed Development Project for Rainfed Areas (NWDPRA), the Ministry of Rural Development's Drought Prone Area Programme (DPAP) under pre-1994 guidelines, collaborative efforts between government and nongovernment (NGO) agencies (the Indo-German Watershed Development Program (IGWDP) and Adarsh Gaon Yojana (AGY), and a number of projects managed solely by NGOs.

The detailed findings of this study are available in Kerr *et al.* (2002). To summarize, the study found that more participatory projects operated by NGOs and in collaboration between NGOs and government agencies were more successful in raising agricultural productivity and conserving natural resources. However, doubts were raised about the ease with which participatory projects could be expanded to a very large scale, as the MORD is currently attempting to do.

The study of Karnataka, Rajasthan and Orissa took a rapid rural appraisal (RRA) approach to examine the approaches of a similar variety of projects run by government and nongovernment agencies, including those with funding from European bilateral agencies. Findings of this study are available in Kolavalli and Kerr (2002). The basic message was that

---

[1]The research was sponsored by the Indian Council of Agricultural Research and the World Bank, and it was conducted by the International Food Policy Research Institute (IFPRI) and the National Centre for Economics and Policy Research (NCAP).

government agencies lack important capacity and motivation to operate participatory projects along the lines of those that NGOs manage. Bureaucratic reforms will be needed to facilitate this transformation.

The remainder of this chapter presents a summary of lessons learned from the two studies described above. Readers interested in the detailed findings on which these lessons are based are referred to Kerr *et al.* (2002) and Kolavalli and Kerr (2002).

Infrastructure development supports watershed development

In the Maharashtra and Andhra Pradesh study, villages with the greatest improvements in performance in agricultural production between the pre- and post-project period were often those with improvements in infrastructure. On average, soil erosion was significantly lower in villages linked by a paved road, and irrigation increased more in villages with high population density. This finding is supported by analysis at the district level by Fan and Hazell (1998), which clearly suggested that improved infrastructure raises agricultural productivity. This would suggest that the growing interest in India in an approach dubbed "watershed plus", in which watershed and infrastructure investments are designed to complement each other, has merit.

Another reason to believe that infrastructure is important is that respondents consistently listed various forms of infrastructural improvements as their top priority for developing their village. There is also a role for participation in infrastructural improvement; in short, people should have a say in what kinds of infrastructure investments are made. This is part of the idea behind the Panchayat Raj legislation for decentralized government. A further point is that people should also be able to choose between watershed and infrastructure investments. In a truly participatory environment in which villagers are equal partners, they should be able to determine whether scarce investment funds should be devoted to watershed development, infrastructure development, or both. It is easy to imagine that some villages must be in greater need of improved infrastructure than watershed development, so there should be flexibility to make this judgement. This is especially so given the small impact of the large amount of funds devoted to watershed development in the past.

Best practices for successful watershed development

The quantitative analysis of Andhra Pradesh and Maharashtra showed that projects managed by NGOs and in collaboration between NGOs and government organizations have performed better than strictly government projects, which in many cases do not show significant improvement over conditions in control villages. The RRA study of Karnataka, Rajasthan and Orissa also showed support for programmes with an NGO component, including those with bilateral funding. While the NGOs have an advantage in that devote more attention and resources to any given village in which they work, their good performance also results from characteristics of their approach that government projects would be able to adopt as well.

Before discussing the best practices for successful watershed development in rainfed areas of India, it is worth reiterating some of the characteristics of these areas that distinguish them from irrigated lands and the most favourable rainfed areas. In irrigated areas, transferring green revolution technology was relatively simple because improved seeds and other inputs were well suited to millions of farms covering huge areas. The new technology was so profitable, with relatively little risk, that farmers were willing to abandon traditional farming systems in favour of new approaches. In less favourable rainfed areas, on the other hand, the success of technical interventions often depends on location-specific biophysical and socio-economic conditions and requires collective action by local people. Farmers pursue complex strategies for producing food and earning their livelihoods. Many apparently attractive production technologies have not been adopted because they are incompatible with existing livelihood systems, thus imposing unacceptable opportunity costs (Walker and Ryan, 1990). Early watershed projects introduced technologies for conservation and production without any input from farmers, all on the basis of trials in experiment stations far from the village and devoid of socio-economic constraints. The lack of sustained maintenance or adoption under these circumstances is not surprising given the difficult conditions prevailing in many rainfed areas.

This background helps explain why people's participation is the key feature of best practices for watershed development. All projects claim to take a participatory approach, but clearly the term "participation" means different things to different people. In the most innovative and successful NGO projects, participation means that local people are full partners in the watershed development programme, with both the authority to determine how the project proceeds and the responsibility to help plan, implement and pay for it. In most government programmes surveyed, on the other hand, "participation" meant convincing local people to go along with the predetermined project design.[2] The findings suggest that full participation is critical to project success, and this should not be surprising given the special characteristics of rainfed areas.

Attitude comes first: project staff consider local people as full partners

The most important and perhaps most challenging - aspect of a truly participatory approach is that project staff must consider local people as equal partners in the watershed development programme. Embracing this attitude is critical to project success because it enables adoption of all the specific participatory measures listed below, and to tapping the imagination and creativity of local people to develop project innovations. Senior staff of the best NGOs share this attitude that they are no better than the people whom they serve, as do some government project officials, especially those in the higher ranks.

Spreading the attitude of equality throughout the watershed profession is not a trivial problem in a system where the traditional management approach is hierarchical and educated professionals have always been led to believe that they are superior to poorly educated programme beneficiaries (Fernandez 1994, GOI & Danida 1996). A change in attitude is critical, however, because of the importance of understanding of local biophysical and socio-economic conditions in watershed areas. Under these circumstances, local knowledge is at least as important as a graduate degree and impressive professional credentials.

While the superior attitude of project staff is especially common in government projects, it can also be found in NGO projects. Some NGO officials think of themselves as saviors of the helpless rural people, a perspective that is not conducive to an equal partnership. Harnessing

---

[2]It is important to keep in mind that these government projects preceded the new, participatory guidelines.

local skills and ideas requires believing that local people are capable of solving their own problems (with assistance in getting organized and tapping funds, public services, and technical assistance). Even in NGOs where senior officials are highly committed to full participation, teaching younger staff members to follow suit may be a challenge. Often the young NGO recruits would have preferred to obtain a government job and, being young and ambitious, they are enamored by the idea of being important and respected. Treating villagers as equals may not come naturally to them (Ajay Mehta, Seva Mandir, personal communication).[3]

Building the bureaucracy's capacity to embrace participatory approaches will require fundamental changes (Thompson,1995). First, people in the bureaucracy need exposure to new ideas and a new attitude about their relationship with villagers, for example through training programmes. However, opportunities and incentives to participate in training are inadequate (Turton *et al.* 1998). Secondly, training will have little or no lasting effect without changes in incentives that cause people in the bureaucracy to feel the need to learn to change the way they work (Thompson, 1995). Senior members of the bureaucracy can lead by example, but they will also require support from changes in incentive systems that encourage their staff to change their attitudes. Some suggestions in this regard are discussed below.

Social organization addresses the needs of each interest group to give them an integral stake in the project's success

The best NGO projects recognize that rural communities are heterogeneous, composed of social groups with diverse, sometimes competing interests. These groups may include people of different castes, land holding status, occupation, gender, etc. Some groups are always more politically powerful than others, who may have little or no say in decisions that affect their well-being. Research in Maharashtra revealed that landless people and women, in particular, often suffered from "successful" watershed development. Sealing off uncultivated lands to regenerate vegetation and reduce soil erosion caused hardships for landless people who relied on these areas for fodder, fuelwood, and other products. Similarly, women had to go further or work harder to collect fodder and fuelwood, and if agricultural productivity rose they had to spend more time in the fields, all in addition to their traditional domestic duties (Pangare and Farrington, 1999).

To try to avoid a possible tradeoff between the objectives of conservation and productivity on the one hand, and equity on the other, some NGOs in Karnataka and Andhra Pradesh organize communities for watershed development by working separately with each interest group they can identify. They help each group become organized and then mediate negotiation between groups, ultimately brokering a watershed development approach in which every interest group stands to gain from overall project success (Fernandez, 1994). This can be described as a building block approach.

Other participatory projects, particularly those in Maharashtra, devote a great deal of effort to social organization but they are less careful to address the interests of each social group. In particular, project plans are approved not on the basis of consensus among interest

[3]Superiority carries benefits, of course; as a result, many professionals are not only uninterested in promoting people's participation, but they are even threatened by it (Fernandez 1994:68; ODA 1997:5).

groups but by a simple vote typically requiring a 70 per cent majority (depending on the project). This approach is easy to implement in Maharashtra with its relatively homogeneous social structure, but often it means that the landless minority has no say in the project design. Typically shepherds have no say in project plans that remove their access to traditional grazing grounds. In some villages the shepherds ignore the grazing bans, undermining the project, while in other villages the grazing ban is enforced and the shepherds suffer. In this case equity and productivity objectives are in conflict.

Under the building block approach, on the other hand, measures could be introduced to give shepherds a stake in protecting grazing lands, for example by granting them rights to other project benefits that depend directly on the grazing ban. The famous Sukhomajri project in Haryana followed this principle, granting irrigation water rights to shepherds who agreed not to graze the catchment area of a small irrigation tank.

Also, it would be easy to devote greater attention to addressing women's interests. Women make up half the farming community, but watershed committees typically include only one woman who is inevitably a token. Reserving half of the membership for women, or working separately with women's groups to enable them to articulate their interests on their own, would be a positive step. For this to work, meetings will have to be organized around women's schedules to reduce interference with housework and child-rearing duties. Technical training for watershed work should also be targeted to women so that they can have greater capacity for decision-making. Working with women as part of social organization efforts would reveal the best opportunities for women to contribute to and benefit from watershed development.

Social organization is so important that it calls for employing project staff who specialize in this area.[4] All NGOs employ staff trained in social organization skills, whereas none of the government projects do. In some government projects agriculture graduates are responsible for organizing villagers even though they do not have the proper training. Typically they are able to do little more than inform villagers that they must form a watershed committee. Other government projects subcontract social organization to NGOs, which is a good idea since they do not have the necessary personnel in-house. But this should only be an interim approach while government organizations develop these skills. Otherwise they will continue to consider people's participation as a discrete activity, such as establishing a committee or informing villagers about the project plan, rather than a process that should underlie every activity.

Evidence from projects visited in Karnataka, Rajasthan and Orissa shows that social organization as conducted by NGOs is relatively inexpensive. Less than 3per cent of the total project funds were spent on awareness building in the Danida (Koraput) project. The per hectare costs of employing NGOs ranged from Rs 500 under SIDA (Pratapgarh) and Danida (Koraput) to about Rs 1050 under SDC (Bidar). These costs are highly affordable considering the potential contribution of thorough social organization to long term project goals.

The major constraint to social organization is not cost but the supply of trained personnel. Currently there are not nearly enough people in the country who are trained in the necessary skills. Further training is a high priority.

---

[4]The need to employ staff with social skills is not unique to India or to developing countries. The Landcare movement in Australia found that recruiting staff members with social skills contributed to improved performance there as well (Campbell 1994).

Project funds flow only after villagers prove they can work collectively

Organizing communities and building social institutions for the long term is a complex task. It requires that each household in a community undertake the effort to work collectively with all the others, including those from other communities or social classes with whom they have never cooperated before. It also requires that every household make sacrifices in a process of give-and-take so that in the end, everyone is willing to go along with the watershed project. Under the building block approach to social organization, which is necessary to establish social institutions to promote long term sustainability, productivity and equity requirements, organizing people is slow and requires patience.

The best projects undertake this difficult process without providing any funds for development activities that require social organization as a prerequisite. They may provide funds for unrelated community needs, such as drinking water facilities or a temple, and they help establish credit groups that help people save money and obtain credit while also learning organizational skills. However, no funds are made available for watershed development until the people have shown that they can work collectively and adhere to agreements to control grazing, maintain investments, and share benefits equitably. They wait for six to eighteen months before investing in watershed development, depending on the project.

Offering funds before such social arrangements are solidified is dangerous because it may lead to superficial social organization that will not be sustained after project funds and staff are withdrawn. In short, field observations suggest that people will quickly establish a watershed committee and demonstrate some token collective action efforts if that is all it takes to qualify for a large injection of development funds. Projects such as the World Bank-assisted IWDP in Rajasthan, the NWDPRA, and some second-rate NGOs operate this way. Some project implementation agencies under the MORD's new guidelines also have pursued this approach (Turton and Farrington, 1998), although this is not the intention of the guidelines. A study by ICRISAT (1996) showed that watershed committees established rapidly under the IWDP in Rajasthan were unable to maintain grazing restrictions once project funds were withdrawn.

The project is designed and technologies are selected in full participation with local people

The best projects do not adhere to rigid, pre-established work plans, which undermine participation because they limit local people's ability to influence the project design. Instead, in participatory projects staff members work hand-in-hand with villagers to develop the project plan according to the local conditions and the needs of the people (Chambers, 1997). For example, local terrain, land use patterns, soil types and availability of raw materials may determine which technical interventions are most suitable and how they should be carried out. Similarly, local social institutions and livelihood strategies may determine the best management strategies for resources such as trees, pastures and water. Flexibility is critical to choose the best approach for a given location.

Closely related to participation in planning the project design is participation in choosing the technology to be introduced on a farmer's field. As mentioned above, early watershed projects prescribed the technology to be used in each project, and farmers often rejected the recommended approaches because of conflicts with existing farming systems. Some current projects like the NWDPRA and World Bank projects still place severe limits on farmers' role in choosing their own technologies, and maintenance of investments made under these projects remains low. Most projects with an NGO component, on the other hand, have taken a much more flexible approach and have better results to show for it.

Technical experts in many watershed programmes are uncomfortable with the idea of allowing farmers to choose which technology to use because they may select something that is suboptimal from a technical perspective. However, farmers' interests are generally consistent with watershed objectives, mainly because conservation and productivity objectives are complementary in the semi-arid tropics. This is because measures that conserve moisture also conserve soil. Also, if farmers do not want a given technology they will not maintain it, in which case it will have no effect at all. For this reason, the most effective technology may be only second-best from a technical perspective (Kerr and Sanghi, 1992).

In any case, a sensible approach for selection of technology is that technical experts from the project staff can teach farmers about a wide variety of improved technologies. Farmers should be encouraged to test and adapt the new ideas, but ultimately they must make the decision of what to adopt on their own plot. Also, projects should begin by investing in project components in which people are most interested, and then expand the scope of the work later, after sustained interaction between project staff and local people brings greater trust and awareness of new ideas (Vaidyanathan, 1989).

Target orientation is relaxed

Government projects traditionally have been planned and evaluated on the basis of achieving physical targets. This approach makes planning and evaluation easy and objective, but it is not conducive to successful watershed development. Watershed development is as much a social problem as a technical one, and social organization takes time. Target orientation makes time scarce, so it discourages project staff from taking the time to resolve conflicts, gain people's input on technology choice, or insist on cost-sharing (Adolph, 1996:4, Thompson 1995:1529). This leads projects to invest funds without adequate assurance of post-project maintenance.

Targets cannot be abandoned completely because they help establish accountability. Without targets, work in any given village might drag on endlessly. Many NGOs "adopt" a village and never leave, severely limiting the area that they can cover. On the other hand, the best NGOs have found a good balance between target orientation and stagnation. They specify in advance the number of years the project will operate in each village, with a social organization phase, an investment phase, and then a pullback phase when responsibility for the project is gradually transferred to the people. Obviously this last stage will be much easier and more successful if the people already bore significant responsibility for the project right from the beginning. The payoffs of participatory planning and cost-sharing will be fully realized when it is time for project staff to withdraw.

The project and the local people share the costs of investment

Cost-sharing between local people and the project is a critical component of successful projects. Not only does cost-sharing stretch scarce project funds but, more importantly, it leads to higher quality work and better maintenance. If farmers are required to help pay for project investments they will insist on accepting only those approaches that they truly want and intend to maintain. Cost-sharing also ensures that villagers will evaluate project inputs on the basis of their long-term productive impacts rather than short term benefits such as employment generation. If farmers do not contribute a substantial portion of the costs, they may lobby for labour-intensive investments that provide work in the dry season even if they do not contribute to long-term development. For this reason, participatory planning without cost-sharing can have a negative effect on achievement of project objectives.

Virtually all projects in India claim to require that farmers contribute to investment costs, cost-sharing in most projects is actually nothing more than a book adjustment, or sleight-of-hand in accounting. In particular, these projects employ farmers to plant trees or build bunds on their fields, paying them an official minimum wage that exceeds the market wage. They require that the farmers contribute 10 per cent of the value of labour, but even after this amount is subtracted, the wage paid by the project is at least as high as the market wage.

In order to ensure that farmers are committed to maintaining work done under a project, their share of the cost must be high enough to remove the possibility of perverse incentives. Higher project subsidies may be justified for some kinds of investments than others; for example, data from Andhra Pradesh and Maharashtra showed that farmers are quick to invest their own funds for soil conservation on flat, highly fertile plots, but less so on sloping plots with poor soils. Since the latter are particularly prone to long term degradation, they should be targeted for project investment funds. Irrigated plots and the best dryland plots, on the other hand, should receive no project subsidies since farmers will invest their own funds there anyway.

In most cases administrative feasibility will require setting fixed subsidy rates for certain activities. Ideally, however, specific subsidy rates should be determined on a location-specific basis in consultation with local people. Project funding should be flexible enough so that funds saved on one activity can be allocated to another; this way project beneficiaries will have an incentive to accept higher cost-sharing requirements. MYRADA and Outreach, two NGOs operating in Karnataka, Andhra Pradesh and other southern states, have experimented with approaches like this. For example, in some cases community members jointly decide the level of subsidy for different kinds of investments, and project funds conserved through cost-sharing are then retained for other development priorities. More recently, both of these organizations have experimented with financing conservation investments entirely on the basis of credit with no subsidy at all. Initial experience appears to be favourable, but it is important to note that this approach became feasible only after a strong thrift group had helped farmers raise their savings and obtain informal credit (Mascarenhas 1998, Fernandez 1998).

Credit should be made more easily available everywhere to help farmers finance land improvement on rainfed plots. Banks have no funds available for this purpose, except under some special schemes that target specific introduced technologies that farmers may not want. This situation needs to be changed.

Many observers and project officials insist that substantial cost-sharing is unattainable, because they have seen first-hand how difficult it is to encourage beneficiaries to devote funds to project activities. But this is no surprise when one considers the villager's perspective of the typical development project, which is target-oriented and follows a pre-established design without local input. Target orientation discourages project staff from pursuing cost-sharing by farmers because it slows down progress, and pre-established project design too often leads to interventions that local people do not want because they are not suited to local conditions. As a result, project "beneficiaries" resist cost-sharing because they know that work will be done anyway and because they doubt that it will bring benefits worth paying for. If traditional physical target orientation is abandoned and local people are allowed to decide how to spend project funds, they will be much more interested in sharing costs. The experience of MYRADA, Outreach and also Chaitanya, an NGO in Andhra Pradesh, demonstrates that this is so.

Finally, cost-sharing in government watershed projects is complicated by the fact that many of them are funded through government employment programmes in which it is illegal to pay less than the minimum wage. At first glance this appears to limit the possibility of serious cost-sharing, but with more careful thinking it is possible to develop creative mechanisms for cost-sharing that are consistent with employment programmes and also have other attractive characteristics. For example, one approach is that a certain amount of money could be allocated in a given village for employment to build watershed structures. If a farmer wanted to obtain some of these resources, he could present an application, or proposal, for what he wanted to do. Once the proposal was agreed to, the farmer would do half of the work with his own resources (or hire labourers to do the work for him), and the employment programme would pay for labourers to complete the job. The farmer would supervise to make sure the work was done correctly, and the labourers would be paid only when the farmer agreed that he was satisfied. The employment funds could be managed by a committee within the village in order to keep track of the funds available and disburse them as different farmers take advantage of the scheme. Such a committee would have to have strong representation by labourers who participate in the programme.

It is important to note that this approach is just one of numerous possibilities. With a little creativity and imagination, a wide variety of innovative approaches can be developed for every aspect of participatory watershed development. The critical point is that projects should be flexible enough to take advantage of such ideas and test them under actual village conditions so that promising approaches can be identified.

The project and the local people jointly manage the budget

An equal partnership between villagers and project staff requires that they have joint control over how the budget is allocated. This is essential for two reasons. First, it reduces the likelihood of graft since project staff become accountable to villagers. Second, it is a critical prerequisite for villagers to have a greater say over project design and decision-making.

In the best NGO projects and the Indo-German project, the budget is fully transparent and villagers jointly decide how the funds are spent. Some other projects have taken the positive step of making the budget transparent to villagers, but they stop short of giving them joint control. This approach retains the benefit of reducing the opportunity for graft. It also encourages villagers to provide input about how funds are allocated, but ultimately their authority to affect the watershed plan is limited. Still, it is far better than the traditional project approach in which even the sarpanch does not know the extent of funds allotted to different activities. This is the practice in the NWDPRA and the Government of Maharashtra's Jal Sandharan project.

Even if projects are required to share budget information with the villagers, they might not do so without pressure. One way to address this is through a publicity campaign to inform all villagers of their rights. This can be done through the mass media as well as through public meetings. NGOs or other independent groups could be hired to create awareness so that local people, armed with information, can assert their rights. (Such a publicity campaign could cover all aspects of project management, including the obligation to satisfy the interests of politically weak groups.)

Field staff are encouraged to take initiative and participate with farmers

Implementing of all the recommendations listed so far will require that field staff have greater authority to make decisions. Field staff are in the best position to identify problems and solutions and to learn how project interventions fare after design and implementation, so senior officials must rely on them to report problems or suggest alternatives. In NGO projects this problem is less serious since the chain of command is small and senior officials spend a lot of time in the field. In government projects, however, this can be more difficult. Senior staff are mainly confined to their offices, and several layers of bureaucracy separate the state-level directors from low-level field staff, acting as a significant barrier to communication. In addition, in the traditional culture of the bureaucracy seniority and superiority are synonymous, so a major shift in attitude will be needed to decentralize authority. High-level officials must lead by example.

Some project staff will be keen to adopt participatory approaches but others may find it difficult. NGOs and bilateral projects try to encourage adoption of participatory methods through close monitoring, but this will not be feasible in large scale government projects because of its high cost. It will be easy to monitor outcomes, such as establishment of watershed committees, but monitoring processes is much more difficult.

Accordingly it is important to develop self-enforcing mechanisms to encourage participation. Two ways to do this are 1) to provide incentives to staff members and 2) to give project participants some leverage to enforce it. An example of an incentive to encourage staff members to work closely with villagers is to require that the watershed plan be presented to the taluka- or district-level office by representatives of different groups in the village rather than the project staff member, yet still hold the staff member accountable for its quality. Requiring that villagers contribute a substantial portion of project costs would also encourage participation since villagers would never pay unless they could have substantial input into project design. Adjusting the current approach to target orientation to allow more time for social organization would also help. A variety of incentives could be developed to encourage workers to participate with project participants; a little imagination will yield numerous ideas.

Approaches to give villagers the means to influence how the field staff operate include giving them joint responsibility for designing the project and managing the budget, undertaking a publicity campaign to make them understand their rights, and providing them with a means to air any grievances about how the project operates.

In general, the bureaucracy needs to undertake a learning process to change attitudes and procedures (Thompson, 1995). "Double-loop" learning is needed in which staff not only review their progress but also ask whether the existing way is the right way to do things (Liebenstein and Maital, 1994). Working in participation with local people will take bureaucracies into uncharted territories in which learning will play a key role. They will need the capacity to assimilate new information and make decisions on new ways of working at all levels. Authority must be decentralized so that field staff members can make decisions appropriate to the circumstances and then learn from their outcomes. Processes will need to be put in place to ensure that individual learning translates into learning by the organization as a whole.

The bureaucracies involved in watershed development tend to be highly rigid. They attempt to control all the actions of field staff as a way to increase effectiveness, but this leads to a narrow, inflexible approach that is actually less effective. Staff at lower levels, even at the state level, are not expected to think, but rather just to do as they are told. Project guidelines are prepared at the central government level, usually by outside experts. Even programme reviews are conducted without involving the people who actually implement the projects, let alone the beneficiaries. It is remarkable the extent to which guidelines drawn up in Delhi or Jaipur can limit implementers and farmers from making sensible decisions. For organizations to be able to learn and to achieve useful outputs, they must rely less on planning, organizing and controlling as the means of increasing effectiveness and focus instead on developing vision and appropriate values (Senge and Sterman, 1992).

One highly innovative approach in a government project was found in the Department of Watersheds in Rajasthan under its first Director, who established a line of communication for every staff member to contact his office directly at any time, with a reply guaranteed within one day. Field staff were strongly encouraged to report problems or suggest new approaches. Unfortunately, even this isolated example of flexibility suffered due to constraints imposed from above. In one case, under the World Bank-assisted IWDP, a field staff member reported to the Director that vetiver grass hedges did not grow well in certain areas where rocky soils limited root penetration. He sought permission to use stone bunds in this area, since farmers had to clear stones from their fields anyway in order to cultivate. But this idea could not be implemented because the responsible office in New Delhi rejected it, insisting on adhering to the approach prescribed by World Bank officials in Washington. (Anirudh Krishna, Director of Watersheds, Govt of Rajasthan, personal communication, 1993.) The project continued to prescribe the use of vetiver grass and other vegetative barriers even in places where they were ineffective.

The village, not the watershed, is the primary unit of social organization, planning and implementation

Since people in rural areas are organized on the basis of villages, participating with them as a community requires using the village as the primary project unit rather than the microwatershed, which would be the logical unit of implementation in a purely technical programme. All of the best projects surveyed were organized around villages rather than microwatersheds; this facilitates implementation of all the participatory approaches discussed here. However, generally the successful projects are able to reconcile the village-based approach with the watershed orientation of the technical plan; they do so in two ways. First, they select villages in which the microwatershed and village boundaries nearly coincide, or in which the microwatershed falls within the village. If a microwatershed is larger than the village, they break it up into sub-units that are treated separately within each village. Second, they do not adhere rigidly to one set of boundaries or another; if the upper reaches of a microwatershed fall outside of the village boundary, or if part of the village falls outside of the microwatershed boundary, they treat it anyway. Just as pragmatic considerations have caused virtually all projects to abandon contour bunds on farmers' fields in favour of "modified contour bunds" (boundary bunds across the slope), the watershed-based approach must yield to a modified watershed approach whose primary orientation is to the village.[5]

Projects operate in villages with favorable social conditions

Before deciding where to implement watershed development, some of the best programmes screen villages to ensure that they possess social conditions that are conducive to successful watershed development. This is particularly important given the extent to which participatory approaches rely on project participants to help manage the project and make it successful. Also, one might argue that how the NGOs and NGO-government collaborative projects screen villages for their work is one of the most important determinants of these projects' success. The best examples of screening villages for favourable social conditions are the Indo-German Project (Farrington and Lobo, 1997) and the Adarsh Gaon Yojana in Maharashtra, which work only in villages that practice shramdan, or voluntary community labour. Shramdan is a good indicator of capability to undertake collective action, which can contribute to watershed project success. In these same projects and also in some NGO projects in Andhra Pradesh and Karnataka, no project investments are made until the villagers have demonstrated that they can successfully control grazing on common lands.

It is important to note, of course, that there is no single critical factor that should be used to screen villages for project participation. Critical social organization skills, and indicators of their presence, may vary by location. For example, projects in Maharashtra have selected shramdan as an important prerequisite, but projects in other states with different customs and traditions may find that other indicators are more important.

Selecting villages with favourable social conditions makes sense in a country where each state contains tens of thousands of villages and the watershed development budget is limited. Given that successful watershed development is a challenge under the best of circumstances, selecting villages with the most favorable conditions is good financial management. Many other investment options remain for villages not selected for watershed development; respondents in the present survey listed roads, schools, buses, telephones, drinking water, public health and many other services as being ripe for improvement. Also, in nonwatershed villages interventions can try to stimulate conditions favourable for watershed development so that it can be introduced later. For example, they can try to help develop social institutions that facilitate collective action, resolve conflicts, and give every community a voice in decision-making.

[5]Kerr and Sanghi (1992) explain why virtually all farmers in the Indian SAT reject strict contour bunds that run through their fields. Although many project guidelines still proclaim that contour bunds must be installed on cultivated lands, field staff know that they can only gain farmers' cooperation if they accede to their preference for bunds aligned to plot boundaries. Watershed officials in even the most technocratic programmes have come to accept this reality.

World Bank-assisted projects that work in large macrowatersheds encompassing numerous villages can pursue the approach outlined here. Work can begin in the villages with the most favourable social conditions, while in other villages steps can be taken to introduce better social organization. Of course, people in those villages can take exposure visits to the areas where work is underway. Also, if people in some villages are simply not prepared to work together in ways that are needed to support a good watershed project, those villages can be excluded from the work.[6]

Government departments operate in coordination with each other

Coordination among state-level government departments is critical to effective operation of large watershed projects, whose activities fall under the domains of numerous departments. Typically these include the departments of soil and water conservation, agriculture, minor irrigation, forestry, rural development, animal husbandry, and sometimes others. Turf wars and incomplete, poorly implemented works are the result when these various departments do not operate in a coordinated manner. Some states, including Maharashtra, Karnataka, Rajasthan and Orissa in the research discussed in this chapter, have established departments of watershed development staffed by members of all the concerned line departments. But experience suggests that this is not enough. In the Maharashtra Department of Water Conservation (Jal Sandharan), interdepartmental coordination was effective at higher levels but not at the field level (Pangare and Gondhalekar, 1998), thus hampering operations on the ground. This is an area in which progress is being made but further encouragement from senior officials is required. The Indo-German Project and the Adarsh Gaon Yojana have handled this issue well; Farrington and Lobo (1997) discuss the intricate approaches taken by the Indo-German Project to iron out interdepartmental administrative complications.

Another problem of administrative coordination in Maharashtra is that Jal Sandharan does not have its own funds. It draws funds from the DPAP and the NWDRPA at the central government level and the Employment Guarantee Scheme (EGS) at the state level. Sometimes project works dependent on EGS funding have been interrupted due to fluctuations in the demand for EGS funds in a given village. This makes effective planning difficult and leads to piecemeal implementation. Funding on a project-by-project basis is okay as long as project officials know in advance what level of funding they will receive.

Monitoring and evaluation must be taken more seriously

The research reported on here suffered from a lack of good data on agricultural productivity and natural resource conditions, but this lack of information has other implications that are much more serious. In particular, it means that government planners lack sufficient data to draw firm conclusions about the returns to different kinds of watershed development investments. Given the vast size of the budget for watershed projects, better information

[6]In principle, the World Bank-assisted IWDP follows such an approach, so it should not be difficult to pursue it again in future projects. However, the IWDP's standards are too lenient and must be reformed in future projects. Delaying the release of project funds, requiring substantial cost-sharing, and generally placing more responsibilities on project participants would discourage them from joining the project unless they are seriously committed to long term natural resource management.

about their performance would go a long way toward more cost-effective government planning. Currently too many funds are allocated on the basis of too little information and, as the findings from this study show, the potential for waste is great.

The data shortage takes two forms: 1) a lack of baseline data against which to compare current conditions, and 2) a lack of monitoring data for easy assessment of current conditions.

Baseline Data

Most projects collect at least a small amount of baseline data while selecting project sites and preparing work plans. In NGO projects, background data cover both agro-climatic and socio-economic issues, while in projects managed by state-level government departments, the data are skewed toward agro-climatic factors. This reflects the technical orientation of most government watershed agencies. Government projects typically conduct detailed soil surveys before commencing work and prepare detailed land use maps. Many NGOs may collect similar data through less formal but equally detailed participatory rural appraisal (PRA) exercises. In both cases, however, typically there is no systematic mechanism for storing the data and making it available for comparison at a later date. Inquiries with government offices revealed that such records are often discarded once the project work came to a close. The reason is that for both government and nongovernment projects, baseline data are usually collected for the purpose of planning, not evaluation.

Monitoring

All government watershed projects keep detailed records of funds spent, structures built, and other physical targets, but such information reveals nothing about impact. It is purely a bureaucratic requirement to limit misuse of funds. Most NGOs also keep records of work done, and again, a small number of the better ones evaluate their own work. It is not known by the authors if any of them collect systematic data on a regular basis or if data are collected only at the time of evaluation. The World Bank's Integrated Watershed Development Project (IWDP) provides a clear example of collection of detailed monitoring and evaluation data; this work is contracted to researchers at state agricultural universities who produce regular, detailed reports on the performance of technical interventions. The MORD and MOA also have guidelines for monitoring and impact evaluation.

Three important problems remain, however. First, it is difficult to obtain the data that have been collected for monitoring. The authors of this chapter tried to obtain such data for this GOI-sponsored study but could not. Second, the data are not organized in a common format across different types of projects, so they are not necessarily useful for comparison between project types. Third, the monitoring procedures under the IWDP and the NWDPRA fail to address socio-economic issues or the implementation process. In the future, monitoring should address process in order to obtain a better understanding of the challenges and impacts of participatory approaches.

Common monitoring and evaluation guidelines are needed

There is a strong need to develop common guidelines for collecting baseline and monitoring data. The difficult question concerns what kind of information should be gathered and at what level. It is best to keep the data set small so collecting and maintaining it do not become a burden. It would be easy for the Ministry of Agriculture and Ministry of Rural Development to issue common guidelines for all projects within their jurisdiction, but generating common guidelines acceptable to multiple ministries and even NGOs would be more difficult. Accordingly, a high level meeting to develop a common framework for data collection should be a high priority. Such a gathering should include not only Ministry officials but also representatives of NGOs and researchers in order to make sure that all parties' priorities are addressed and that a workable, usable system is developed. A tiny proportion of the vast watershed budget in each ministry could then be set aside for collecting and maintaining such data in a representative sample of all kinds of watershed projects throughout the country. A common interministerial cell could be responsible for monitoring watershed projects. Arrangements could be made to gather data from all kinds of projects, including those of NGOs.

Simple baseline data for each project could be collected at the village and household level and then updated every year. They would have to include a balance of socio-economic and biophysical data. Collecting biophysical data can be time-consuming and prone to error, so there is a need for indicators that are as simple and as possible to collect. For example, the National Remote Sensing Agency or its state-level counterparts could be contracted to provide satellite photos during critical periods of the year. At the start of each dry season, for example, satellite images taken at regular intervals would provide information on the rate of recession of vegetative cover and water bodies after rain has stopped falling. This would be a powerful indicator of moisture conservation and afforestation efforts. This technique could also provide information about the status of severe, highly visible soil erosion problems. Data from the satellite images could be transferred to computerized databases so that monitoring could be done quite effectively. This system would be much less prone to error than the kind of ground-level data gathering approach conducted in this study.

This approach has been undertaken sporadically for evaluating various watershed programmes. For example, remote sensing images of a DPAP project site in Anantapur district of Andhra Pradesh clearly showed signs of successful water harvesting (S.P. Tucker, Dept of Rural Development, Andhra Pradesh, personal communication). However, remote sensing for evaluation is not undertaken on a regular basis. It needs to be institutionalized to contribute to a better understanding of the real impact of watershed development investments.

Aerial photos (from planes rather than satellites) are used in many countries to monitor natural resource conditions and plan or evaluate watershed development projects. Aerial photos could also be effective in India, but security laws currently restrict their use. Perhaps under a government initiative they would be made available.

The monitoring system currently used by the World Bank-assisted IWDP provide some useful ideas about collecting biophysical data, but they are weak on the socio-economic aspects. Better input from social scientists will be required as part of a balanced monitoring evaluation effort. Also, greater effort will be needed to incorporate the views of people in villages, since they are the ones most directly affected by the watershed projects and have numerous insights about their operation.

CONCLUSION: A CALL FOR CAUTION

The findings of the analysis show that participatory watershed projects managed by NGOs have made a significant contribution to agricultural productivity and natural resource conservation in the study areas. More technocratic, top-down government projects, on the other hand, have fared less well. In fact, for many performance indicators the government projects did not perform any better than control villages with no project.

Collaborative projects between NGOs and government agencies have performed particularly well, and this appears to bode well for efforts to expand participatory approaches to a large scale. However, it is important to acknowledge that the NGO-government collaborative projects analyzed in this study have benefited from favourable treatment that cannot be extended on large scale. For example, as mentioned above, all of their villages had been the site of previous watershed projects (as had almost all other projects in Maharashtra), and in most of their sites an experienced NGO had already been active in the village for several years. Moreover, as these were high profile projects subject to relatively frequent visits from high ranking government officials, project staff worked particularly hard and development funds for all kinds of activities were allocated on a priority basis. Such special treatment will not be possible as these projects continue to expand, so it is premature to draw conclusions about the potential for scaling up based on the findings presented here. However, these comments are not meant to detract from the good performance of these projects; resources should be allocated to experiment further with government-NGO collaborative projects and any other efforts to introduce more participatory approaches to government-funded projects.

The major lesson to be learned from this study is that most government watershed development investments have yielded disappointing results given the vast resources allocated to date. Lessons learned from early projects have been put to good use in more participatory approaches on a relatively small scale, but expanding them to a large scale remains uncharted territory. The new MORD-MOA Common Guidelines represent an attempt to scale up participatory approaches, but progress has been sporadic.

A strong argument can be made that watershed investments should slow down, focusing on experimenting with innovative participatory approaches, until there is sufficient capacity among government staff to work in a more decentralized, participatory way. However, given that large watershed budgets have already been put in place, the focus should be to use project funds to encourage such government reform. The MORD, for example, does this by disbursing funds only when state and district governments show that they are making progress in adopting more participatory approaches. If it can help encourage bureaucratic reform it will represent an important spillover benefit that will offset slow progress in the actual watershed development objectives.

References

Adolph, B. (1997). Scaling up participatory approaches to watershed management: challenges and opportunities. Summary proceedings of the regional workshop, 8 and 9 Jan 1997, Krishi Vigyan Kendra, Zahirabad, Andhra Pradesh, India. Integrated Systems Project Report Series No. 7. Patancheru 502 324, Andhra Pradesh, India: International Crop Research Institute for the Semi-Arid Tropics, and 70593 Stuttgart, Germany: University of Hohenheim.

Campbell, C. A. (1994). Landcare: Communities shaping the land and the future. St. Leonards, NSW, Australia: Allen and Unwin.

Chambers, Robert. Whose Reality Counts? Putting the Last First. London: Intermediate Technology.

Fan, S. and P. Hazell (1998). Role of infrastructure in production growth and poverty reduction in Indian rainfed agriculture. Washington, D.C.: International Food Policy Research Institute.

Farrington, John, and Crispino Lobo (1997). Scaling up participatory watershed development in India: Lessons from the Indo-German watershed development programme. Natural Resource Perspectives 17. London: Overseas Development Institute.

Fernandez, Aloysius (1998). The Myrada Experience: Will People Take Loans for Treatment Measures on Private Lands in a Micro Watershed? Bangalore: Myrada.

Fernandez, Aloysius (1994). The Myrada experience: The interventions for a voluntary agency in the emergence and growth of people's institutions for sustained and equitable management of micro-watersheds. Bangalore: Myrada.

Government of India and Danida (1996). Review-cum-Appraisal report, Karnataka Watershed Development Project, Karnataka. New Delhi: Ministry of Foreign Affairs and Danida.

Kerr, John, Ganesh Pangare and Vasudha Lokur Pangare (2002). (Forthcoming). An Evaluation of Watershed Development Projects in India. Research Report. Washington: IFPRI.

Kerr, John and N. K. Sanghi (1992). Indigenous soil and water conservation in India's semi-arid tropics. Sustainable Agriculture Programme. Gatekeeper Series Paper No. 34. London: International Institute for Environment and Development.

Kolavalli, Shashi and John Kerr (2002). Mainstreaming Participatory Watershed Development in India. Development and Change. Volume 33, No. 2 (April).

Liebenstein, Harvey and Shlomo Maital (1994). The organizational foundations of X-inefficiency: A game theoretic interpretations of Argyris' model of Organizational learning. *Journal of Economic Behavior and Organizations*, **23**: 251-268.

Mascarenhas, J. (1998). Organisational and human resource development aspects of enhancing cooperation between people and institutions. Outreach Series Paper 10. Bangalore: Outreach.

ODA (1997). ODA Karnataka Watershed Development Project: Project document (final) Volume 1. Bangalore and New Delhi: Government of Karnataka and Overseas Development Administration, British Development Cooperation Office, British High Commission.

Pangare, G. and S. Gondhalekar (1998). Evolution of the watershed development programme in Maharashtra: A holistic view. Paper presented at the National Workshop on Watershed Approaches for Managing Degraded Land in India: Challenges for the 21st Century. April 27-29, Department of Wastelands Development, New Delhi.

Pangare, Vasudha Lokur and John Farrington (1999). Strengthening the Participation of Women in Watershed Management. Chapter 4, pp.118-127, in Farrington, John, Cathryn Turton and A. J. James. Participatory Watershed Development: Challenges for the Twenty-First Century. New Delhi: Oxford University Press.

Senge, Peter M. and John D. Sterman (1992). Systems Thinking and Organizational Learning: Acting Locally and Thinking Globally in the Organization of the Future. *In:* Kochan, Thomas A and Useem Michael (eds.) Transforming Organizations. New York; Oxford; Toronto and Melbourne: Oxford University Press, 353-70.

Thompson, John (1995). Participatory Approaches in Government Bureaucracies: Facilitating the Process of Institutional Change. *World Development*; **23**(9): 1521-54.

Turton, Cathryn, John Coulter, Anil Shah and John Farrington (1998). Participatory watershed development in India: Impact of the new guidelines. Overseas Development Institute, London.

Vaidyanathan (1989). Integrated watershed development: some major issues. Foundation day lecture. New Delhi: Society for Promotion of Wasteland development.

Walker, T. and J. Ryan (1990). Village and household economies in India's semi-arid tropics. Baltimore: Johns Hopkins.

Watershed Management: Issues and Policies for 21st Century pp. 228-234

Appropriate Policy and Institutional Arrangements for Efficient Management of Rainfed Watersheds in 21st Century

G. Sastry, Y. V. R. Reddy and H. P. Singh

INTRODUCTION

The agricultural sector contributes about 28 per cent of GDP sustaining more than two-thirds of the country's population. The primary task before this sector is fulfilling the food needs of a growing population. Food grain production has kept ahead (2.5 per cent p.a.) of population growth (2.1 per cent p.a), thereby maintaining food security. The demand for food grains is estimated to be about 232-235 million tonnes by 2015 and 260-264 million tonnes by 2030. Such an exploitation is likely to result in degradation of natural resources and loss in land productivity. Land capability based farmer demand driven Watershed Management is one strategy that increases and sustains land productivity on long-term basis without seriously damaging natural resource base. Focusing on the water resource development, emphasis has been laid on sustainability issues with the concept of whole watershed planning. Such an initiative underpins a shift in agricultural policy and a strong commitment in farmers of watershed management to increase and sustain the biomass production. Watershed Development Approach to rainfed farming encompassed in the WARASA (Watershed Areas' Rainfed Agricultural Systems Approach) - Jan Sahbhagita Guidelines is a vehicle for achieving the twin objectives of enhancing production while simultaneously preserving the natural resource base. The thrust in the watershed approach is a low-cost; location specific technologies, which are practical and feasible, that can provide opportunity for local innovations rather than capital and chemical intensive farming.

Hence a "Common Approach for Watershed Development" was jointly formulated and adopted by the Ministry of Agriculture and the Ministry of Rural Development, Government of India (2000) incorporating the strength of their earlier first generation based watershed programmes. The restructured watershed programme provides for decentralization of procedures, flexibility in choice of technology and provisions for active involvement of watershed community in planning, execution, monitoring and evaluation so as to make it sustainable. Financial support for a Watershed Management Programme is finite and therefore withdrawal by the PIA at a particular stage is definite. Therefore, sustenance of WSM after withdrawal is essential to assure optimal land productivity on a sustainable basis.

GUIDING PRINCIPLES

Sequences of activities and their operational modalities would vary from situation to situation. Hence the Guidelines are flexible so that the desired modification could be considered at different levels. The main elements of the guiding principles are: conservation of natural resources, integrated development of natural as well as social resources, *in situ* moisture conservation, sustainable farming system, adoption of ridge to valley approach, due emphasis on production enhancement activities for land owners and livelihood support for landless families, democratic decentralization in decision making, transparency in transactions, mobilization of the community at the village level, direct funding to the community, emphasis on "Government" participation in "Community's" plans, contributory approach to empower the community, building upon indigenous innovations, initiatives and ideas; equity for resource-poor families and empowerment of women, moving away from subsidy oriented development to self reliant development with monitoring and evaluation, convergence of activities/schemes of Government and non-governmental organizations etc.

WITHDRAWAL STRATEGY

Research in integrated watershed development programmes in different agro-ecological regions over the past two decades clearly demonstrated increased and sustained land productivity till the PIAs functioned. However, data with regard to sustainability after withdrawal by PIA's is scanty and needs to be collected and analyzed for formulating suitable withdrawal strategies.

Though natural resources were improved and quality of resources of farmers were enhanced through watershed development programmes, there was no strategy to maintain the technological inputs such as structures, systems etc. leading to damages after withdrawal by PIA. Hence, there is an urgent need to evolve suitable withdrawal strategies to maintain and sustain watershed management activities particularly after withdrawal of finance.

PEOPLE'S PARTICIPATION

Participation by stakeholders is mandatory for the success of the watershed management programmes particularly in rainfed eco-system. During 1980s, the emphasis was laid in implementing watershed management interventions but associating the farmers/villagers in planning, implementing and monitoring the programmes was considered not so important.

According to the new guidelines, framed by the Ministry of Agriculture/Ministry of Rural Development, stakeholders should be involved from preparation of watershed management plans for execution and then to maintain watershed infrastructure / activities after the financial/ administrative withdrawal from the watershed by PIA. Therefore, capacity building of the stakeholders should be undertaken to improve the skills, knowledge etc. of the programme and maintenance.

Key elements in participatory approach:

- Farmers' empowerment and ownership.
- Peoples' friendly, process based and physically target oriented programmes.
- The approach should fit into the farmers' pace of life rather than the convenience of donors.
- Mobilization of community as UGs & SHGs, Youth Clubs, Women Groups etc.
- Assured and quick benefit generation.
- Poverty alleviation through capital and income generation.
- Distributional equity among men and women; all social groups, classes and castes.

- Farmers' capacity building and farmer centered planning.
- Farmer managed funding.
- Farmer centered monitoring and evaluation.
- Creation of a "Corpus Fund".

METHODOLOGY

Twenty watersheds were considered in which the integrated watershed management practices were implemented and PIA's withdrawn at the end of the project period. The watersheds so selected were supported by NWDPRA, MRD, ICAR, International Donors and NGO's. The authors interviewed both PIA's and stakeholders for various success levels. The data were collected from treated and a control watershed. Where a control watershed was not available, the pretreated condition of the treated watershed was adapted as control. The data so collected were analyzed and results presented.

ANALYSIS OF SELECTED WATERSHED CASE STUDIES

Empowering the women as leaders of the working User Groups for various watershed interventions with a concept of "saving while earning" through executing land and nonland based works. Savings increase the level of confidence and empower women to participate in decision making. Since there is an in-built profit margin in any schedule of rates followed by Government Departments, and same was getting creamed-off by middle men / contractors. By executing works through Self-Help Groups ensures that the profit goes to the Self-Help Group only. The payments are made according to the quantum of work done by the group through a simple estimate understandable by the people. The practice of muster role was abolished and the total payment for the work done by the group is measured, entered in the measurement book and payment is made through a crossed cheque in the name of Self-Help Group. The group then withdraws the money through a group resolution for making payments to the members based on work done by each member as recorded by Self-Help Group itself and the rest is saved as Group Corpus Fund, which is used for internal lending on interest as decided by the group again. They are encouraged to save almost to the extent of 50 per cent of payment made to them as Corpus fund in the group. The practice of setting targets for works has been dispensed. The villagers themselves conduct their own PRA exercise, arrive at fixing the developmental activities to be taken up in the watershed, as per their absorption capacity. The works are then allocated to the self-help groups, mainly women. Regular training programmes were conducted for capacity building and for proper motivation of the women groups in executing the works and save while earning through executing the works to increase their credit base, ensured future maintenance and sustainability of the work done. Workshops were organized for all the women self-help groups where they shared their experiences. Such an activity motivated everybody to perform even better than what they have been contributing and provided feed back for tying-up loose ends.

At Tiptur (BAIF managed by a Non-Governmental Organization), Karnataka, community mobilization through meditation camps to alleviate from social problems in the initial phase followed by implementation of land/watershed management interventions was followed. Graded bunds were constructed along with farmer based farm ponds with 1.0 ha catchment area and with a capacity of 250 cum. Farmers applied harvested water as irrigation and desilted these farm ponds themselves because it is within their means. Again, at the tail end a check dam was constructed to harvest the remaining water also to raise fish. Tree based farming was well received by farmer group. In a similar watershed at Ralegan Siddhi, community mobilization and organization along with much needed water for irrigation ensured subsequent participation, increased land productivity and sustainability, as evidenced by Table 1.

Table 1. Impact of the NGO managed watershed, Ralegan Siddhi

Item	Pre w/s Dev. Prog.	Current status	Increasing rate
Cropped area (double cropping) (Ha)	630	956	52 per cent
Cropping intensity, per cent	98	164	–
Inter cropping area, Ha	–	65	>100per cent
Use of improved seed, Ha	50	860	17 times
Use of chemical fertilizer, tonnes	8	83	10 times
Irrigation area, Ha	56	447	8 times
No. of wells	34	103	3 times
No. of community wells	–	5	> 100per cent
Average yield of village (As per cropped area) (lakh)	6.7	128.2	19 times
Per capita yield (Rs.) (As per cropped area)	445.6	6465.7	14.5 times

At Fakot watershed (Dehradun), also financed by Ford Foundation (implemented in 1979 and withdrawn in 1986), runoff and soil loss data were recorded with due Soil Conservation interventions. The data show that the soil loss reduced compared to the pre-treated watershed conditions. Also the dependency on forest for fodder and fuel reduced due to integrated watershed management and sustained even after withdrawal by the PIA. As withdrawal strategy the implementing Agency i.e. CSWCRTI, Dehradun continued its association with the farmers/beneficiaries with occasional visits to interact with the farmers.

Table 2. Indicators of sustainable development of natural resources and nonarable lands

Indicator	Pre-project period	At the end of the project	After financial withdrawal
Runoff, per cent	42	15	13
Soil loss,			
Tonnes/ha/ year	11.1	2.7	2.0
Dependency on forest			
Fodder	60	40	20
Fuel	80	70	40

The self-help groups were created to get over the limitations of rural banking system. The groups functioned by exerting cohesive moral pressure on its members to keep them from straying. It was noticed that the recovery of loans advanced by SHGs is far better than the recovery in conventional banking system, eg. Adilabad, Ramanathapuram and Ralegan Siddhi and several other watersheds. SHGs of landless villagers took up productive non-land based activities. One of formidable challenges of watershed development is to create in it a stake for the landless and marginally placed farmers.

The data collected from the watersheds were analyzed and presented in Fig 1. It may be seen from the figure that cropping intensity increased by 25 - 50 per cent in case of 45 per cent of the watershed studies; BC Ratio registered between 1.0 - 2.0 in respect of 60 per cent of the watershed; migration from watershed villages reduced from 20 - 50 per cent in respect of 30 per cent of watersheds. However, watershed management activities sustained for more than 10 years in case of 20 per cent and 35 per cent of watersheds sustained between 5 to 10 years indicating viability of watershed management programme in rain fed ecosystem. Augmentation of water resources through *in situ* moisture conservation and water harvesting on watershed basis was responsible for sustenance of watershed programmes.

WATERSHED MANAGEMENT WITHOUT WITHDRAWAL STRATEGY

Watershed management projects, where no specific withdrawal strategy has been noticed, the watershed interventions have practically disappeared due to farmers unwillingness to continue maintenance of these structures. Interestingly, the farmer based Farm Ponds in one of the watersheds was not desilted and allowed to continue to siltation. Similarly, in another watershed the bunding and terracing was not maintained for their proper size and shape. This has resulted restoration of pre-treated (degraded) status.

RECOMMENDED WITHDRAWAL STRATEGIES

- Intent to withdraw from Watershed programme may be let known to the stake holders at least two years before proposed withdrawal. Withdrawal should be in a phased manner.The watershed activity in which farmers have gained capability may be withdrawn first.The farmers must be made aware of finiteness of time frame and financial support and that the stakeholders have to take over one day.
- Constitute strong and committed WCs /WAs / SHGs / UGs to undertake watershed management programmes. Associate these Groups to maintain watershed infrastructure during the project life. Towards last one-quarter of life of the project, the maintenance component should rest with these groups only. Structures such as embankments, checkdams and gully plugs etc should be supervised regularly and repaired as and when it is noticed, to save on cost.
- Capacity building at all levels of stake-holders to protect and maintain infrastructure created. This involves training of watershed committee members, visits to other successful watersheds and interaction with the farmers of successful watershed programmes from other areas. Such a capacity building increases their level of confidence.
- Create institutions that would address equity and gender issues and care for the watershed assets after withdrawal. Institutions are required to strengthen backward (such as technology, credit, seeds and other inputs, linkages with Line Departments from the beginning) and forward linkages (make remunerative prices, grading, quality, storage, processing and transport).
- Develop effective leadership that would involve transparency in decision making and in fund utilization from the beginning and post project assets maintenance should be transferred to this leadership. Ensure, by and large, decision making participatory

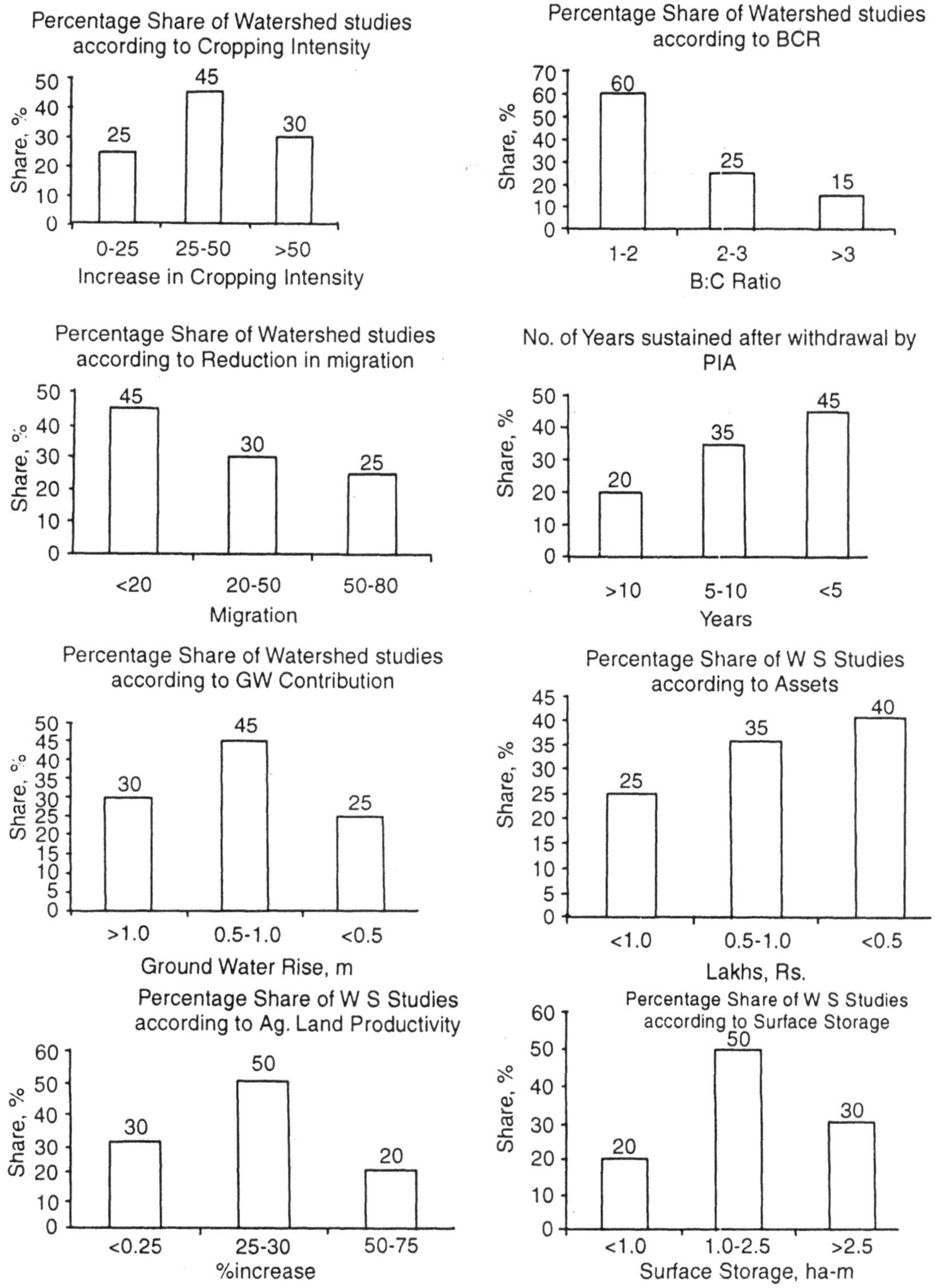
Percentage Share of Watershed studies according to Cropping Intensity
Share, %
0-25
25-50
>50
Increase in Cropping Intensity
Percentage Share of Watershed studies according to BCR
Share, %
1-2
2-3
>3
B:C Ratio
Percentage Share of Watershed studies according to Reduction in migration
Share, %
<20
20-50
50-80
Migration
No. of Years sustained after withdrawal by PIA
Share, %
>10
5-10
<5
Years
Percentage Share of Watershed studies according to GW Contribution
Share, %
>1.0
0.5-1.0
<0.5
Ground Water Rise, m
Percentage Share of W S Studies according to Assets
Share, %
<1.0
0.5-1.0
<0.5
Lakhs, Rs.
Percentage Share of W S Studies according to Ag. Land Productivity
Share, %
<0.25
25-30
50-75
%increase
Percentage Share of W S Studies according to Surface Storage
Share, %
<1.0
1.0-2.5
>2.5
Surface Storage, ha-m

Fig. 1

and capable of mobilizing the resources from the beginning of the project implementation process. The interest accrued from corpus fund should be used for maintenance and the fund should not be touched, as such.

References

Agnihotri Y., Mittal S. P. and Arya, S..L. (1989). An Economic Perspective of Watershed Management Project in a Shivalik Foothill Village. *Indian Journal of Soil Conservation,* **17**(2).

Mishra, P. R. *et al* (1980). Operational Research Project on watershed development for sediment, Drought, and flood control. Tech. Bulletin, CSWCRTI, Research Center, Chandigarh.

Mallik J. K. (1997). Growth of Agriculture in Independent India: 50 Years and After. Reserve Bank of India Occasional Papers. Vol 18, 2 & 3 Special Issue (June-September).

Watershed Management: Issues and Policies for 21st Century pp. 235-240

Strategy for Development of Watersheds in Tamil Nadu

K. Sivasubramaniyan

Water, population control and education are the important elements for India's economic development. Tamil Nadu's population is 62.1 million (as per 2001 Census) as against the all-India of 1027 million. The land area of Tamil Nadu and India is 12.99 million hectares (MHA) and 304.87 MHA respectively (1996-97). These figures show that the land area and population of Tamil Nadu account for 4 per cent and 6 per cent of all-India (for other details see Table 1).

Table 1. Land utilisation in Tamil Nadu and India, 1996-97 (Mha)

Category	Tamil Nadu	Per cent	All-India	Per cent
Total geographical area	12.99	100.0	304.8**	100.0
Forests	2.14	16.5	68.8	22.6
Barren and uncultivable	0.48	3.7	19.1	6.3
Land under non-agril. uses	1.94	14.9	22.4	7.3
Arable land	8.43	64.9	194.5	66.5
Cultivable waste	0.34	2.6	13.9	4.6
Permanent pastures and other grazing lands	0.13	1.0	11.0	3.6
Land under miscellaneous tree crops and groves	0.23	1.8	3.6	1.2
Current and other fallows	2.25	17.3	23.2	7.6
Net sown area	5.49	42.2	142.8	46.9
Gross sown area	6.46	49.7	189.5	62.2
Net irrigated area	2.81	51.2*	55.1	38.6*
Gross irrigated area	3.35	51.9*	73.3	38.7*

Note: *Refers irrigated to sown area. **Refer, as per reporting area.
Source: CMIE, Agriculture, November 2000. Season and Crop Report, Tamil Nadu, 1996-97.

The net sown area (NSA) of Tamil Nadu was 6.04 MHA in the early 1960s which declined to 5.34 MHA in the mid-1990s, as corresponding figures for all-India was 135.91 MHA and 142.22 MHA. The Net irrigated area (NIA) in Tamil Nadu and India during the early 1960s and the mid-1990s shows that NIA in Tamil Nadu has increased only marginally from 2.47 MHA to 2.63 MHA but at all-India level this has doubled from 25.54 MHA to 53.51 MHA. This clearly indicates that any further increase in both NSA and NIA in Tamil Nadu is very difficult and the only possibility is to raise the production by adopting land saving technologies as well as increasing the water use efficiency to reach higher productivity both in irrigated and rainfed lands. As the yield from irrigated land is more than that from rainfed land more land should be brought under irrigation.

The productivity of rainfed land in Tamil Nadu is only a third of that of irrigated land. Available data indicate that there has been no significant improvement in the yield from rainfed land, whereas the yield from irrigated land has been showing a significant rising trend. As the major portion of the cultivated land is still rainfed, substantial increase in productivity of rainfed land is essential to both raise agricultural production and improve the living conditions of the rural population whose mainstay is agriculture. Soil conservation through watershed development and dry farming are integral to agricultural development. However, resources allotted for these are small with little impact compared to irrigation investments (GOTN 1997).

Agricultural sector in India contributes 28per cent of GNP and accounts for 65per cent of overall employment. Nearly three-fourths of India's population living in rural areas still depend upon this sector for their livelihood. Although India has achieved self-sufficiency in foodgrains production by increasing it from a mere 51 million tonnes in 1950-51 to about 209 million tonnes in 1999-2000, compared to other developing countries, the level of productivity attained in India in both irrigated and unirrigated lands is much less. This has been one of the major reasons for the poor living standards of Indian farmers. Although several factors are responsible for the low productivity of agriculture, the primary factor is the inadequate and unsustainable irrigation facilities created and managed by both the government and local communities in the country. Even though, the irrigated area of the country has increased more than two-fold over the past five decades the rainfed land still accounts for 62 per cent of the cultivated land and contributes nearly 44 per cent of agricultural output (Ninth Five Year Plan: 1997-2002).

With persistent increase in population and steep decline in per capita water supply, meeting the increasing demand for foodgrains in the coming decades may turn out to be a major problem. Since a major portion (about 80 per cent) of available water is used for agriculture further augmentation of supply is possible only through conserving and developing small watersheds wherever feasible. After independence the contribution of 'small water bodies' such as tanks, springs and diversion networks has been declining all over the country. This decline is considerably more in the southern states especially in Tamil Nadu. Preservation and development of these water bodies will help not only to increase surface irrigation potential but also to raise groundwater recharge. The same principle is also applicable for the development of watersheds which are smaller in size, and capacity compared to tanks and other water bodies.

Though the water available to us on earth today is not different in quantity from what it was in the past, the supply of water in terms of per capita availability is much less now due to growing population. For instance, in 1991 India's annual per capita water availability was estimated as 2214 cubic metres (CuM) against the global average of 9231 CuM (Ninth Five-Year Plan: 1997-2002). The corresponding figures for 1996 are 2167 CuM and 8338 CuM (World Development Report: 1999-2000)[1]. The latest estimates show that in 2000, the per capita water availability was 1244 CuM in India as against the global average of 7122 CuM (Human Development Report, 2000)[2]. This clearly points out the necessity for India to conserve and develop water resources in all possible ways. One of the important and effective methods to utilize the available rainwater is country-wide development of watersheds in a consistent manner.

[1]The total supply includes river flows from other countries.

[2]The total supply includes only annual internal renewable water resources.

CONCEPT OF WATERSHED DEVELOPMENT

A watershed is a line of separation between water-flows to different rivers or basins or seas. Watershed development is nothing but developing micro level water harvesting structures like check dams, percolation ponds and other rainwater absorbing mechanisms. All these enhance soil fertility, enrich moisture regime and promote ecological balance through conservation of forests. So far India has not had any effective programme to harvest rainwater except in some regions where it is done in an ad hoc manner. In order to promote this strategy all traditional water harvesting systems should be developed and converted into effective watershed development mechanisms.

The basic idea behind watershed development is to promote the economic well-being of the village community. This could be done by slowly developing a village microplan through "watershed approach" which includes conservation of land and water resources, production of inputs supply, and capacity building. Development of resource users organisations for managing these watershed activities in an ecologically sound and economically viable manner is necessary. But most of the impact studies on watershed programmes reveal that the programme starts and ends without having any linkage or integration with economic growth (V. K.Misra, 2001).

STATUS OF WATERSHED DEVELOPMENT IN INDIA

In 1986, National Watershed Development Programme was launched for rainfed areas in 16 states. The programme was modified in the light of the suggestions made by the Committee of Secretaries constituted to review the programme. Thus, in 1991 revised guidelines for identifation and formation of watersheds were issued. The new guidelines gave importance for planning and preparation of a complete plan of the watershed. The programme was also made more people-oriented and participation of beneficiaries was encouraged. The technical bias of the programme was reduced and it was directed more towards providing economic benefits. The coverage of watershed was increased and inherent monitoring and evaluation component was included (Ratna Reddy, 2000).

IMPACT OF WATERSHED PROGRAMMES

The necessity and importance of watershed programmes are now felt by both the government and the people. As a result, a number of research studies were undertaken to examine the efficacy of the programmes from different perspectives. All these studies not only proved the economic viability of the watershed programme but also underlined the fact that it is the only alternative to the development of rainfed agriculture in India. In fact, some studies have even shown that watershed development programmes score over existing intervention programmes in terms of employment generation and natural capital regeneration in moderately degraded regions (Chopra and Kadakodi, 1993).

In order to understand the overall effect and importance of watershed programmes the difference between watershed management and watershed technology as indicated by researchers needs to be explained. While watershed development is associated more with technology, watershed management is more idealistic. While the success of watershed development as a technology is well established, the idea of watershed management is proving to be the main bottleneck for the widespread success of the programme. Understanding the interplay between land, water and people is equally important in studying watersheds. Therefore, watershed management is more than just the cost-benefit analysis of investments. In short, watershed development is essentially a community-based one, its adoption and success depend largely on the cooperation of people, and the potential beneficiaries of the programme (Ratna Reddy, 2000).

It is important to note that some of the watershed development programmes which have been functioning for the last few years were evaluated by the Centre for Science and Environment. Its observations are: (*i*)In the north Indian states of Gujarat, Madhya Pradesh and Rajasthan, "villages covered under the watershed development programmes managed the year 2000 drought better than those without such programmes, (*ii*) In Rajasthan, due to water harvesting activities undertaken during the year 1993-98, the grain stock in the village around Mandlikpur had gone up and the average income of the village had doubled. The surplus created helped them sail through the successive three years of drought smoothly and (*iii*) In Neemi village, in Jaipur district of Rajasthan, a prosperity has been reached due to watershed development through tanks. As it is observed that "after the tank filled up with water, the villagers started keeping animals. The production of milk has increased and as a sign of changing times people have started keeping cows instead of goats." The village also sustains five colonies of agriculturists from other states along with its own 100 families and 6000 cattle (CSE, 2001).

NECESSITY FOR RESEARCH ORIENTED WATERSHED DEVELOPMENT

Generally, dryland areas are characterised by greater agro-climatic diversity than the traditional irrigated land areas. Hence, the former requires location-specific research for evolving viable technologies and practices. But, so far dryland areas have not received the priority they deserve in agricultural research. For instance, Green Revolution was characterised by land saving but water-using technologies, whereas dryland areas need water saving enterprises and practices which optimise output per unit of scarce water. In the absence of such technologies, the farmers in dryland areas go in for water intensive crops like paddy when water resources are conserved, giving rise to conflicts on water and scarcity of drinking water (Hanumantha Rao, 2000). Experience shows that there is scope for increasing productivity of both irrigated and rainfed lands, the former has received considerable attention from the days of the Green Revolution and the latter has been neglected. Hence, due attention needs to be paid to promote, sustain and stabilise watershed programmes in different parts of the country where water scarcity prevails even moderately or marginally.

As a matter of fact, the watershed development projects implemented so far and now under implementation cover only 10 per cent of the dryland area. These may serve at the most as pilot projects to demonstrate the validity of certain development strategies and to promote their replication in a large measure which is yet to be realised. Actually, three decades after we launched the Green Revolution, the country is preparing itself for a second revolution in the form of watershed development on the farm front. However, this will be with regard to those areas which are rainfed and where incidentally more depletion of soil fertility and loss of moisture are taking place. These areas have great potential for and are a challenge to future agriculture in the country, particularly in Tamil Nadu. It assumes added importance in Tamil Nadu since it has already exhausted 95 per cent of surface water and around 75 per cent of groundwater, there is necessity to improve the position of water resources in a sustainable basis. Tamil Nadu has, according to revenue records, nearly 2 million hectares of degraded and wasteland and about 4.2 million hectares of dryland, part of which is now under degradation. Together, these constitute nearly 50 per cent of the geographical area of Tamil Nadu.

STRATEGIES TO BE ADOPTED FOR WATERSHED PROGRAMMES

- Minor irrigation tanks to be improved to restore their capacity, and development of small water-harvesting structures such as farm ponds, check dams and percolation ponds should be constructed wherever possible in a massive scale to utilise the maximum quantum of rain water.
- Registered watershed association has to be formed in each watershed for its sustainable development. People's participation right from the stage of project planning, preparation, project completion and post project management would be the hallmark for the effective functioning of watershed association. Further, internal and independent external evaluations, periodic monitoring are important for successful implementation of watershed programmes.
- Non-Government organisations participation should be promoted to develop watershed programmes in different areas.
- To conserve all sources of water—rain, river, ground, estuarine and treated sewage water and use them efficiently and in a conjunctive manner.
- Promote a state-wide movement for watershed management and foster the revitalisation of water harvesting, saving and sharing traditions, including practices like Kudimaramathu.
- In order to know the specific areas for watershed development, preparation of the watershed Atlas for entire Tamil Nadu state should be given prime importance.
- All subsidiary rural development programmes should be integrated under one single head in the name of 'watershed and wasteland development programme' and preference be given for restoration and creation of watersheds in the rural areas.

CONCLUSION

For a long time now dry farming regions in India have been neglected on the policy front. This neglect is mainly due to the concentration of public resources in the well-endowed regions for meeting the country's food requirements, through irrigation development and green revolution technologies, which are complementary. But it has been realised of late that the optimum productivity levels in these well-endowed regions have already reached their potential and meeting future demand is limited. Moreover, further increase in area under irrigation is not only limited but also expensive.

Given the necessary shift in the policy framework in respect of agricultural research, extension and supply of inputs like seeds, infrastructure development, price and credit policies, dryland areas hold considerable promise for the development of agriculture. In order to develop dryland areas soil moisture conservation strategy should be widely adopted. Without enhancing or preserving the soil moisture conditions we cannot increase productivity of dry land crops. Hence due importance should be given both in the (government's) planning process and also in the research circles. In order to achieve this objective an unanimous decision to be taken to declare a year say 'year 2002' as a "watershed development year" of the nation and intensive programmes/activities should be undertaken in favour of achieving this objective. Finally, it is pertinent to add the recent statement made by Anil Agarwal, "there is no village in India which cannot meet its basic cooking and drinking needs through rainwater harvesting. It does not matter how much rain you get, if you don't capture it you can still be short of water." Given the fact that India is one of the most well-endowed nations in the world in terms of average rainfall, there is no reason why it should suffer from drought (CSE, 2001).

References

Centre for Monitoring Indian Economy (CMIE), *Agriculture*, November 2000.

Centre for Science and Environment, *Catch Water*, Vol.3, No.3, June 2001.

Chopra, Kanchan and Gopal K. Kadekodi (1993). 'Watershed Development: A Contrast with NREP/JRY/ *Economic and Political Weekly*, June 26.

GOI, Ninth Five Year Plan, 1997-2002, Vol. II, Thematic Issues and Sectoral Programmes, Planning Commission, New Delhi.

GOTN, Report of the High Level Committee on Development of Wastelands, Watershed Development and Implementation of Poverty Alleviation Programmes, Department of Agriculture, Chennai, May 1997.

GOTN, Season and Crop Report of Tamil Nadu, 1996-97, Directorate of Economica and Statistics, 1998.

Hanumantha Rao, C. H. (2000). Watershed Development in India: Recent Experience and Emerging Issues, *Wastelands News*, August-October.

Human Development Report 2000, UNDP, Oxford University Press.

Misra, V. K. (2001). Watershed Wonders: Inter-disciplinary Integration and Institutional Development for Making Watershed Development a People's Movement. *Wastelands News*, February-April.

Reddy, V. Ratna (2000). Sustainable Watershed Management: Institutional Approach. *Economic and Political Weekly*, September 16.

World Development Report 1999-2000. Oxford University Press.

Watershed Management: Issues and Policies for 21st Century pp. 241-252

Participatory Watershed Management and Development Scenario: Future Threats and Thrusts on Irrigation Projects—Approach on Rights and Risks

K. R. Sivaraman

INTRODUCTION

There is no life on earth without water, our most important resources apart from air and land. During the past three centuries, the amount of water withdrawn from freshwater resources has increased by a factor of 35, world population by a factor of 8. With the present world population of more than 6 billion still growing at a rate of about 90 million per year, and with their legitimate expectations of higher standards of living, global water is expected to rise by a further 2-3 per cent annually in the decades ahead. But freshwater resources are limited and unevenly distributed. In the high consumption countries with rich resources and a highly developed technical infrastructure, the many ways of conserving, recycling and reusing water may more or less suffice to curb further growth in supply. In many other regions, however, water availability is critical to any further development above the present unsatisfactory low level, and even to the mere survival of existing communities or to meet the continuously growing demand originating from the rapid increase of their population. In these regions man cannot forego the contribution to be made by dams and reservoirs to the harnessing of water resources. Demand for water is steadily increasing throughout the world. Water is becoming an increasingly scarce commodity worldwide. There are vast water-stressed areas where people survive with little or no water. Yet, in some parts of the world huge quantities of water are wasted. To assuage this scarcity, there is growing pressure to harness and utilize surface water sources, like rivers. But the extent to which we should interfere with river systems is a subject of some debate.

The demand for water is an increasing phenomenon due to population growth, extensive farming due to green revolution, industrial growth, urbanisation and other factors. The demand for potable drinking water is more prominent rather than the availability, which become polluted due to various means. Water and water quality conservation is very much in need. All major and minor irrigation projects were constructed throughout the world wherever possible. Yet there are deficit major part of the world. Government yet to over come this aspect. Is it a problem to deal with the people themselves? There is one thing common in both mega cities and remote villages in India, and why not, the same problem in most of the countries in the world, the water scarcity? Where we have gone wrong in formulating the right water policy? For finding solution, which suits our sub continent, this paper deals with the threats and thrusts on various irrigation projects, and right approach by implementing Participatory Irrigation Management and Participatory Watershed Management and Development.

WATER AND ITS UNIQUE FEATURES

Water is a renewable resource and a mineral. That renewable is accomplished by natural and artificial processes. We are at our best when we aid and abet natural processes in such manner as by control of run-off and evaporation through reforestation and cover crops, by storage, by groundwater recharge and by fitting our use to the yields which can be replenished naturally. The paradox of the world's water resources is that while it is generously abundant in its totality, and sometimes back we have not reached the point of social and technological development either to live where the water is or to bring the water easily to the places where we are.

Now technology has improved and we are able to bring the water from the source to its use where people need to live and settle. Of an approximated 160 million cubic kilometers (40 million cubic miles) of water on earth, not more than 0.5 per cent is readily accessible for man's use. Of the total, 97 per cent is in the ocean and salty seas; 2.25 per cent is in snow and ice, with some usable on melting. Of the 3 per cent of the total rated as fresh, some is frozen and much is in ground strata over 750 m (2,500') below the surface. Only one tenth of the total fresh water is estimated to be in ground water strata less than 750 m below the surface. Of that source, a safe withdrawal, which could not deplete or deteriorate it by saline water incursion is placed at 2 × 1013m³/year (0.50 × 1010 million gallon/year), but three-fourth of this falls on the saline oceans and seas. Thus 1.25 × 1014 m³/year (3 × 1010 million gallon/year) precipitates on land. This is distributed as rather rapid evaporation and plant transpiration to the atmosphere, percolation to replenish the ground water and run-off to streams, lakes and artificial impoundment's. The run-off to our most accessible sources is about 4 × 1013m³/year (1 × 1010 million gallon/year). Such global estimates indicate the limit of readily available fresh water which is about 6 × 1013m³/year (1.50 × 1010 million gallon / year). That is the sum of run-off and safe groundwater withdrawals. The necessity of rational management to meet rising usage by more people, more manufacturing, more irrigation becomes more evident when the questions of where it is and where it is needed are joined.

A LOOK AT DAMS

One of the most efficient ways to manage water resources for human needs is by the construction of dams that create reservoirs for the storage and future distribution. Currently there are about 45,000 dams higher than 15 meters throughout the world. While some are more than 2,000 years old, about 73 per cent have been built in the last 50 years. The reservoirs formed by these dams store some 3,600 km³ of usable water. Across the world, one-third of the land area is well watered and two-thirds arid and semiarid. The same fractions can be applied to the Indian sub-continent and as well as to Tamilnadu, especially to the eastern coastal plain also.

A difficult task

The World Commission on Dams (WCD) task was exceedingly difficult. There are about 45,000 large dams in the world. Although the first hydel dam was built in 1980, by 1949, the world had 5,000 large dams. By the end of the 20th Century, this figure had increased to 45,000 in more than 140 countries. Of these 22,000 are in China alone, a country that had only 22 large dams in 1949. The other four in the top five large dam nations are the United States with 6,390, India with 4,000 and Spain and Japan which varies from 1,000 to 1,200 each. An estimated 160 - 320 new large dams are built each year. Between 1989 and 1993, an average of four million people were displaced annually by 300 large dams.

According to the WCD, an estimated 40 to 80 million people have been displaced by large dams. The commission estimates that "For millions of people on all continents, displacement has essentially occurred through official coercion". The report also records that "little or no meaningful participation of affected people in the planning and implementation of dam projects including resettlement and rehabilitation has taken place". In addition, large dams have affected 60 per cent of the world's rivers, often ecosystems irreversibly. The WCD has concluded that on the whole large dams have had a negative impact on ecosystems, most of which cannot be mitigated through corrective measures. The record of large dams in flood control has also been mixed—they have helped in some instances and, in others, made communities more vulnerable.

In sum, large dams seems to present more problems than solutions. Although they have played an important role in irrigation and increasing food production, they have also resulted in widespread water logging and salinity in some area—38 per cent in the Indus and Gangetic basin, for instance. There have been large cost cover over-runs, partly due to inflation and partly because all the costs were not factored in at the outset. For instance, in a substantial number of the dams studied by the WCD there had been huge underestimation of the number of people to be displaced and the cost benefit ratio has been boosted, to implement the project.

Different approach

Instead of assessing the value of these dams on the traditional cost benefit basis, the WCD suggested a totally different approach: the rights and risks approach towards dam development. "Reconciling competing needs and entitlements is the single most important factor in understanding the conflicts associated with development projects and programmes —particularly large-scale interventions such as dams. The approach developed by the Commission of recognizing rights and assessing risks (particularly rights and risks) in the planning and projects cycles offers a means to apply these core values to decision making about water and energy resource management", states the WCD report at its outset.

Dams and water—India

There are about 4,291 large dams in India covering 9 per cent of the World Dam Population, counting 695 currently under construction. Of these 2,256 dams were built in the peak period between 1971 to 1990, during Green Revolution. About three-quarters (73 per cent) of the 3,596 completed dams are situated in three western agriculture states. Large dams in India are constructed and owned by the respective State Governments as a whole. The majority of Indian dams have been built for irrigation and some multipurpose. India has a variable monsoon climate and is subject to cyclical drought. The country has a large network of rivers with an estimated annual flow of 1,900 km^3, much of which occurs from June to September. Total withdrawals of water are estimated to be 750 km^3 currently, and have been projected to increase 1,050 km^3 by 2025. India's constitution provides primary power to the states to develop water resources within their boundaries, and certain powers to the

Union Government to regulate and develop inter-state rivers to the extent declared by Parliamentary Law.

Current trends in sectors

Irrigation: Irrigated area increased from 22.60 million hectares in 1951 to about 89.60 million hectares by 1996. While large dams contributed to growth, almost two-thirds of irrigated potential was created through minor irrigation projects. There is a gap between irrigated potential and irrigation utilisation of about 8.80 million hectares. Yields in irrigated areas are on average twice that of dry land (rain-fed) agriculture.

Power: The share of hydro of total installed capacity in India declined from 50 per cent in 1963 to 25 per cent in 1998 (21,981 MW). An estimated 2/3 of the hydropower capacity is attributed to storage projects, one-third to run-of-the river schemes. The Central Electricity Authority assessed the country's aggregate technically feasible hydro potential at 94,000 MW consisting of a mix of small, micro-hydro, and large dams.

Water Supply: An estimated 90 per cent of the country's drinking water supply comes from groundwater. Drinking water requirements of urban areas where groundwater resources are not utilised are met from the reservoirs of irrigation and multipurpose projects.

Flood Management: Embankments rather than dams have been used as the primary flood control strategy with other flood management measures. Flood adaptation rather than total flood control and adoption of water and soil conservation measures are gaining currency. Few dams are primary purpose flood control.

Other Factors: The Government forecasts a total estimated storage potential from dams of 384 BCM by 2025.

Social Dimension: The estimates of number of people displaced by the large dams in India range from 21 million to 40 million. While since 1976 various State Governments have issued directives and enacted policies and legislation for resettlement, application has been inconsistent.

Environment Dimensions: Forestland submerged for reservoir construction was estimated at 5,00,000 hectares in 1980. Compensatory reforestation has been applied to areas and ecosystems. According to the regional offices of Ministry of Environment and Forests (MOEF), in the majority of cases, project authorities did not comply with the conditions specified for dam approvals.

Areas of Controversy around Large Dams: India has a very active debate on water and energy resource development and specifically on the role of large dams. The recent drought in the northwest and south has also prompted a far-reaching discussion of fundamental approaches to water management and use in the country, including the role of all levels of government and local communities, in addition to the full range of social, environmental and development issues debated on the role of large dams, specific aspects of the debate in India include:

- The feasibility, practicality and scope for various means of water storage (groundwater recharge, water harvesting, small and large dams).
- The role of traditional and large-scale alternatives for water management and supply including the extent to which options are complementary and mutually exclusive.
- The proportion of the budget, costs and delays in large dam projects in the main agricultural states.
- The scale and magnitude of involuntary displacement arising from large dams.
- The capacity and resource availability and political will for effective mitigation of social and environmental impacts of large dams.
- Perspectives on unresolved social and environmental issues from part dam projects.
- Adequacy of policy, institutional, and financial support for performance improvement of dam and non-dam alternatives.

Rethinking the issue

But if this is really done, it is more than likely that many projects would either be modified or even rejected. Clearly, such a contingency does not sit well with government that find it easier to big build and centralize management rather than allow a more democratic, decentralized system that reduces their control. It is therefore not entirely surprising that India does not follow the above WCD framework. In the long run, the attitude of funders of these projects will have an impact on the number of large dams that will be constructed in developing countries.

This will clearly have to be accompanied by a push for alternatives to energy and water needs and a strategy to deal with the problems that dams that are on the drawing board, or half built, are throwing up. For the water and energy needs of populations around the world are as real as the problems that large dams have thrown up. Sustainable solutions are still urgently needed and on a scale that can match the extent of the need. The result is switching it over medium or minor irrigation systems, especially tank irrigation systems. The role of Participatory Irrigation Management (PIM) and Participatory Watershed Management & Development (PWM) will overcome certain issue, and solve the water requirement problem to a certain extent.

DAMS AND WATER: TAMIL NADU

Tamilnadu is being an agrarian state, special attention need to be paid for the development of groundwater and surface water resources, as it is essentially be required for irrigation, industrial, domestic and livestock needs. The total geographical area is being 1,29,901 sq.km. having a population of 1.02 billion as per 2001 Census. The total number of major reservoirs is 61 and also 39000 big and small tanks. The annual normal rainfall of Tamil Nadu is 945.70 mm and the total irrigated area is about 47 lakh hectares, irrigated through reservoirs, canals, tanks, tubewells, borewells, open wells, filter points and other sources. The estimated surface water potential is around 30 million ac. ft., whereas the estimated ground water potential is about 11.50 million ac.ft. excluding the saline one. The utilization of surface water exceeds 90 to 93 per cent, whereas the utilization of groundwater is about 62 per cent. The per capita water resources of Tamil Nadu is 4224.50 cubic meters which steely falls short of the all

India figure of 24645 cubic meter. Administratively this state has been divided into 30 districts, which are sub divided into taluks and blocks.

This State has a long coast on the east stretching and slightly on the south west for nearly 920 kms from Pulicat Lake on the North and Capecomerine (Kanyakumari) on the south and south west respectively. Generally the slopes are gentle from the foothills on the western ghats towards east, south east and south to meet the sea at Bay of Bengal, Indian Ocean and Arabian Sea. The geographical terrain pattern of the state can be broadly classified to coastal region, the central plateau and western hilly terrain.

Water balance

The surface water and groundwater potential for all river basins (except Cauvery) is 13015 Mcm and 15346 Mcm. Thus the total water resources works out to 25361 Mcm. The total demand for all the sectors except Cauvery basin was 31821 Mcm in 1994 and will be 30220 Mcm in 2044. Most of the river basins are deficit basins, vide report published by IWS, Chennai.

Planning tools and decision support system

The design, analysis and management of water resources systems are very complex because of the unique nature of the resource, water, the large scale of the system, the non-linearisation of the relationship depicting interactions with the systems and of the system with the environment, the stochastic nature of inputs and outputs, the constraints on resources, the horizon of analysis and the time lag in certain aspects of system behaviour. Analysis of such system is intended to aid decision-making. Planning of water resources of a river basin can be achieved by system analysis. The application of systems method such as mathematical optimization and simulation can significantly aid in the definition, evaluation and selection of water resources investment, designs and policies.

Even though the present availability of water resources, both surface water and groundwater resources are more than 30 per cent short than to meet the all aspects of requirement, only by effective conservation and management, it is being meeted out. There are some short-term measures and long term measures to be adopted to overcome this situation.

Short Term Plan

- Equitable distribution of irrigated water supply by better management.
- Improving the performance of the existing irrigation system by suitable structural and on farm measures.
- Micro irrigation - drip and sprinkler irrigation.
- Crop rotation, raising garden crops, orchards, and other less water consumption crops.
- Encouraging less water consuming industries in water scarcity area.

Long Term Plan

- Irrigation efficiency should be stepped up to 60 per cent by modernization.
- Modernizing and desilting Irrigation Tanks, by the traditional Kudimaramathhu schemes.

- Industrial and Municipal wastewater should be treated and reused.
- The concept, namely the value of the produce obtained per unit of water consumed should be introduced.
- The industries should be encouraged to implement Rain Water Harvesting Techniques, by the construction of ponds within their campus.
- Conjunctive use of Surface water and Groundwater, wherever possible and feasible can be implemented and encouraged by doing institutional financing.
- Rain Water Harvesting Techniques and surface water savings, by preventing the water letting into the sea during flood season.
- Training and Public awareness.
- Artificial Recharge Techniques.
- Construction of check dams, bandharas, and percolation ponds.
- Environmental Planning.
- Forming Farmers Participatory Irrigation Management.
- Creating awareness about better Watershed Management & Development.

WATERSHED MANAGEMENT AND PARTICIPATORY WATERSHED MANAGEMENT (PWM)

Resource management on watershed basis is given a renewed thrust. Conservation of water assumes significance in view of the growing population and simultaneous increase in water needs of the people. Spasmodic rainfall adds a challenge to the efforts of proper utilization of the resource. Vagaries of nature may apparently seem to cloud the results of artificial recharge structures. A bad rainfall year for example after adopting conservation practices in an area effectively nullifies shortage in storage of reservoirs. Analysis of long term water level fluctuations combined with rainfall data, crop yields, etc on the same time scale will give meaningful and statistically significant results. A comparison with such results in different areas with different treatments of watersheds will show which of the methods are most suitable for a given agro-climatic condition. Hence, lot of perseverance and patience is required before the fruits of conservation can be seen.

The diversification in the cropping pattern as well as introduction of high yielding varieties of crops has greatly helped to attain self-sufficiency in food. Application of fertilizers and pesticides in the field has become common practice to farmers, which is in one way or other, polluting the groundwater gradually. Technical and scientific findings of water management must reach the farmers for adoption. Further saline water having an EC value more than 5,000 micro siemens/cm can be used for raising crops by drip irrigation, provided the SAR value is not exceeding 5 to 7. (Higher SAR ratio water will reduce the yield per acre). The fertilizers are mixed and used with the saline water. In these techniques waste and fallow land can be developed and converted into orchards and garden lands. To conserve soil moisture, asphalt spraying and adding pressmate/coir mulching techniques are being done on the ground surface to avoid evaporation losses. Though our economy may not permit such costly irrigation practices, it is quite possible and within our reach to adopt water management methods to conserve waters. Thus the management of surface water and groundwater sources in an integrated manner is possible and equitable distribution can be assured.

The department in this regard is actively pursuing studies on water conservation, watershed management, artificial recharge and conjunctive use of surface and groundwater techniques. Apart from the main task of monitoring of Surface water\Groundwater network activities, site specific studies may also be undertaken which will help in proper planning of the groundwater resource. This is also a useful evaluation of the watershed activity and a proper socio-economic evaluation can be attempted.

Participatory Watershed Management is the best method to maximize the irrigation potential available to irrigate maximum area for long term strategy. One of the best techniques is conjunctive use of surface water and groundwater resources.

CONJUNCTIVE USE OF SURFACE AND GROUNDWATER

The concept of conjunctive use is taken to imply the coordinated and harmonious development of the two sources with the sole purpose of maximizing agricultural production. It is not only the joint use of water resources of more than one type but also their exploitation through efficient use in techno-economic terms by asking advantage of the interaction between the several sources and the impact of development of one source or another. The interaction between surface and ground water includes contribution of baseflow from ground water, the recharge to ground water from surface water, artificial recharge of surface water to ground water, indirect recharge by ground water through augmentation tube wells or borewells or dug wells into canals, recharge from irrigation system to ground water.

In various irrigation commands serious concerns are the harmful effects of water logging and conversion of arable land into salt infested user land. As a follow up of the action plan of conjunctive use of surface and ground water is to be undertaken with the following objectives:

- Evaluation of water resources in respect to quantification of surface and groundwater resources available in space and time so as to determine the water balance.
- Identification of the critical areas of water logging and soil salinity as well as scarcity area within the command.
- Matching the demand of various sectors with the available water resources and to evolve a strategy to meet the projected demands of future.
- Mathematical modeling studies to simulate hydrogeological situation, general various scenarios to evolve strategies for optimum development of groundwater resource.
- To evolve a suitable plan for controlling the problem of rising water levels and also depletion of water levels.
- To test the sustainability of the present irrigation with respect to conjunctive use of water resources and suggest improvement for future.
- To evaluate the economic aspect of ground water development plan with respect to cost benefit ratio, internal rate of return and pay book period etc.

Advantages and disadvantages of conjunctive use

Advantages of Conjunctive Use

There are several advantages for adopting Conjunctive Use Techniques in the water resources utilization, which are as follows:

- Ground water exploitation lowers the water table and reduces water-logging problem.
- Use of ground water helps reduce peak demands for irrigation.
- Smaller surface/ground water storage can provide for water requirements during drier periods.
- Increased water supply without much investment in existing pumping device.
- Seepage from canals adds to ground water thereby providing scope for recharging ground water.
- Supplemental supplies from ground water ensure proper irrigation scheduling.
- Surface and sub-surface out flows are minimized, causing reduction in peak runoff and flood discharge.

However, there are some constrains in Conjunctive Use Planning:

- Greater power consumption, more pumping and from greater depths. (in case of borewells or tubewells only).
- Possibility of deterioration in groundwater quality due to influx of salts leached down from soil, by application of fertilizers and pesticides, which is very meagre, and can be curtailed by changing cropping pattern.
- More complex and more costly option and control of conjunctive use and artificial recharge can be augmented and overcome by better farmers PIM.

The fundamental principle of conjunctive use planning is that there must be full recognition of the meaning conjunctive use. It means the complementary use of the many natural sub resources of both system surface and ground water to allow more cost-effective development than could be obtained even with the optimum plans for independent development. Thus there is great scope for time-wise regulation of the natural fresh water resource under conjunctive use.

Some other principles for the guidance of the actual planning processes are as follows:

Storage capacity analysis, water transport facilities, energy production and use, recharge capacity analysis, commodity of source of water, system time-response characteristics, maintaining control of water developed for use, water logging and root zone salinity balance, basin salinity balance, interactions with existing groundwater structures, and reliability enhancement.

Tamilnadu is in the advanced stage in the utilization of surface and groundwater. The sources available are limited, which should be utilized to the maximum possible extent by the integrated and the conjunctive use of surface and groundwater. The following three methods can generally be adopted in Tamilnadu:

- Diverting the surface water from the areas where the storage capacity is adequate for two crops to the adjoining areas where there is no regular water supply for irrigation.
- Where there is storage capacity for only one crop, the groundwater available in the ayacut area has to be pumped and supplied to the adjoining elevated areas for raising one crop, as the groundwater level will rise due to recharge by applied water, which may enable to stabilize the second crop also by use of groundwater.

- Where the surface water is now supplied by rotation, periods of supply of surface water can be broken so that groundwater can be used in intervening periods.

Through such conjunctive use of surface and groundwater, in the command areas of the waterbodies located in Tamilnadu, the following can be achieved:

- It is an integrated approach for the use of both the resources fully, to imply the coordinated and harmonious development of the two sources with the sole purpose of maximizing agricultural production. Areas, which have no assured water supply, can be provided for atleast one crop.
- In areas where one crop is now raised, there is a possibility for raising two crops.
- It may also be possible to raise a third summer crop in the reservoir ayacut areas, which will consume less quantity of water by suitable crop rotation.
- It may also be possible to use simple contrivances like windmill and or by employing solar energy for groundwater extraction which is in short supply.
- Extensive measures have to be devised to popularize animal power, which for so many reasons has gone out of usage, and obsolete.

Thus by minimum expenditure and optimal utilization of available water, the State's income and agricultural production can be maximized. Besides, the equitable distribution of the available water will result in the development of backward areas, giving large-scale employment to the rural rustic and the educated alike. The pilot study conducted during 1982 - 1984 in Krishnagiri reservoir area was fruitful and successful. By adopting the same techniques, project report for such Conjunctive Use Studies were prepared for the entire Sivagangai and Pudukkottai Districts and proposal were sent to Government for approval.

NATIONAL WATER POLICY AND PIM

The National Water Policy (1987) of Government of India suggested the progressive involvement of farmers in the management of irrigation systems particularly in water distribution and collection of water rates. Separate guidelines have also been issued on farmers participation in irrigation management. The State Government have followed suit either by enacting or by amending the irrigation Acts to the requirements of Participatory Irrigation Management (PIM). The broad outlines set for PIM activities are:

- Setting up of Water User Agencies (WUAs) on hydrological basis (mini watershed basis) with the enrolment of atleast 51per cent of the farmers as members. Registration of the WUAs either under Cooperative Societies Act or Registration of Societies Act.
- Signing a MoU between the WUAs and WRO and getting the system control over to the WUAs. WUAs assuming the responsibility of obtaining required water for their jurisdiction and distributing it among the members judiciously, i.e. shouldering the responsibility for the management of system under their jurisdiction.
- WUAs shouldering the responsibility of collecting the water cess from the members, remitting the same to the government (WRO) and availing the incentives offered thereon.

- WUAs carrying out the maintenance works in canals and related infrastructures within their jurisdiction like the traditional old Kudimaramattu type.

On the face of it, the programme is simple and straight forward. But, the reports on the performance of the WUAs formed earlier under the obligation of the rules indicate certain constraints and lack of modification and power.

- Lack of policy and legal support from the government.
- Lack of authority and power to the WUAs.
- Lack of financial support to WUAs.
- Lack of cooperation and support from the irrigation agencies.

All the WUAs under PIM are formed by the efforts of the officials of PWD, WRO and sometimes with the help of reputed NGOs. The efforts and dedication put forth for this cause are beyond suspicion. But, even the best WUAs formed under the most favourable conditions fail due to various reasons. There cannot be another best example for an ideal WUA for PIM. There should be a strong government backing, strong local leadership and an ideal and strong change in cropping pattern depends upon monsoon changes.

CONCLUSION

- PIM can only help in reducing the cost of food production to some extent, and thereby the cost benefit analysis should not be made simply with the values of food production, and should be analysed with the user demand and also the total cost involve in import of food grains, time and place. To make the irrigation system more efficient and responsive it may be necessary that GMIS—Government Managed Irrigation System be transferred to FMIS—Farmer's Managed Irrigation System, by forming WUAs viz. PIM and PWM, which alone would be responsible to cater all the needs for good and effective irrigation management.
- The practice of fixing numerical targets for the creation of PIM in conjunct with PWM prescribing uniform procedures may be avoided since farmers organisation is only social services.
- The farmers may be progressively trained and exposed to the day to day achievement in scientific farming activities by conducting programming, taking them to educational tours in other parts of the country, different methods of irrigation practices by creating awareness, and by conducting training programmes. "Turnover" of the irrigation systems is not the end of the PIM process, but it is the beginning of the watershed management.
- Realization of the necessity of judicious use of water alone will make the farmers organizations to whole-heartedly cooperate with the irrigation strategy in the management of the system.
- Regular interactions and mutual trust between the WUAs and the WRO and a transparent schedule shall alone be able to achieve the goals of PIM.
- Ultimately it must be the effective participation of the government organisation, which are responsible for scientific implementation of Watershed Management Techniques and farmers shall be a part of the process to make the PIM a success.

References

A guide for estimating irrigation water requirements, Technical Series No: 2,July 1971, Water Management Division, Ministry of Agriculture. (Dept. of Agriculture), New Delhi - 110 001.

Groundwater Manual, Bureau of Reclamation, U.S. Department of Interior, 1977. John Wiley & Sons, New York, 1981. A Water Resources Technical Publication.

Patric A. Domenico. Watershed Management Techniques. Mc Graw Hill Book Company Inc., New York 1972.

Survey of the Environment 2001, a publication of The Hindu, Chennai.

The Madras Agricultural Journal, Vol. 59, No: 2, Special Issue on Groundwater and Wells, Agricultural College and Research Institute, Coimbatore - 641 003.

Training Course on Groundwater Techniques (First Course - Sep. - Oct. 1970) - Lecture Notes not published, UNDP Groundwater Project, Thanjavur, Tamilnadu.

Watershed Management: Issues and Policies for 21st Century pp. 253-258

Challenges in Watershed Development: Case Studies

R. K. Sivanappan

INTRODUCTION

Watershed development and management is an integration of technology within the natural boundary of a drainage area for optimal development of land, water and plant resources to meet the basic minimum needs of the people in a sustained manner. The poor in the rural areas who are struggling for survival can not be expected to pay heed to the conservation strategies and environment, unless their daily needs of food, fibre and fuel are met with. Still a more urgent need is for assured and full employment for all these people. Integrated watershed development and management is not only the most effective solutions to many of the problems mentioned above, but also the effective solution to many other common problems like drought, floods etc. It includes the integration of many scattered programmes like soil conservation, afforestation, minor irrigation, fodder development and other development activities in a well prepared micro watershed project. Based on study of climate, land, water and plant resources on the one hand and man, animal resources on the other, this offers hope for bringing about sustained natural resources development. It also provides solution to many environmental problems like soil erosion, siltation in tanks and reservoirs, improper land use, land degradation, lowering of ground water table etc. Once these are solved, the overall productivity, income of the family and employment opportunity in the villages could be increased and thereby the living conditions of the rural population can be enhanced.

WATERSHED

A watershed is an area from which runoff, resulting from precipitation flows past a single point into a large stream river, lake or an ocean. In other words a watershed is that area in which all the precipitation converges and drains past a particular point. The term watershed, catchment area or drainage basin can be used interchangably. A watershed may be only a few hectare as in the case of small ponds/checkdams or hundred of square kilometers as in the case of rivers or big reservoirs. For convenience, watersheds are classified in terms of size into: (*i*) Basin (*ii*) Catchment (*iii*) Sub catchment (*iv*) Watershed (*v*) Sub watershed and (*vi*) Mini and Micro watersheds.

Each watershed is an independent hydrological unit, any modification of the land use in the watershed will be reflected on the water as well as in the sediment yield of the watershed. Watershed has become an acceptable unit of planning for optimum use and conservation of soil and water resources. A watershed is a hydrological unit which produce water as an end product by interaction of rainfall and watershed factors. The quality and quantity of water produced by a watershed is the index of amount and intensity of rainfall and the nature of watershed management. In some watersheds, the aim may be to produce maximum quantity of water distributed throughout the year. In the another watershed, the priority may be to reduce the peak rates as runoff for minimizing the soil erosion and floods. Yet in another the concern may be for increasing infiltration so as to improve the soil moisture regime for maximum production of crops/trees. Therefore, different measures/activities to be introduced in order to achieve the objectives of different watershed. The area of the watershed, may be of any size, but of the viable watershed may be about 500 - 2000 hectares. The area in the watershed may be completely forest or it can be of only agricultural land or combination of both. Since large areas, both forest and agricultural lands are degraded, the priority should be given to the high priority watershed, while taking up watershed development programme. This can be identified from the aerial/satellite/remote sensing maps.

Concept of watershed development

The concept of integrated watershed development is that the development and management of the resources in the watershed so as to achieve higher production that can be sustained without causing any deterioration in the resource base or causing no ecological imbalances. This calls for the formulation and implementation of a package of programmes of action for optimum resource use in the watershed without adversely affecting the soil and water base or life supporting system. The concept assume more importance in the context of planning for sustained development. The watershed development aim at preventing watershed degradation that results from the interaction of physiographic features, eliminate unscientific land use, inappropriate cropping pattern/tree selection, soil erosion thereby improving and sustaining productivity of resources leading to higher income and living standard for the inhabitants in the watershed area. Therefore, it involves restoration of the ecosystem and protecting and utilizing the locally available resources within a watershed to achieve sustainable development.

The development activities are going on in the country from 1951 onwards, but all the works were taken up in a haphazard way fixing the area and funds as the target. Though the works were completed and amount spent, the impact of the development was not felt by the people and these are also not sustainable. In fact, the farmers were not happy with soil and water conservation works carried by the department in earlier years since it was not treated well on watershed basis in a saturated manner. The concept of treatment/development on watershed basis was accepted by the Govt. in eighties and it is catching up well in all places only now. The impact of the development works are felt by the people since the entire watershed is fully treated after proper planning. It is the policy and principle of the Central and State Governments that all the development works related to land, water, and other natural resources management must be taken on watershed basis. The various programmes under operation such as (*i*) Drought Prone Area Programme (DPAP), (*ii*) Desert Development Programme (DDP), (*iii*) Integrated Watershed Development Programme (IWDP), (*iv*) National Watershed Development Programme for Rainfed Agriculture (NWDPRA), (*v*) Tamil Nadu Social Forestry Programme (*vi*) and, Tamil Nadu Afforestation Programme (TAP) are having the basic objectives of land and water resources management for sustainable production and environmental protection. Hence all the activities/works should be tackled on watershed basis.

Objectives of integrated/comprehensive watershed development

The objectives of the watershed development and management are to (*i*) utilize the available land to its maximum productivity by adopting various/suitable measures according to the land capability and without any environmental degradation, (*ii*) maximize productivity per unit area, per unit time and per unit of water to meet the food, fodder and fuel requirements of the people living in the watershed, (*iii*) conserve as much rain water as possible in the place where its falls, increasing the ground water level and maintaining it for sustainability and stabilise the flow of water for long period, (*iv*) prevent soil erosion and reduce sediment production by means of suitable soil and water conservation structures, (*v*) control the damaging runoff and moderate floods in the downstream areas, (*vi*) drain the excess water safely and avoiding gully formation, (*vii*) maximize the water storage capacity in the watershed, both in the soil and constructing many water storage structures, (*viii*) improve the infrastructural facilities in the watershed and (*ix*) increase the level of income and status of the people living in the watershed.

Components/treatments in the watershed

The following are some of the ways and means by which the watershed should be treated in order to achieve the objectives of the watershed development. The treatments may be different soil conditions, topography, rainfall pattern etc. Based on the above factors, watershed development plan can be prepared and implemented with the cooperation/involvement of the people/farmers in the watershed by different disciplines joined together.

The major activities include the following

Water resources development: These activities include construction of earthen embankments/bunds, construction of check dams(temporary and permanent), construction of percolation tanks and providing drainage to remove excess rain water.

Erosion control devices: Contour and staggered trenching, gully control structures, construction of silt detention tanks, providing vegetative and stone barriers and retaining walls

Land development (Agricultural land): Contour ploughing, land leveling, contour bunding/graded bunding/compartmental bunding and terracing in steep slopes.

Agricultural development: Improved dry land farming techniques, suitable crop selection, agro forestry techniques and introducing horticultural crops.

In situ moisture conservation techniques for agricultural crops include micro catchment, basin/furrow farming, broad bed and furrows, tie riding/random tie ridges and water spreading. For tree crops the activities such as water spreading, saucer basin/crescent bund, 'V' ditch forming, tie ridging, catch pits and deep pitting.

The most important factors affect the watershed behavior and need to be studied for development are size and shape of watershed, topography, soil characteristics, precipitation, land use, vegetative cover including trees and animal population.

The variety of data necessary for planning a watershed development project will depend on the nature and purpose of the activity. Broadly speaking, the data necessary for planning watershed development can be grouped into three categories such as (*i*) hydrological data, (*ii*) soil and land use data and (*iii*) economic and social data. Implementing the project requires expertise in a variety of fields like Agriculture, Horticulture, Forestry, Soil and Water Conservation, Hydrology, Socio economic and Community Development. Therefore, adequate training of personnel in the above subjects will have to be provided to ensure the success of this new and complex activity in the watershed development.

CASE STUDIES

The Government of Tamil Nadu is implementing various watershed development programmes in the State under various schemes—namely DPAP, WGDP, Wasteland Development, etc. About Rs.75 - 100 crores are spent every year for this type of works. In spite of large amount spent in the last 15-20 years, there is not much impact noticed by the farmers/people. The main reasons for this are that the works are not carried out in a strict watershed basis and also not carried out all activities (Different disciplines) in the same watershed in a holistic way. Further the money spent (allotted) is not enough to saturate the required Soil and Water Conservation and Water Harvesting measures. However, there are some success stories in watershed development in Tamil Nadu which can be taken as a model for the future development in the state. They are (*i*) Seepalakotai watershed in Lakshman Nagar, Theni Dt under Western Ghat Development Programme (WGDP) and (*ii*) Alankulam Taluk, many watersheds in Tirunelveli Dt. under DANIDA aided comprehensive watershed development programme.

The various works executed and the impact of these projects are detailed below:

(*a*) Seepalkotai Watershed - Lakshman Nagar in Theni Dt. under Western Ghat Development Programme

Lakshman Nagar, a small village in Theni district, Tamil Nadu which is situated at a distance of 10 km east of Chinnamanur and 28 km south east of Theni, is the foot hills in between Varisainadu and Kambam valley in the western ghats. The annual rainfall is about 700-750 mm in the area. There are about 40 wells in the Seepalakottai watershed and the watershed area is about 350 ha. Since it is a very interior village, the farmers are having about 10-12 acres of dryland with wells and the remaining lands are under rainfed conditions.

Thiru L.Muthusamy and his four brothers live in this village. When the contour bund was laid by the Agricultural Engineering Department, ten years back, it helped Muthusamy and his brother to bring their lands under cultivation. Each brother has 10 acres of land with one well. They used to raise irrigated cholam with the water available in the well and take a single crop in about 1-2 acres. In dryland groundnut/cholam/redgram are normally cultivated. When the Agricultural Engineering Department took up the implementation of Western Ghats Development Programme, four check dams were constructed in a deep jungle stream. Initially knowing the benefits of percolation ponds, the brothers gave 2 acres of land to construct one percolation pond. After reaping its benefits, they gave another 2 acres to construct another percolation pond in the watershed.

Before implementation of the soil conservation measures/structures the wells have water lifting to irrigate only for 1½ to 2 hours. So they resorted to deepening of their wells every summer spending about Rs.10,000 to Rs.15,000. The average depth of the well is below

100-120 and in summer there may not be any water in the wells. They used to get about five bags of cholam per acre. Hence to supplement the income, they worked as agricultural labour also. After the contour bunds, check dams and percolation ponds are constructed in the watershed area, the brothers cultivated banana, coconut, mango and guava under irrigation in 15 acres and keep another 15 acres with ploughing to cultivate cotton and vegetables totaling 30 acres under irrigation. In the remaining 15 acres or so dryland Cholam is cultivated with an yield of 7 bags/acre. This is the case with all the farmers in the villages. This could be possible with the increase in the water table in the wells and soil moisture in the land. The wells have enough water now to pump for about 6 hours a day with electric motor. They do not go as labourers now. The water table is about 60′ from ground level throughout the year as compared to 100 - 120 ft. before implementation of the watershed development programme. The farmers in the village provided houses to the farm workers in their farm to do farming throughout the year. All the farmers in the village say that after implementing Western Ghat Development Programme in the watershed, the rain water is retained and hence water is available in plenty. The silt accumulated in the check dams/ percolation ponds are removed by the farmers every year and use it in their lands as fertilizers.

There was no drinking water in their wells during summer before the implementation of the watershed programme. Now the village has a water tank with taps and they get water throughout the year. This is the impact of watershed development works carried out in the watershed by the Western Ghats Development Programme involving the farmers.

Lakshman Nagar is having enough water and food at present and the farmers are happy and they help many poor people who come for job to their village. Hence this is another Ralegoen Siddhi of South.

(*b*) Alankulam Taluk - Many Watersheds in Tirunelveli Dt. under DANIDA aided Comprehensive Watershed Development Programme

The experience of Alankulam comprehensive development programme is different model from Lakshman Nagar watershed in Theni Dt. The DANIDA aided comprehensive watershed and development programme is in operation for nearly 10 years in the drought prone Dt. From the experience gained in the Phase I (first 3 years) the project component and approaches were modified in the Phase II (last 7 years) and this has helped in developing agro forestry and self employment in the watershed. The emphasis is forming Village Development Association (VDA) involving all the villagers in planning, implementation and to take care of sustainability. Further there are project committees with farmers representative for the smooth functioning to take care of the micro level problems in the watershed. The community organisers—a member of the local N.G.O are helping the VDA's in various activities. The various activities contemplated in the selected watersheds are:

- Soil and Water Conservation: Under this, check dams, Contour bunds, Vegetative Barrier on contour, repairs and maintenance of existing ponds/Tanks, percolation tanks are the various types of works implemented in consultation with VDA/farmers.
- Provide a bore well (upto 75 m depth) with a hand pump to give water for the trees planted in the watershed (Fruit and other trees). One borewell for every 10-12.5 ha or to cater the needs of about 10 farms.
- Creation of agro forestry in drylands after land and water development.

- Helping farmers to have inter cropping including fodder crops.
- Establish research and demonstration farms in the farmers field so that farmers can gain knowledge to select the crops/trees for their land.
- Training to the farmers in agricultural. Technology and also take them to the successful farms to gain practical knowledge.
- Training in growing nursery and skills to the farmers/landless people for their livelihood.
- Organising seminars/workshops to improve knowledge.

With deep involvement of the field staff and full cooperation of the farmers/VDA, the project was successful in the entire Alankulam Taluk in many watersheds. It was reported by the farmers that they are able to get an income of Rs.5000-7500 per acre from their dry/rainfed land which was barren and not producing anything before the watershed development programme was introduced.

The main reasons for the success are: (*i*) Involvemnent of fatmers through VDA's (people's participation), (*ii*) Holistic approach in implementing the programme in the watershed, (*iii*) Involvement and cooperation of different disciplines working in the project namely Agrl. Engg, Agronomist, Horticultralist, Veterinary and Forest specialist and (*iv*) The flexibility in the project to enable them to modify any programme if the farmers think otherwise.

The success of the project has reflected the policies/systems of the Government in other similar type of development. The lessons learnt are: (*i*) forming VDA and giving training, (*ii*) Involvement of local NGO's, (*iii*) Integrated approach involving various disciplines, farmers/NGO's, (iv) Providing borewell with hand pumps for every 10 farmers or 12 hectares, (*v*) providing training to the farmers as well as mid level staff working in the project including study tour etc.

CONCLUSION

The watershed development project can be successful only if we emphasise 'people's participation in all the activities of the project including planning, implementation, monitoring and evaluation and maintenance thereby sow the seeds of success in the programme. We must also carefully analyse three aspects of cost, return and risks as these are very important in every project. In order to achieve the goal and to fulfil the aspiration of the people, we should have inter-disciplinary experts in watershed management as indicated earlier. There are experiences like Ralegoan Siddhi and other watersheds in the country. However, the experiences of Lakshman Nagar and Alankulam models can be replicated in all watershed development programmes which will solve water and food problems of the state.

Watershed Management: Issues and Policies for 21st Century pp. 259-264

Watershed Development in India: Experience of NABARD

V. Puhazhendhi and V. Tagat

INTRODUCTION

Scientific management of land is the key aspect of sustainable agriculture, as land being the fixed resource and subject to degradation due to exploitative use. Integrated watershed management has been accepted as the most rational approach in preventing deterioration of ecosystem, restoration of degraded lands and improving the overall productivity for sustained use. Approximately 170 million hectares of land in India are classified as degraded land, roughly half of which falls in undulated semi-arid areas where rainfed farming is practiced, that indicate the existence of vast potential for future expansion in irrigation management.

In the past, watershed development has been promoted solely out of the central grants and subsidies. Of late, stress has been laid on sharing the costs with the watershed communities through people's participation. The trade-off between private returns and social returns and also between short run and long run returns are more pronounced in case of watershed development programmes. NABARD has been implementing a number of micro watershed projects and involved in evolving participatory watershed development approach that can be replicated over wide areas. An attempt has been made in this paper to highlight and examine the issues involved in watershed development. The experiences of Indo-German Watershed Development Programme (IGWDP) implemented in Maharashtra have been used as case study to strengthen the arguments wherever feasible.

INDO-GERMAN WATERSHED DEVELOPMENT PROGRAMME

The IGWDP has been operationalised by NABARD in Maharashtra, India since 1992 through the bilateral aid from the Federal Republic of Germany (FRG) routed through Kreditanstalt feur Wideranfban (KfW) for mainline phase and through Deutsche Gesellschaft feur Technische Zusammenarbeit (GTZ) to the Watershed Organisation Trust (WOTR) for the capacity building phase. There are two phases in the programme, "Capacity Building Phase"(CBP) and "Full Implementation Phase" (FIP). The CBP supported by GTZ-WOTR is basically building the capabilities of village communities and NGOs to plan and implement and supervise the watershed activities through a "learning by doing" approach. On successful completion of CBP, these projects move into the FIP supported by KfW-NABARD.

The ex ante impact of the programme are expected mainly related to increasing groundwater levels in the areas, changes in cropping pattern, reduced soil erosion, reduction in seasonal migration of people, creating employment opportunities for landless and increasing women's participation. The strategy followed by NABARD is to provide for all the common costs and joint costs by routing the KfW funds for the purpose and encouraging people's participation and NGOs in cost sharing, implementation and maintenance. Further channelising institutional credit for supporting the agricultural, animal husbandry and non-farm activities, has also been appropriately included in the programme.

The NABARD-IGWDP evolved out of the sporadic successes achieved by a few NGOs, who were able to demonstrate that spectacular results could be obtained if technology walked hand in hand with the aspirations of the people whose welfare it was meant to secure. The principal concern on which the NABARD initially focused its attention was that participatory watershed development should be replicable over wide areas. The actual programme formulation was preceded by a series of consultations between NGOs, KfW representatives and NABARD. Since the partners in the programme had their area of operations stretching from international levels to the village level, the creation of confluence of interest within and across these levels was the first major milestone in the evolutionary process.

The distinctive mark of the programme has been the ability to evolve a participatory delivery system through involvement of watershed communities. The programme has successfully evolved a technically sound, but participatory watershed planning methodology, a strategy of coherent transition from capacity building to full-scale implementation of micro/village level watershed projects, a framework for field level collaboration among NGOs, community based organisations and government's line departments, a framework for participatory co-operation between international and national organisations, development of a set of procedures for hassle-free implementation of the projects and creation of durable institutions for capacity building and programmed implementation of projects.

The programme is currently covering 92,000 hectares of private and other land in 20 districts of Maharashtra, involving 57 NGOs working in 95 watersheds. In terms of financial achievement, about Rs. 554 million has been disbursed under the programme. The programme period of 4 years had exceeded in most of the projects. The area treated under the programme was around 63827 hectare with the average grant of Rs. 5812 per hectare. The cost of investment could be broadly grouped into three categories (*i*) soil conservation, (*ii*) water conservation and (*iii*) cost of management and training. Among different components, area treatment has been given priority over drainage line treatment.

EMERGING ISSUES

Peoples' participation in watershed management is viewed as a dynamic group process in which all members contribute to the attainment of overall objective, share the benefits, exchange information and knowledge and follow the set of rules for common interest. Need for people participation assumes greater significance when there is common resource management, which is an essential component of watershed development. With commitment and conviction, people can discipline themselves for their betterment. Villagers need to contribute their stake/share (voluntary labour etc.) and feel the ownership of the programme. The experiences of NABARD-IGWDP implementation have identified the following issues/rationale for people participation in the Watershed Development Programme.

- Reduction in development cost to the Government or NGO and ease in planning, implementing and monitoring.
- Increase and fair distribution in the perceived and actual benefits to people participating in the programme.

- Increase in awareness and decrease in perpetual dependence of people on the government/NGOs and thereby making the programme self-sustained.
- Gradual empowerment of the society and economically disadvantaged people.

Various strategies for encouraging people's participation in watershed development have been evolved over time. The earlier strategies depended on the top down approach and people's organisations like panchayats were involved in implementation and monitoring. Now the focus has shifted to involving the people right from the planning stage by organising them in the form of informal watershed development societies. This approach to a greater extent reduces the conflicts between the ecological (technical) sustainability and social sustainability. One of the grass root level institutional arrangements that are found to be effective in ensuring people's participation is Village Watershed Committee (VWC). The village community is also managing the funds through the VWC for total transparency and cost effectiveness.

EMERGING STRATEGIES

The emerging strategies for ensuring people's participation in watershed development programme are:

- Formation of small homogeneous groups, involve them from planning stage, entrust the responsibility of management and maintenance of the assets created under the watershed project and establishment of appropriate federal structure.
- Greater role for women at every stage.
- Educating all the participants through their own members about the tangible benefits, which can accrue if the common goals are collectively achieved and the social benefit expected to be derived due to watershed management.
- Clear rule for bearing costs jointly and distribution of benefits from common property resources and equitably.
- Identification of conflicts over the use of common resources and settled through people's involvement.
- While a micro-watershed is identified by its physical characteristics, it is the peoples' commitment that decides for inclusion or otherwise in the programme. It is in effect a "self-selection" process.
- People to demonstrate their willingness/eagerness to undertake the watershed project by visiting nearby watershed for experience/exposure and agreeing to contribute 4 days of *shramdan* (voluntary labour) initially without any support from the programme.
- Villagers agree to bring down the cattle population within the carrying capacity of the land; ban free grazing and free felling and eschewing crops, which require more amount of irrigation.
- A pilot project - 50 to 100 ha - is taken up for treatment measures where the competence and willingness of the community/villagers and the ability of the NGO to mobilise and work with the villagers is put to acid test. This project lasts for 12 - 18 months period and affords the right opportunity for learning while doing. Technical skills required in soil and water management is demonstrated, practised and refined in the field itself.

- VWC, a people organisation with adequate representation of women is formed for planning, implementing and eventual maintenance of the treatment measures/ structures. VWC is responsible to Gram Sabha or entire village community.
- Village community is asked to partner the process of watershed development through their contribution of one-day-a-week free labour contribution. As an incentive for their participation, the programme envisages ploughing back 50 per cent of free labour contribution to the maintenance funds of VWC.
- The treatment measures follow a "ridge to valley approach". The "net-planning" approach adopted demands survey of each of the plots in the watershed and suggest appropriate technical measures for conservation and improvement in consultation with the farmer and his family.
- Efforts are made to encourage VWC to think of ways and means to involve the landless in project activities and design appropriate systems of benefits arising from common property resources.
- The women in the community, besides being represented in the VWC are encouraged to undertake project related activities like raising a nursery, kitchen gardens, in addition to inculcating the habit of thrift and funds management / rotation amongst themselves. Besides, women promotion/development activities are undertaken within a ceiling of 5 per cent of project measures.
- The village community and NGOs are allowed to implement the full-scale watershed project only on successful completion of CBP. This strategy has to a large extent eliminated the incidence of failure at a later stage.
- The village community in association with NGOs manage the entire fund meant for project measures. The management cost of NGOs is funded separately.
- A maintenance fund is created out of peoples' contribution, ploughing back 50 per cent of voluntary labour and also an end-of- the- project incentive for future repairs and maintenance of structures.
- The village community is made aware of the exit requirements of NGOs towards the later part of the project to allow VWC to take charge of future maintenance, development and possibilities.

With adequate monitoring and follow up, broad strategies during planning and implementation the IGWDP, watershed projects have produced impacts in the lives of people, which were hitherto unheard of. The following are a few of the striking impacts of the projects.

- The water tables have risen in project villages and there has been increase in number of wells with perennial water.
- From the status of drinking water being drawn through tankers to the village, transformation has made the village a milk surplus village.
- Shift in cropping pattern towards high value crops and increase in cropping intensity. There has been increased production of high value crops and increase in foodgrain production (dryland millets).
- Soil erosion has been arrested and the green cover in the watershed has considerably increased.

- Seasonal migration of people from village reduced to a significant extent. Even "reverse-migration" reported in respect of people doing marginal jobs in cities.
- The landless have new jobs: fishing in community ponds, farm labour, off-farm activities etc.
- Social disparities have reduced; women have a greater say in development and school attendance has increased.
- Institutional partners, i.e. VWCs, NGOs, WOTR, NABARD etc. are richer in their knowledge, experience and capabilities for participatory watershed development. Even one VWC has decided to function as NGO to facilitate watershed project in a nearby village.

The major learning from the experience of implementing more than 80 projects under IGWDP and also the in-house evaluation and mid-term reviews conducted, can be summarised as under:

- Peoples' participation is an essential component of watershed development. With commitment and conviction people can discipline themselves for their betterment.
- Villagers need to contribute their stake/share (voluntary labour etc.) and feel the ownership of the project.
- Creating peoples' organisations like VWC for planning, implementing and maintenance is a must for success and sustainability.
- Having a pilot phase of proof testing the abilities and motivation of community and NGO during the CBP phase insures against most future possibilities of failures.
- Women and landless need to be integrated into the project design appropriately.
- Technical considerations and skills for soil and water management, based on "ridge-to-valley"principle can be understood by villagers through learning-by-doing, demonstration/exposure visits etc.
- The village community should manage the funds for project measures through the VWC for total transparency and cost effectiveness. However, timely availability of funds in adequate measures in advance is a critical requirement for watershed projects.

WATERSHED DEVELOPMENT FUND

With the successful sustainable models of participatory watershed development being implemented in Maharashtra spreading across the government circles, a Watershed Development Fund (WDF) was created with NABARD with an amount of Rs. 200 crore equally contributed by GOI and NABARD. While two third of the Fund will be utilised for loans to State Governments for watershed development; the remaining fund will be used for grant based projects through NGOs as also for promotional and capacity building efforts. The objectives of the proposed Fund are to spread the message of participatory watershed development. The Fund is being utilised to create the necessary framework conditions to replicate and consolidate the isolated successful initiatives under different programmes in the government, semi-government and NGO sectors. Thereby, all the actors involved—watershed community, Central and State government departments, banks, agricultural research institutions, NGOs and NABARD can act in concert to make a breakthrough impact in participatory watershed development. The WDF is proposed to be operationalised in close coordination with the Central and State Ministries and Departments as a continuum of their efforts but with a distinct identity and contribution from NABARD and the Government of India, financially and managerially. The fund will be utilised to implement watershed development programmes in 100 priority districts in 3 years spread over 14 states. During the year 1999-2000, six States—Andhra Pradesh, Madhya Pradesh, Orissa, Uttar Pradesh, Maharashtra and Gujarat were targeted to initiate works. In the year 2000-2001, Bihar, Haryana, Himachal Pradesh, Jammu & Kashmir, Karnataka, Rajasthan, Tamil Nadu and West Bengal are targeted to initiate works. Three State Governments have already signed the MOUs for participating under WDF and seven more have expressed their willingness to join. The officials of State Governments, Project Implementing Agencies, NGOs from project states as also officials of NABARD from the project states are being exposed to IGWDP - Maharashtra model of watershed development through exposure visits, training and workshops.

LOOKING AHEAD

Encouraged by the successful implementation of IGWDP in Maharashtra, KfW has expressed its desire to partner replication of the IGWDP model in two more major states of India viz. Andhra Pradesh and Gujarat. NABARD has successfully completed 30 projects under the Indo-German Watershed Development Programme and some more are nearing completion. These projects are emerging as 'role models' for replication in other parts of the country. The completed watershed projects are to be sustained with appropriate post-development management aspects. The experiences highlights the crucial role of capacity building including appropriate training for the village community, increased contribution from farmers and a more participatory planning process. Further, effective soil and water conservation measures especially during drought period and adequate and timely credit support from the institutional sources would sustain and stabilise the watershed programme.

The success of the programme, however, is further dependent on educating the farmers and in providing extension services to consolidate the benefits of watershed development by diversifying the agricultural activities. This involves taking up new/commercial field crops, horticulture activities, vegetable cultivation, floriculture, dairy, stall-fed goat rearing, fishery, etc. The Village Watershed Committees (VWCs), constituted in the watershed projects, are being empowered to develop a network for marketing their produce to realise the best possible prices. In some of the projects, farmers are being encouraged to organise themselves into co-operative societies. Some co-operative dairy societies have also been formed for better marketing link up to realise higher prices for the milk.

Watershed Management: Issues and Policies for 21st Century pp. 265-274

Joint Forest Management and Watershed Development Programme in Tamil Nadu: An Experience in TAP

C. K. Sreedharan

INTRODUCTION

Earth is a "Water-Oxygen" planet and both of them are threatened now. Water is the most critical factor for all kinds of living beings and bio-mass production in rainfed areas. Way back in 1977 a Syrian delegate remarked during the World Conference on Water Resources held in Argentina that "the day is not far off when a drop of water will cost more than a drop of oil". The world is fast approaching the threshold of such a tragedy. To prevent the threshold becoming a catastrophe, we need urgent and scientific water management. Drought hits more people in the world than any other natural disaster. Asia is the most disaster prone continent and India tops the Asian list. Due to enormous demographic, industrial and agricultural pressures the ground water availability has entered a critical phase globally and in more threatening proportions in developing countries like India, which have huge human and livestock population.

WATER MANAGEMENT SCENARIO

With the degradation and deterioration of catchment areas the run off results in heavy loss of surface water, which remains to be largely untapped. Ground water has several advantages over surface water. It can be developed quickly, *in situ*, near the place of use. Surface water requires suitable sites and structures for harvesting and storage. Ground water is also not susceptible to evapotranspiration, seepage and pollution. However, the ground water resources have been exploited to the hilt. Demographic industrial, and agricultural pressure have become manifold. New river valley reservoirs have run into economic and ecological problems. The onus has gradually shifted to ground water.

There has been a systematic degradation and deterioration of forests, which are the primary catchments to many streams, lakes and waterholes. The surface vegetation has become denuded. The vegetation present acts as sponge soaking water into the ground. The roots bind the soil and thereby ensuring it is not lost in the run off. The degradation or deterioration of any one feature of the watershed has a compounding and cyclic effect on all others. Without vegetation, water is not harvested to its maximum potential within the catchment. This leads to less recharge of ground water and less bio-mass or vegetation above the ground completing the cyclic effect. The coastal areas suffer inward incursion of sea water due to reducing amounts of water flow within the water table.

NEED FOR INTEGRATED WATERSHED MANAGEMENT

The need of the hour is an integrated and sustainable watershed development programme. The watershed management should be poised to check the following chain of events:

(*i*) Rain water run off without having time for seeping into soil, (*ii*) Soil erosion leading to degradation of soil health and fertility, (*iii*) Reduction in storage capacity of water bodies due to increasing silting and bed load, (*iv*) Less water availability in both surface and underground water resource and (*v*) Resultant barren lands sans productivity and consequent poverty.

To arrest the cyclic nature of degradation and reversal of events, the approach has to be seen in wide angle. The integrated approach should address to the bio-physical (soil, water and forest conservation) and socio-economic (human communities and their social and economic problems) factors equally.

The package of this integrated programme should address the critical issues like afforestation for catchment improvement, *in situ* soil and moisture conservation (both vegetative and engineering), economic water management practices in dryland agriculture, location specific agronomic practices, rain water harvesting through cement checkdams and percolation ponds in the downstream locations to convert surface flow into subsurface storage through ground water recharge, socio-economic development of the watershed community—both individual oriented and community oriented inputs, multi sectoral linkage for involvement of line Departments like Agriculture, Agri-engineering, Horticulture, Animal Husbandry, Public Health, Social Welfare, Rural Roads, Public Works Department etc. This will facilitate dovetailing of both technology and programmes in the Project Area.

The above integrated approach will go a long way in making the watershed development programme efficient and sustainable in the long run. Watershed approach would facilitate the containment of deterioration of natural resources and develop the total resources in such a way that it results in improvement of quality of inhabitants ultimately.

THE NEED FOR PARTICIPATORY APPROACH

Watershed Development programme should always be community centric because the life of the local community is directly involved. Water comes on the top of village development agenda. Therefore the utility of accumulated wisdom and skill of the local watershed community should not be lost. Their inherent knowledge about drainage pattern, location of recharge structures on lineaments etc. will be of immense use in designing appropriate treatment for efficient watershed management. People's participation has historical background. During Chola period there were instances of maintaining waterways and waterholes by people themselves in what was known as "Kudi Maramathu". In modern times the developmental planning strictly regimented its approach on technology and administrative nuances than placing confidence on the people to whom the ultimate benefits are intended. People are by and large willing and open towards development endeavours. Their practical indigenous knowledge is precious. In recent times several stories are emerging with shining examples of peoples participation. Starting from Anna Hazare model in Ralegaon Sidhi much water has flowed under the bridge. Several participatory programmes have come to stay. Only problem is consolidating the achievements and sustaining them forever.

EXPERIENCE IN TAMIL NADU AFFORESTATION PROJECT

Forests are the major catchments to several macro and micro watersheds. Over the years due to severe demographic, cattle and economic pressures, natural forests have been gradually reduced to the stage of degradation. The depletion of vegetation has affected the carrying capacity of rivers, reservoirs, and small waterholes. Continuous run-off has resulted in erosion of fertile top soil, sedimentation, fall in groundwater table etc. Most of these degraded forests, which are located all over the State, are in catchment area leading to severe water scarcity in about 3000 Villages adjoining them.

With an immediate task of restoration of the degraded forests a massive afforestation project has been conceptualised. People's participation through Joint Forest Management is the basis for implementation of the Project. The main thrust is on eco-restoration and biological upgradation of the degraded forests. For this purpose a degraded forest area with 250 ha. is selected on watershed basis. These areas have a crown density of less than 0.4. Degraded forests are high priority watersheds. The following works are undertaken with people's participation.

- Afforestation by gap planting.
- *In situ* soil and moisture conservation through gully plugging, vegetative barrier, contour trenches, semicircular bunds, contour walls etc.
- Water harvesting by construction of cement checkdams and percolation ponds.
- Rehabilitation of forest dependents and other poorer sections through alternate income generation activities.
- Community development works to enhance the quality of life of local community.
- Multi sectoral linkage to intensify the developmental activities in the programme villages.

A brief overview about the project is as follows:

An overview of Tamil Nadu Afforestation Project

Degradation of forest due to unlimited biotic interference and the consequent impact on the quality of life of rural inhabitants and the ecology of the State as a whole was recognized as one of the priority problems facing Tamilnadu, which necessitated tangible and time bound action. As a result Tamilnadu Afforestation Project funded by Japan Bank for International Cooperation (JBIC) with an outlay of Rs.499.20 crores is being implemented from 1997-98.

Objectives of Tamil Nadu Afforestation Project

Tamilnadu Afforestation Project has the following major objectives.

- Improving the productivity of forests by creating heterogeneity of tree species and through intensive soil conservation and water harvesting measures.
- Preserve areas, which are rich in biodiversity.
- Motivate people to adopt tree growing culture to create resources and assets for meeting the requirement of fuelwood, small timber, and fodder in the shortest possible time.
- In consonance with these concepts optimize the use of land resources on scientific basis and provide employment opportunities in rural areas.
- Enhancing the capabilities of Tamilnadu Forest Department by establishment of Geographical Information System, various training programmes and reorientation in Management of Natural Resources through Joint Forest Planning and Management.

Target

Tamil Nadu Afforestation Project envisages the following target for its forestry programme for five years (1997-98 to 2001-2002).

1.	Local Bio-diversity conservation through JFM		2,75,000 ha.
2.	Buffer zone activities		1,000 Villages
3.	Incentive for local tree cultivation		250 lakhs (seedlings)
4.	Conservation programme	Block :	1,00,500 ha.
		Linear :	500 kms.
5.	Production programme	Block :	730 ha.
		Linear :	1,000 kms.
6.	Conservation of species		2,176 ha.

Project strategy

Protective measures, enforcement and enactments have failed to stem the decline in degradation of forests as a result of apathy of the local population and absence of any precise plan for enlisting their cooperation. Joint Forest Management (vide National Forest Policy 1988) emerged as an effective tool for forest development and planning. The local population have been treated as "Stake Holders" and "Partners" in Joint Forest Management. Therefore Joint Forest Management is the hub of Tamilnadu Afforestation Project.

The project has focussed its aim squarely on institutionalizing the Joint Forest Management for sustainable and long-term natural resource management. Local population plays an active role in the implementation of the project in a transparent manner. Significantly, to maximize the realization of benefit of the project, the approach to implementation is done on "Watershed basis". With a holistic approach equal focus and emphasis is laid on forestry, development of buffer lands and the socio-economic upliftment of the village. The above strategy is designed to optimise the impact of ecological restoration and biological upgradation of forests, socio-economic development of interface villages, and soil and moisture conservation and water harvesting for hydrological improvement of the watershed.

The project is implemented in the following sequential manner with the active participation of the stakeholders.

- Selection of watershed.
- NGO participation for awareness creation and mobilization.
- Institution building (development of VFC)
- P.L.A., resource inventory, identification of beneficiaries and prioritization of work and micro planning.
- Execution of forestry development works, watershed development through soil and moisture conservation and water harvesting.
- Introducing alternate employment to forest dependents and weaker sections.
- Community development works.
- Inter sectoral linkage for dovetailing development.
- Benefit sharing.

Project achievement

Table 1. Number of village forest councils functioning

1997-1998	199
1998-1999	200
1999-2000	200
2000-2001	200
2001-2002	200
TOTAL	999

Table 2. Total number of village forest council members

Year	Members		Total
	Male	Female	
1997-1998	40653	32974	73627
1998-1999	41799	39732	81531
1999-2000	38832	34537	73369
2000-2001	35691	31425	67116
2001-2002	32930	30274	63204
TOTAL	189905	168942	358847

Table 3. Area under Joint Forest Management

1997-1998	60892 ha.
1998-1999	88240 ha.
1999-2000	74057 ha.
2000-2001	75150 ha.
TOTAL	298339 ha.

Table 4. Agro-forestry and homestead planting through incentive for local tree cultivation scheme

1997-1998	49. 62 lakhs
1998-1999	49.75 lakhs
1999-2000	50.00 lakhs
2000-2001	50.00 lakhs
TOTAL	199.37 lakhs

Table 5. Local employment generation

(in lakh Mandays)

Year	Male	Female	Total
1997-1998	24.67	27.08	51.75
1998-1999	60.97	40.66	101.63
1999-2000	61.30	40.87	102.17
2000-2001	60.35	40.23	100.58
TOTAL	207.29	148.84	356.13

Alternate employment

Alternate employment has been provided for forest dependents and rural poor in the project village through several income generation activities. These measures are designed to increase the earning capacity and wean the forest dependents away from the forests. Women who are engaged in firewood collection and grazing are greatly benefited by these activities.

Community development

All programme villages tackled under TAP are poor with least civic amenities like transport, protected drinking water, education etc. The project plans to give a facelift to these villages by taking up several development activities according to the wishes of the local villagers. Construction of overhead tanks, drilling borewells, construction of community halls, providing infrastructure for schools, electrification, installation of TV sets, construction of fair price shops, libraries, sanitation facilities are some of the activities done.

IMPACT OF WATERSHED DEVELOPMENT

Impact of watershed development programme on the geo hydrological environment was studied by an internal evaluation. Sixty watersheds developed from 1997-98 to 1999-2000 at the rate of 10 per cent sampling was selected to study the impact. Baseline data on the watershed was already collected while the project was implemented. Another survey was conducted to record the present status. A comparative analysis was made to study the impact.

The following parameters were studied that include (*i*) Water table level (in 5 to 10 observation wells situated within one to two km) from RF boundary, (*ii*) Rainfall at baseline and subsequent years, (*iii*) Correlation between rainfall and water table, (*iv*) Changes in total area under cultivation, (*v*) Increase in area under irrigated cultivation, (*vi*) Cropping pattern and (*vii*) Intensity of paddy cultivation.

A brief summary of the findings with appropriate data is given below for reference.

Hydrological impact

Lack of water is the most important factor leading to the decrease in agricultural productivity. Agriculture being the primary occupation in the rural economy, singularly affects the economy and social equilibrium when it fails. The selected watersheds under Tamil Nadu Afforestation Project suffer from low water table levels and agriculture was severely affected. Drought was prevailing in these tracts punctuated by a gradual reduction in rainfall. The rainfall pattern observed in the selected watersheds is illustrated below.

Table 6. Pattern of rainfall in selected watersheds

Project Year	No. of sample Watersheds	Rainfall (mm)			
		1997	1998	1999	2000
1997-98	20	1032	944	800	1030
1998-99	20	1035	920	756	943
1999-2000	20	984	994	738	800

It is seen from the table that during 1997-98 watersheds received 8.52 per cent and 22.40 per cent deficit rainfall in subsequent two years. In 1998-99 watersheds received 11per cent and 27 per cent deficient rainfall in subsequent two years. The data during 1999-2000 watersheds received 1per cent surplus rainfall in first year and 25 per cent deficit in the second year. Due to the these factors the agriculture was touching all time low in productivity and drought was intimidating these fragile economics. One of the important aspects of watershed development is drought-proofing. The following project package aims to redeem the drought conditions: (*i*) *In situ* soil and moisture conservation in the catchment area by vegetative and engineering methods; (*ii*) Water harvesting by construction of cement checkdams to serve as micro reservoirs and (*iii*) Construction of percolation ponds for recharge of ground water table.

The downstream lands especially those in the vicinity of "Zone of influence" are benefited by (*i*) Prevention of sedimentation through run off, (*ii*) Arrest of flooding by impounding the rain water, (*iii*) Increase in water table levels observed in agricultural wells and (*iv*) Increased availability of water both for irrigation and drinking. There has been an appreciable increase in the water table levels due to various water harvesting measures undertaken in the watersheds. 5 to 10 wells within the "Zone of influence" were selected for hydrological monitoring and monthly data on water table level were recorded. An analysis of the data on ground water table indicates that there is an increase in the level of water table in spite of subsequent drought years.

Table 7. Impact on watertable, 1997-98

Number of Watersheds observed	Average water-table level at baseline (1997) (m)	Average water table at the end of		
		I Year (1998) (m)	II Year (1999) (m)	III Year (2000) (m)
20	9.15	8.49	7.85	7.80

Table 8. Impact on watertable, 1998-99

Number of Watersheds observed	Average water table level at baseline (1998) (m)	Average water table at the end of	
		I Year (1999) (m)	II Year (2000) (m)
20	7.80	6.95	6.90

Table 9. Impact on watertable, 1999-2000

Number of Watersheds observed	Average water table level at baseline (1999) (m)	Average water table at the end of I Year (2000) (m)
20	11.20	10.77

Water table level is the depth at which water is available from the ground level. (i.e.,) depth to water.

Correlation with rainfall

The increase in watertable level when compared with the rainfall makes an interesting reading. While the rainfall has been showing a steady decrease, the water table has been showing gradual increase. This points to the fact that rainwater harvesting in the catchment area has effectively percolated to the ground and in addition the quantity of water harvested through checkdams and percolation ponds has actified the dormant springs and aquifiers. The comparative analysis of rainfall and water table is given below.

Table10. Rainfall and water table

(Per cent)

Changes in rainfall		Changes in water table level
	1997-98 Watersheds	
1998 - (-) 8.52		
1999 - (-) 22.4		(+) 7.20
		(+) 14.20
	1998-99 Watersheds	
1999 - (-) 11		
2000 - (-) 27		11
		11.50
	1999-2000 Watersheds	
2000 - (+) 1		(+) 3.8

The above data unambiguously proves that inspite of relatively poor rainfall the water table has been raising gradually which is mainly due to the water harvesting measures undertaken in the watershed. Wherever appreciable increase is not noticed the water table has not gone down and has been stabilized. The increase in the ground water table is the most tangible benefit enjoyed by the downstream dwellers. Thus drought situation due to deficient rainfall was managed to a great extent by water harvesting in the watershed. Consequently the impact is visibly felt in the agrarian front in area under cultivation both in rainfed and irrigated condition and cropping intensity and individual cropping extent. Data were collected from cultivators in the watershed for analysis. The analysis shows that (*i*) there is an increase in total operational area (both rainfed and irrigated) and the total area under cultivation has gone up from 6404.95 ha. to 6842.39 ha. showing an increase of 7 per cent, (*ii*) there is an increase in intensity of cropping especially in paddy and (*iii*) increase in area under cultivation has been observed in cash crops like paddy, sugarcane, and groundnut.

Impact on agriculture

The impact of water harvesting measures on the agricultural pattern in the downstream lands has been visibly felt. The drought stricken scenario has changed due to increased availability of water which is the single contributory factor leading to rejuvenation of agriculture.

Area Under Cultivation

There is an increase in overall extent under cultivation as tabulated below.

Table 11 shows that the total extent under cultivation has increased despite deficiency in rainfall. The impact is profoundly felt in irrigated cultivation, which has registered an impressive increase mainly due to increased availability of ground water in the downstream area.

Table 11. Total operational area under cultivation

Year	Total area cultivated		Percentage of increase	Irrigated cultivation		Percentage of increase
	Total at Baseline	Total at Present		At Baseline	At Present	
1997-98	2061.49	2117.04	2.71	1212.82	1423.31	17.4
1998-99	1612.26	1753.61	8.74	794.09	914.00	15.11
1999-2000	2731.20	2971.74	8.78	1656.78	1826.19	10.39
	6404.95	6842.39	6.84	3663.69	4163.50	13.70

Cropping Pattern

When viewed from the cropping pattern angle there is a significant shift in preference to cash crops like paddy, sugarcane and groundnut. Crop wise analysis reveals the following trends.

Table 12. Impact on cropping pattern

Name of the crop	Area under cultivation at Baseline	Area under cultivation now	Percentage of increase
		1997-98 Watersheds	
Paddy	773.98	944.52	22
Sugarcane	36.32	79.04	119
Groundnut	501.80	536.79	7
		1998-99 watersheds	
Paddy	952.93	1135.59	19
Sugarcane	166.75	204.73	23
Groundnut	395.88	426.43	8
		1999-2000 Watersheds	
Paddy	1057.40	1241.20	17
Sugarcane	355.98	366.49	3
Groundnut	528.00	584.00	11

The above analysis points towards the fact that farming practice has gained the much needed momentum.

Cultivation of paddy is the yardstick for measuring improvement in agriculture. Area under paddy cultivation has increased by 16 per cent as it is now cultivated in 3321.31.0 ha. compared to 2784.31.0 ha. during base year. Another important observation emerging from the study of agricultural pattern in the watershed is the intensity of paddy cultivation, which is the mainstay of food production in Tamil Nadu as seen from the following table.

Table 13. Changes in area under rice cultivation

Project Year	No. of water sheds studied	Paddy cultivation intensity number of villages			
		Cultivating Twice		Cultivating Thrice	
		At baseline	At present	At baseline	At present
1997-98	20	8	12	0	1
1998-99	20	5	16	0	0
1999-2000	20	13	11*	0	6
Total	60	26	39	0	6

*two time cultivation changed to three time cultivation.

The above analysis of the impact of water harvesting and hydrological improvement in agriculture brings to the forefront the rejuvenation of village economy, which enormously ensures security of the economic environment. Thus complementarity between agriculture and forests has been well established.

SUGGESTED VISIONS FOR FUTURE

Integrated approach is more advantageous than sectoral approach. Therefore, along with watershed development with afforestation and soil and moisture conservation works the following issues need concentrated and concerted approach.

- Conservation of natural resources in a watershed development programme is best ensured if local communities are directly involved in the management and receive guaranteed benefit as a result of their conservation and rational utilisation.
- Prudent agronomic and efficient water management practices in downstream command areas. Since water is the most critical factor for all kinds of living beings and bio-mass production in rainfed areas, prudent water management practices are of paramount importance.
- Integrated farming practice by linking micro enterprises like Dairying, Animal Husbandry, Poultry, Pisciculture etc., will enhance the income generation of farmsteads during non rainy seasons reducing the demand for water.
- Promotion of agroforestry in the arable drylands and non-arable arid lands to increase the biomass production outside the forest area to protect the catchment tree cover.
- Pro-active participation of the watershed community in all facets of the programme giving a sense of belonging to the works undertaken.
- Recognition of local knowledge and wisdom instead of too much reliance on technology.
- Socio-economic package to mitigate poverty of watershed communities.
- Constant monitoring and evaluation to keep track of the development process.
- Continued support to the programme till sustainability is achieved instead of abrupt exit policy which negates and nullifies all the hard earned gains.

References

Anon (1996). Tamilnadu Afforestation Project Document, Tamil Nadu Forest Department.

Anon (2000). Imprints of JFM in Tamil Nadu - An impact study, Tamil Nadu Forest Department.

Dhruva Narayana (1993). Soil and Water Conservation Research in India, 437-447

Michaelsen (1991). Participatory approaches in watershed management, Planning. *Unasylva*, **42**(194): 3-7.

Watershed Management: Issues and Policies for 21st Century pp. 275-289

Sustainable Watershed Development: Case Study of a Successful Model

Philip Abraham

INTRODUCTION

The aim of a development project is to improve the quality of life. Strong efforts made during the project period often produce good results. However, the benefits do not last long unless care is taken to build sustainability into such development projects. In agriculture dominated economies like India, where two thirds of the cropped area is dependent on rainfall, efficient use of natural resources is necessary for sustainable socio-economic, environmental and social development. To a large extent, this sustainability depends on the management of land, water and vegetative resources that provide life-supporting systems for human beings and animals. Watershed management has now become the new paradigm for natural resources management and conservation.

According to Ministry of Agriculture:

- India has 2 per cent of the world's land, 4 per cent of its fresh water resources but carries 18 per cent of the World's population and 10 per cent of its cattle.
- India has geographical area of 329 million hectares of which 47 per cent cultivated, 23 per cent under forest, 7 per cent under non-agricultural uses, remaining 23 per cent falls under the categories of barren and un-cultivable land, culturable wasteland, permanent pastures and fallow lands.
- Per capita availability of land declined from 0.89 hectares in the early 1950s to 0.37 hectares in the mid 1990s and is expected to reduce further to 0.19 hectares by 2020.
- India produces more than 200 million tonnes of foodgrain from 142 million hectares of cultivated land. Irrigated areas account for 37 per cent of crop-land and contribute 55 per cent of food grain production whereas rainfed areas account for 63per cent of cultivated land but contribute 45 per cent of food grain production.
- Over 5.3 billion tonnes of top soil is lost every year due to soil erosion. Nature takes about a thousand years to form one cm. of soil.
- The average soil loss is estimated to be over 16 tonnes/hectares/year.
- Soil loss varies from less than 1 tonne/hectare/year in dense forests to more than 80 tonnes/hectare/year in the Shivalik Hills.

WATERSHED MANAGEMENT

Watershed management is one of the important development programmes aimed at improving land use on a sustainable basis in dry areas, thereby improving the rural prosperity.

Watershed management in India has been defined as rational utilization of land and water resources for optimum and sustained production with minimum hazards to soil and water. Conservation meaning thereby, proper land use, protecting land from all forms of degradations, building and maintaining soil fertility, conserving water for agricultural use, proper management of water for drainage, flood protection, sediment reduction and increasing the productivity of land etc.,

Watershed management approach is basically driven by the concern of balancing exercises of man made interventions on natural resources, as part of the livelihood process. The model of watershed development is an agreeable measure to achieve 'sustainable development'. As 70 per cent of our cultivated land fall under the rainfed areas, watershed management is one of the critical factors for improving agricultural production. Loss of vegetative cover, followed by soil degradation through erosion, has resulted in scarcity for water as well as soil nutrients. Restoring ecological balance and increasing productivity in such regions necessitate undertaking all developmental activities on a watershed basis. Watershed management is a holistic approach to bring about development of integrated farming systems on watershed basis which aims at optimizing use of land, water and vegetation in our area to alleviate drought, moderate flood, prevent soil erosion, improve water availability and increase fodder, fuel and agricultural production on sustainable basis. The range of issues relevant to integrated watershed management includes environmental issues, crop and livestock production, social and cultural concerns and infrastructure planning. Scientific watershed approach looks at the various means of conservation as well as an optimum pattern of exploitation of land resources for sustainable agricultural land use, defined in its comprehensive sense.

Focus on people

Although watershed management is often defined bio-physically, the watershed problems are often due to human disturbances in their efforts to earn their livelihood. The goal of watershed development activities are to improve the quality of life of people. Hence, people must be the focal point of watershed management programmes and innovation should be planned with their need in mind.

COMPREHENSIVE WATERSHED DEVELOPMENT PROJECT (CWDP)

Comprehensive Watershed Development Project, Tirunelveli is being implemented in Tirunelveli and Thoothukudi Districts of Tamil Nadu from 1990, as a bilateral project of Govt. of India and Royal Kingdom of Denmark and implemented by Agricultural Engineering Department of Government of Tamil Nadu. Phase I of the project was implemented from the year 1990 in 34 villages of Tirunelveli and Thoothukudi districts for 3 years. The main activity of the project was to raise shelterbelts and develop interbelt areas to arrest wind erosion. An area of 17,860 ha. of theriland and wasteland were brought under cultivation by controlling erosion. On successful completion of Phase I, the Royal Kingdom of Denmark sanctioned the Phase II from August 1994 for a period of 7 years to cover an area of 42,500 ha. in 120 villages with the total financial outlay of 41.2 crores. A sum of Rs.31.77 crores will be shared by Danida and Rs.9.95 crores by Government of Tamil Nadu. During Phase II, the main activity is Agroforestry, involving active participation of farmers. The project has been extended upto March 2003 based on the recommendation of the Review Mission.

Development objective

To enable the land users in priority watershed to practise dry land agriculture, range management, horticulture, sylvipasture and forestry including conservation and use of natural resources on a sustainable basis.

Major objectives

- To enable relevant authorities including extension staff to support farmers and other land users, in priority villages to identify, prioritize, plan and implement suitable land use systems through farmers association.
- To demonstrate and popularize different land use systems aimed at sustainable production and increased income in priority areas.
- To control wind and water erosion in priority area.
- To create short and long term employment opportunities thereby increasing the household income in the priority areas.

The main components of the project are agroforestry, silvipasture, shelterbelt, interbelt development, research and demonstration, training and extension etc. These components are carried out through Village Development Associations (VDA).

Planning and implementation strategy

Phase I: The success and sustainability of any project largely depends on people centred planning and implementation strategy with sufficient scope for flexibility. In Phase I, the project was initially designed to control wind erosion to cater to the problems arising out of the strong wind that gushes through the gaps in the Western Ghat. Therefore, the main activity of the project was to plant shelterbelt and break the wind speed and develop the land for useful purposes and develop interbelt areas through economic, horticultural trees. In this process people's participation in the project was not given much importance and was limited to obtaining consent letters. There were extension activities to inform the people. It was mostly target oriented Government programme based on top-down approach.

Change of strategy in phase II

The second phase of the project witnessed a major change in the components of the project. The shift in emphasis was from shelterbelt and interbelt development to agro-forestry. People's participation was given paramount importance. People were involved in planning, implementation, monitoring of activities. This necessitated changes in the existing strategies and introduction of new strategies.

The project adopted the following strategies

- Collaboration with reputed and experienced NGOs for imparting training to the target groups and for enhancing the activities of the community organizations.
- Emphasis on the participatory watershed planning with a view to facilitate the farmers and other target groups to own-up the project activities.
- Provision of one-time grant to landless families in order to enable them to undertake non-farm income generation activities.

- Flexibility in planning and implementation with an emphasis on quality standards rather than quantity of work achieved.
- Conducting socio-economic survey based on social mapping involving all the project staff and people.
- Involving the landowners in the shelterbelt and inter-belt development; channelising the benefits mostly to the landowners, giving due consideration to women and landless.
- Continuance of multi-disciplinary approach in implementing the project activities.
- Giving responsibility of raising nursery to the farmers.
- Training on participation and handling of group management to farmers with a view to foster group approach to watershed management.
- Training on PRA, community organization and watershed development with a view to enable the project staff, especially the technical staff to "work with the people".
- Appropriate financial support to the Village Development Associations for take-off and sustenance.
- Provision for staff training and development.
- Research and demonstration, as needed.
- Several committees for decision making, coordination and monitoring.

Process of people's participation in CWDP

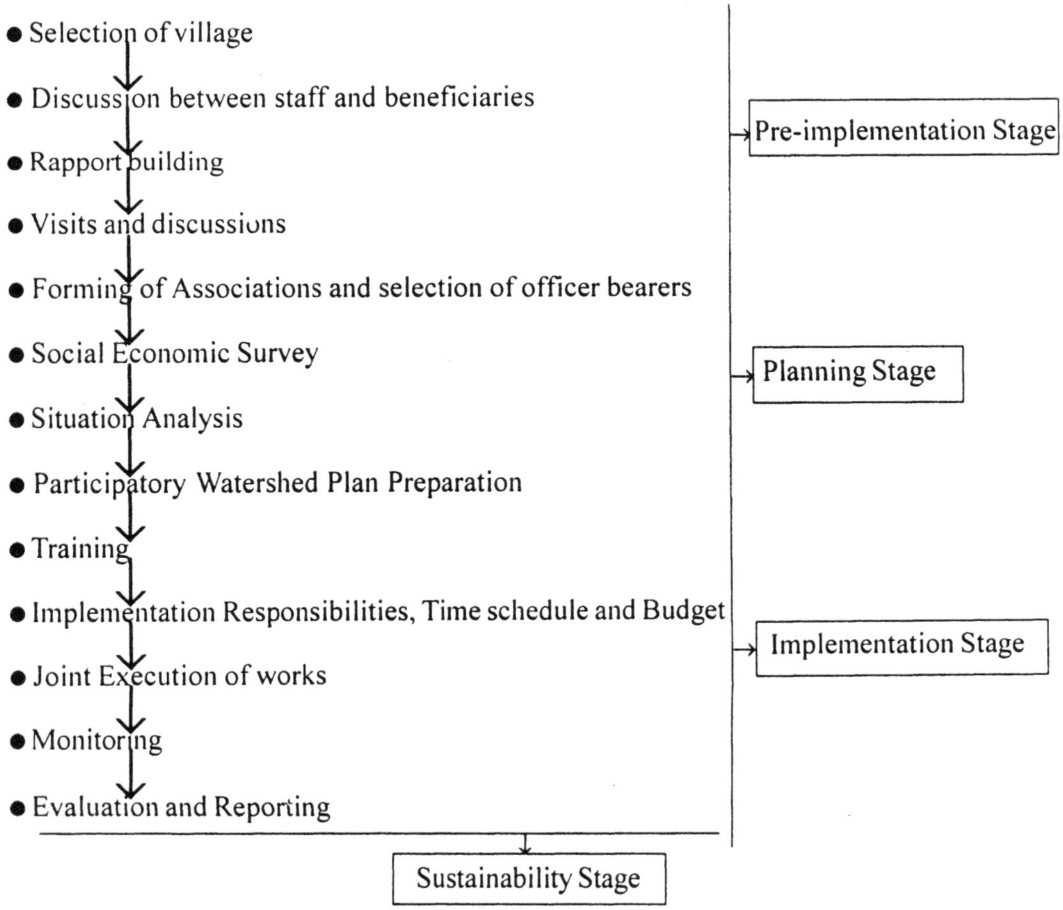

Villages were selected through a rapid survey on wind velocity, vegetative cover, water erosion, potential area for treatment and social indicators such as land ownership, absentee land owners, potential problems etc. After selecting the village, the project staff along with community organizers visited the village and identified local leadership. Project details and the problems of the area are discussed. A Village Development Association (VDA) is formed. This is registered under Societies Registration Act. Participatory Rural Appraisal (PRA) techniques such as resource mapping, social mapping, wealth ranking, tree preferential ranking, transect walk and focused group discussion are conducted to identify their local resources and needs. Based on the data and information gathered, participatory watershed plan is prepared in which the extent of land proposed by farmers, extent of land in which agroforestry is to be raised, tree species to be planted, borewells to be drilled, land development activities to be carried out, assistance to landless, and responsibilities of the association in implementing the plan etc. are finalized. Then activities envisaged are jointly executed. Periodic review at various levels, emphasizing quality and process evaluation, are carried out regularly. The project staff continuously monitor and evaluate the activities of the project through VDAs. The periodical Joint Review Mission which consists of members from GOI, GOTN and Danida have played a major role in project monitoring.

Main activities of CWDP

Integrated Agroforestry

The wasteland that are lying idle can be developed by planting useful and productive trees. These need to be drought resistant also. Seventy per cent of the trees planted in CWDP are cashew and the remaining are neem, vagai, tamarind, mango, pomegranate and other horticultural trees. One borewell is provided for every 12.5 hectares. The borewell is fitted with hand pumps and pot watering is carried out at the rate of 15 litres of water per tree for a fortnight during non-rainy days. The survival rate is more than 85 per cent after 3 years. Watering the trees is a collaborative effort between the owner of the trees and the project on a 50 : 50 basis.

In the first year a disc ploughing is done before pitting. Nursery raised seedling of about 3 months old are planted during monsoon period. Pitting and planting is the responsibility of the farmer, while the seedlings are supplied free. Neem cake is supplied by project free. In the first year and second year, farmers are encouraged to take up rainfed crops on the watershed. Limited quantity of seeds are supplied free. Fodder development is another area where trainings and inputs are given. Agroforestry is integrated with soil and moisture conservation, livestock, Village Development Association (VDA), forest, horticulture and annual cropping. Convergence of services at the village level and sustainability coupled with short-term, medium and long-term benefit to farmers are also kept in mind. Simultaneous efforts at capacity building also takes place through various training programmes.

An analysis of this main component of CWDP throws up several important aspects.

Transfer of Technical Knowledge

Agroforestry facilitates transfer of knowledge and skill to the farming community. This is a real unseen sustainability—the technique and skill of nursery raising. Although it is only one farmer who is raising nursery, after a proper and formal training, the entire village sees and understands it, as the nursery activities are taking place in the same village. The various aspects include, the criteria for the selection of the nursery site, seed collection, seed treatment, mixing of red-earth, sand and farm yard manure, filling up of polythene bags, watering and planting. In activities like planting, pitting, manuring, alignment etc. learning is by doing. Through a formal farmers training, plant protection, plant care and pest controls are explained.

The concept of use of controlled watering for tree species was unknown in the area before CWDP came in. Farmers knew only of open wells and energized borewells. In CWDP, borewells fitted with hand-pumps, are used for pot watering the tree species. In several cases this water is used for drinking purposes and has been a boon to the villagers. A formal training given to the farmers on repair and maintenance of borewells, also act as great help, as they need not depend on any outside assistance. A tool kit is also supplied for repair work.

The impact on gender equity is not to be ignored. Women are given priority in all training programmes. The pot watering of trees is done by women. These labour charges go to the women and especially the children. As is the case always, better finance with women also means more say in the decision-making at home. This situation contributes heavily towards the overall family and village development, especially the education and nutritional status of children.

Livestock has always been part of agriculture and ever since man started farming activities. In CWDP, health camps for livestock is always coupled with a training to the farmers on animal care, fodder development and breeding techniques and practices. In a highly drought prone, low soil depth area, livestock especially cows and goat, is very often the only means of survival. Market is not at all a problem.

The soil conservation and planting of trees contribute a lot towards ground water replenishment. This helps in garden land farming as wells would yield water for longer duration than usual. Trees help not only seepage of water through its root zones but also prevent moisture loss due to the shade it provides.

Moisture and Soil Conservation

Due to wind erosion for several years, the top soil has been completely eroded. The soil depth for planting and agricultural activities has become very shallow. Several soil and moisture conservation activities have been undertaken to prevent such erosion. Initially, the main component of the project was to plant shelterbelt. Later in Phase II the project has realized the importance and more than 50 per cent of the budget is allocated to spend on soil and moisture conservation activities. In order to break the handpan to facilitate infiltration and proper root growth disc ploughing/strip ploughing is undertaken in the agroforestry areas. To recharge the ground water, small sized ponds like percolation ponds, kuttais, deepening of ponds, desilting of existing ponds and to control erosion check dams, gabion structures are constructed across the waterways. For *in stitu* moisture conservation, vegetative hedges, *in situ* bunding, strengthening of existing bund, compartmental bunding, contour bunds are constructed.

"Catch water where it falls", being the principle, ploughing is the primary moisture conservation activity. Field compartmental funds are the next. During the training, farmers are informed about the need and importance of moisture and soil conservation. Soil conservation in the project area is very important as wind speed goes up to 60 to 70 kms per hour during peak season. By involving the people in these activities, from the planning stage itself, through a PRA exercise of transect walk, people have realized the need, importance and usefulness of these activities. Presently farmers are taking up such activities on their own.

Capacity Building

Capacity building both to the staff and people are crucial for sustainability. If activities are implemented without sufficient information to the people, it will never be sustainable. The staff need to be oriented and exposed to principles and techniques of community organization. Training needs assessment survey was conducted for the staff and training plan prepared and implemented. There are several training programmes for the people such as farmers training, nursery training, VDA office bearers training, landless training, SHG training, artisans training, gender training etc. These trainings are conducted in the village itself. Syllabus is prepared beforehand and a course material in local language is handed over to the trainees. The resource persons are from the project itself or outsourced experts. Several camps are held every year for livestock and training on animal and bird care and fodder development.

Since the resource persons come from various departments, establishing of linkages with these, is also an off-shoot of capacity building. The various subsidies, schemes and partnerships available with the departments are availed by farmers.

Assistance to Landless

The poorest of the poor in the village are the landless and they can not be left out. Chances of their being left out in a land-based activity are high. In every VDA about 10 landless are advised to select any livelihood development activity of their choice, the project contribution not exceeding Rs.1500/- per family. The VDA organizes this. Most often (about 50 per cent) go for goat or cow and the rest for small business activities.

The Workshops

CWDP, has several workshops. One that is conducted every year is the Annual Evaluation and Planning Workshop. These workshops are conducted normally outside the project area. The participants include a cross section from the policy makers in Danida and Government, to the representatives of farmers. The members of staff are also represented on a cross-section basis. These workshops provide a forum for exchange of ideas, concepts. Thus planning becomes more realistic. This has been a major strength of CWDP. It has been a strength because, we have been able to identify areas that require correction and corrections were made then and there.

Farmer's Day

An interesting development in the history of CWDP was the celebration of farmers day for the last three years. The objective was:

- Mutual sharing of experience and knowledge with a view to create the attitude of ownership of different components of the project.

- Making the outsiders aware of nature and importance of participatory project.
- Enabling the staff to see and exchange ideas with the stakeholders from various sub-divisions.

The important activities organized during the day included exhibition, sports, informal interaction, public meeting addressed by dignitaries, medical check up and so on. The farmers day in the initial year was organized with the project funds. In the subsequent years VDAs met part of the expenses. The ultimate aim was to see that VDAs themselves organize the annual show without depending on the project for finance. This celebration really has through the staff and people more closer.

Newsletter

To disseminate the technical know-how to VDA members a quarterly newsletter is being published by the project. This helps the VDAs to share their experiences/success stories with other VDAs.

The Committees

On a regular basis several committees are functioning at different levels. These committees meet on a pre-fixed date and time and minutes of all meetings are recorded. These committees are VDA committee, Section Level committee, Sub-divisional committee, District Technical committee and State Level committees. Monitoring and process evaluation take place at these meetings. Very often communication gaps are the reasons for failures of several projects. These committees provide a forum for continuing communication between various stake-holders.

Wind Erosion Control

CWDP was initiated with the prime objective of controlling the wind speed and developing the land affected by it. Change in the nature of activities did not affect achieving this prime objective. Today this is one particular area called theri land where vast areas used to be covered with red deposited soil, the problem is completely arrested. It is the demonstration effect that comes to the fore when it was demonstrated to the people that the problem arose due to the cutting of trees and the solution is only planting of trees. Several thousand hectares of theri land are today cultivable not only through the effort of the project, but also due to the participation and awareness of the people.

Research and Demonstration

As part of the ongoing activities at CWDP, Research and Demonstration occupy an important place. In collaboration with Agricultural. College & Research Institute, TNAU, Killikulam field researches are conducted. There is a continuous process documentation of these. The themes for the research include, spacing, different methods of watering, manuring, drought resistance etc. Demonstration plots are established in order to demonstrate to the farmers, the species that are introduced new in the area and also to demonstrate the result of right agricultural practices. Trainees are taken to these spots for practical exposure. Seeing is believing. Before CWDP, cashew was not a commonly planted horticultural tree in the area. Presently in CWDP area alone cover about 25000 hectares of land with cashew. Other farmers are also developing their lands with cashew. Introduction of right drought resistant species in the area has been an important achievement of CWDP.

Sustainability of VDAs

The activities initiated in the area can be sustainable only if there is a village forum that is sustainable. In CWDP, the VDA is seen and developed more as an organization responsible for initiating and maintaining development activities in the village than just watershed activities. The bye-laws of the VDA provided for this.

- Carrying out project activities in a transparent manner.
- Holistic development approach.
- Sharing the benefits of the project among all stakeholders.
- Offer micro-credit to its members.
- Catering to the basic developmental needs of the village.
- Utilizing their own resources for the village upliftment.
- Serving as a focal point for co-operation with any other service provider or Government Department.
- Sustaining and replicating the project activities in the village.

The watershed development plan for the village is prepared in a participatory manner. People are fully involved in planning, implementing and monitoring the activities of the project.

An initial grant of Rs.5000 (Rupees Five Thousand only) is given as a one time grant to each VDA to meet the initial infra structural expenses. Additionally, seed money is given for VDA as an incentive to match the contributions by members. The maximum amount provided per Association is Rs.7,000/- over a period of three years. This fund ensures organizational sustainability. The seed money along with regular subscriptions from members will grow and help the VDA to revolve the fund for internal lending among members. The members of VDA are meeting once in a month (minimum) at a mutually agreed place, date and time. At the meeting, all the matters related to the project are discussed in a transparent manner. Resolutions are passed with mutual consent. When necessary, matters are taken up with higher authorities for benefit of the village.

Micro credit beneficiaries are identified and loan disbursed with easy accessibility. The service area banks, to which VDAs are linked, are willing to advance loans to the VDA at lower interest rates. This credit facility from financial institutions is availed by VDA and recycled to its members in a micro credit form with easy accessibility. The interest rates to be charged are mutually agreed with members and VDA. Landless persons are identified and assisted by providing income-generating assets. The pro-active members of the grass root institution such as VDA help in the sustainability of the project activities and its spread to neighbouring areas. The monthly meeting serves as a forum for its members to voice their need and seek solution.

Economic sustainability

Getting freed from the clutches of the private money lenders has come as a blessing to the VDA members. Micro-credit at easy terms and accessibility is one of the most attractive activities of the VDA. The service area banks to which VDA is linked are also giving credit at the stipulated rate to the Associations to provide micro credit to its members. This financial assistance enables members to undertake income-generating activities and increase their income status. Recoveries of loans by VDAs is 100 per cent. This excellent rate of recovery has improved the availability of funds with VDA. This popular lending activity also provides for economic sustainability of the VDA. Thus the VDAs are enabled to continue even after the project is withdrawn, sustain the development work in the village and continue to deal with the objectives mentioned earlier.

Case of Radhamudayarkulam village

It is worth the effort to study one VDA as a case out of nearly 175 VDAs. A Participatory Impact Evaluation carried out found that about 75 per cent of VDAs are sustainable. The details of one of such is given below in order to provide an idea of a sustainable VDA.

(*a*) *Short Background*

Radhamudayarkulam village is one of the oldest villages in Tirunelveli district in Tamil Nadu. Agriculture is the main source of their livelihood. It is dependent on rainfall which is scanty. Average land holding size is 4 acres. A large percentage of land is lying waste. This is due to reasons such as unreliable rainfall, inadequate resources to dig wells and develop the land etc. The quality of land for tree or horticultural production is good. There are 84 members in the VDA of which 56 are males and 28 females. The village covers a total area of 314 hectares and CWDP has covered an extent of 138 hectares of dryland by planting trees like cashew(*Anacardium occidentale*), neem(*Azardirachta indica*), tamarind (*Tamarindus indica*), vagai (*Albizia lebbeck*) etc. At a distance of 4 kms the residents are able to get bus transport. There are electricity and telephone connections to the village.

(*b*) *Initiation and Growth*

In April 1996, a team of staff from Comprehensive Watershed Development Project (CWDP) consisting of an Assistant Engineer from Agricultural Engineering Department and a Community Organiser from an NGO came to the village. They introduced the project and its activities. It was also explained that the villagers need to organize themselves into a Village Development Association (VDA). This was followed by several visits by the community organizer. The VDA was officially started on May 15, 1996 when the office bearers were elected and bye-laws were adopted.

That was the beginning. Immediately there was a flurry of activities. These included training for the office bearers of VDA on record keeping, training on nursery raising, identification of land for development, PRA meetings, training on how to plant, protect and maintain trees, geophysical survey to identify points for drilling of the borewells etc. Nov-Dec. each year is the rainy season. During this period, horticultural species such as cashew, pomegranate, mangoes and economic trees like neem, vagai, tamarind etc. were planted. Meanwhile, regular meetings of the VDA started. Monthly subscriptions were collected from members and in all meetings, a statement of account was read out. From CWDP, the VDA received Rs.3,000/- towards the initial expenses of the VDA and a matching grant of equivalent amount of the subscription. From this core fund, members started taking loans.

(c) Economic Sustainability of the VDA

Prior to the initiation of the VDA the sole source of credit to the entire village was the private money lender. He used to charge upto 60 per cent rate of interest. Such costly credit led villagers from bad to worse economic scenarios. After the VDA started giving loans to its members at an interest rate of 24 per cent, the scenario completely changed. The table below indicates the growth of this loan fund and its utilization. Most of the loans are utilized for agricultural purposes.

Year 1998-99	1999-2000	2000-2001
Rs.1,04,201	Rs.1,32,961	Rs.1,90,435

Recovery of loans is 100 per cent. This has led to growth of lending activities. It has also made the VDA a sustainable organization. Thus, VDA provided an assured source of micro credit in the village. Economic confidence of the villagers started brimming. At this stage, a commercial bank came forward and offered Rs.10,000/- as loan at an interest rate of 12 per cent. The VDA has since repaid the loan. The return from the investments in the dryland has started. The cashew trees have started yielding. The return from the timber will still take a longer period. Today, the VDA has sufficient funds to lend to its members and is not interested in taking another loan from the bank.

(d) Organizational Sustainability of the VDA

The development activities initiated in the village can go on and be sustainable only if the organization, namely the VDA, is sustainable. The essential factors that are expected to help the VDA to continue activities even after the project is phased out are the following.

Legal Status: The VDA is a registered body with proper rules and regulations, under the Societies Registration Act. As per registration rules, the VDA must hold a General Body meeting every year and present the audited statement of accounts and an annual report. The VDA has also met regularly every month and taken decisions by mutual consent. This makes the VDA transparent and cohesive. Strict adherence to the bye-laws ensures proper organizational sustainability.

Maintenance of Proper Records: The VDA maintains several records and registers such as loan register, membership register, cash book etc. This contributes towards organizational sustainability. In the case of Radhamudyarkulam VDA, all records are written by members themselves and not by a paid animator or by the community organizer. By doing it themselves, members are able to comprehend and make improvements/corrections as required. All the records are maintained as prescribed.

Integrated Development Approach: The VDA provides a forum for all the members to express their needs and problems relating to infrastructure. VDA has served as the right platform to make representation to higher levels of Government about needs of the village in this regard. This VDA is also interested in obtaining better return on agricultural produce. For example it intends to process the cashew nuts and assist members to obtain a higher price. Led by a committed leadership, armed by a legal status and proper records, supported by cohesive action, the VDA is striving to serve its members in better ways. Leadership development has resulted in the village members being elected to the local bodies (Panchayats). The VDA has become a focal point for total village development and has initiated several activities towards this end. The VDA has constructed a drinking water system exclusively for school children. They have provided cooking vessels for the preparation of mid day meals for school. Infra-structural facilities in the form of table, chair and cub board has been provided by the VDA to the school. In addition, they are clearing the overhead drinking water tank periodically, constructed by Rural Development Department. The VDA office bearers are in close contact with elected people's representatives in order to avail more basic amenities to the village. Presently one of the felt needs of the village is establishment of a processing unit for the cashew nuts produced in the vast area in and around the village.

Documentation of development experience and participatory impact evaluation

CWDP has a life span and is expected to phase out by 2003. Before phasing out, it was important that the various experiences that are in CWDP are documented and disseminated to the world outside. Therefore, a document comprising the development process and experiences have been prepared under the leadership of Dr. R.K.Sivanappan. As part of the exercise, a website is created at GOTN server and a video cassette also prepared. Another Participatory Impact Evaluation study was conducted under the leadership of Dr. Alok K. Sikka

As part of the study, the sustainability of VDAs was estimated using a rating index.

Rating Index	Percentage
0-1 (embryonic)	–
1-2 (emerging)	12.50
2-3 (growing)	12.50
3-4 (well developed)	50.00
4-5 (mature)	25.00
Total	100

The analysis indicates that 75 per cent of the VDAs are well developed or active and are doing an appreciable job in organizing, developing and sustaining the project activities.

Threats to sustainability

There are some shortcomings, which threaten the sustainability of VDAs. Some of them are:

- Formation of VDA is an evolutionary process. Some VDAs were formed in a hurry without adequate spade work. Hence, the members of such VDAs are yet to fully understand the philosophy and scope of the VDAs.
- Savings in many VDAs have not picked up. Such VDAs are found to be interested only in taking seed money from the project.
- Members of a few VDAs did not evince much interest when the project was phased out.

Recommendations for sustainability

Taking into consideration the experiences at CWDP, there are several lessons that can be of use to other watershed development activities elsewhere. The recommendations given below also include improvements on the present practices at CWDP

People Centred Focus and Approach

Starting from the pre-planning stage, the centre of development should be the people. Very often the focus is on check dams, trees, livestock, ploughing, land development activities etc. Basing on such activities a budget is prepared and activities are initiated. It is during the process of implementation that the people get involved. By this time people are called in to contribute, to participate and to take care of the structures for maintenance. In effect a people's group is formed to maintain the activities. This process must stop. The structures and activities are for the people. Therefore, any Govt. or any other Civil Society Organisation (CSO) activity must first start with the people.

Bottom-up Planning and Targets

There has always been difficulties experienced by people regarding the targets already fixed the need based/bottoms-up planning approach. The Govt. and CSOs need to have an approved physical target and budgets before talking to the people. At the same time, development activities need to be based on local problems and priorities and must take into consideration existing resources. Kerala has attempted to face this problem by introducing the village based planning process for the Five Year Plans. There are differing opinions on the success or otherwise of the attempt. However, it is definitely a starting point. Unless we try, how do we learn. There are several arguments as to why bottoms-up planning is not possible. Without getting into arguments one has to agree to the fact that it is the right thing to do. It is the beneficiary or the person(s) who faces the problem and knows about its implications and solutions

Ownership by Local Bodies

Decentralized decision making is essential for the development of a country like India which is very large both geographically and demographically. The 73rd Consititutional Amendment has provided the opportunity to implement this. The Panchayat Raj institutions should not only be given responsibility to implement development programme, including watershed, but also given the resources or powers to collect resources. The line departments ultimately should become "Consultants" to the PRIs. It is essential for such a business-like situation to develop, to get best service. It will also help Govt. Departments to become financially more self-relevant. There is a lot of difference between people themselves carrying out an activity and some one else doing it on behalf of the people. This is called "ownership" by people. Ownership can be achieved only when people understand the technicalities, methods, cost, organizational matters, maintenance and uses and problems relating to watershed activities. Therefore, complete transfer of information on all these factors is essential, if the activities are to be owned by people or local bodies.

Role of CSOs

Civil Society Organisations have their own specific role to play in the watershed activities due to their practices such as quick decision-making, less paper work, expertise in community organization etc. They also act as a pressure group to the PRIs. Present policies of the Govt. also encourages the participation of NGOs in Govt. programmes. NABARD and even commercial banks have several schemes that are being channeled through NGOs.(CSOs). people also have a better rapport with NGOs.

Sustainable VDAs and Self Helf Groups (SHGs)

How to make a Community Based Organisation, by whatever name it may be known, such as Village Development Association (VDA), Grama Sabha, Watershed Committees, user groups, informal groups, groups for common property maintenance, SHGs etc. can be made sustainable, is a question that very often has not been addressed. In order to make such groups sustainable there are several factors that are essential. These are: complete information to the people of the activity, effective leadership, legal status (registration), accurate and transparent keeping of accounts, regular meetings for effective communication, financial sources, linkage with Govt. Departments and commercial and financial institutions, holistic approach, linkage with PRIs, support from CSOs etc.

Monitoring and Communication

There needs to be an inbuilt system of monitoring. This is different from physical and financial target reviews. Monitoring quality is important. Monitoring with the people by external agencies is essential. This also calls for ongoing and productive communication between people and external agents. Such communication also helps capacity building. Such monitoring and communication meetings need to build into the project at the planning stage itself and should be held at the village itself and not a Govt. office or NGO office.

Gender/Benefit to Farmers

The active involvement and understanding of the watershed activities by women is essential for its sustainability. Any human being would first ask the question "What do I benefit from this?" when an action is proposed. This benefit may be direct or indirect. The benefit from a watershed activity is often long term. However, people want immediate short-term and long term benefits. Therefore, while planning watershed activities it is important to provide entry point activities and micro-credit facilities (eg. Loans by VDAs, SHGs etc.) immediately. Watershed plans must provide for some short-term benefits such as cash crops, vegetables etc. The long term benefits could be from horticultural trees, timber, appreciation of land value etc.

Flexibility (Change and Dynamism)

Total flexibility at the village level, within certain parameters of budget and guidelines is essential. The situation in adjacent village could be much different. There should be scope for changes as and when a change is noticed. Only when there is scope for corrective measures can there be dynamism. Concurrent evaluation that is participatory can facilitate change and dynamism in watershed projects.

Convergence and Programme Integration

Presently the line departments function separately and parallely. This must change. Various services of the Govt. must convergence at the Panchayat level. This can bring in integration. There are examples of a watershed programme in which once one department comes and does the ploughing, then another department come and does land development and watershed construction activities. After sometime another department comes and plant horticultural plants. Finally the livestock department is not involved and the farmers loose their grazing area and no fodder grass or trees gets planted. Result—break in the villagers life. Livestock is bread and butter of village life and the watershed project comes and breaks that also down.

Capacity Building of the Staff and Farmers

It is often assumed that whether Govt. or NGOs, staff know every thing. In CWDP, one of the first activities in the first phase was training to the staff on principles and practices of community organization. Exposure and field visits do much more than class room learning. The capacity building to the staff should not only be on community organization but on technical matters as well. Updating and upgradation of knowledge and skill of staff is essential. The same principles apply to farmers also. In the case of farmers, handing over training materials to them, methodology, practical experience exposure, technical and organizational matters etc. are important matters to be noted.

Farmers Contribution and Participation

These are terms that are often used wrongly. All watershed activities are people's activities and it is the external agencies that go and participate in the people's programme. Not *vice versa*. In the same manner it is the external agency that is contributing to the village development. Nothing should be given free. Only when there is sweat of the farmer he/she will feel the importance of it. Remember the farmers will shed sweat only when he/she feels it will benefit him/her.

Information Technology

Watershed activities should not get behind times. The advantage of internet is that information is available at easy accessibility anywhere at any time. There is a need to have localized watershed websites in local language. In another six months time probably internet will be available through cable. Such easy availability of information and its sharing will help farmers to implement watershed activities in a better way.

CONCLUSION

Time has come to stop implementing watershed activities and begin implementing sustainable watershed activities. The psychological satisfaction of carrying out watershed activities will not be of any use to the farmer unless it is sustainable and continue to benefit the farmers even after the external agents withdrawal. The most important factor to remember is, that farmers should "own" watershed activities. Only then it will be sustainable.

Watershed Management: Issues and Policies for 21st Century pp. 290-293

ASSEFA's Experience in Watershed Management

P. Thangaiah and Raja Mohamed

ASSEFA (Association for Sarva Seva Farms) is inspired by the Gandhian philosophy of Sarvodaya "Welfare of all" and is the second phase of Bhoodan-Gramdan Movement—a non-violent socio-economic land gift movement initiated by Acharya Vinoba Bhave. ASSEFA started in 1969, is presently engaged in Integrated Rural Development activities covering Agriculture, Animal Husbandry, Rural Industries, Health , Education and Cultural activities. ASSEFA's activities are spread over in six states in India covering 67 blocks and 3500 villages.

ASSEFA's strategy is mainly working with the people and empowering rural communities to sustain and continue the programmes on a long term basis. ASSEFA believes in Community based movements through Community based organizations, like Gram Sabhas and Women Groups besides various activity groups at various levels. ASSEFA has promoted parallel organizations for intensified activities in specific fields.

- The Nidhi Foundations at the village level and Federation of Nidhi Foundations at the area level called (Sarva Jana Seva Kosh (Peoples' Bank) for social credit activities.
- The Sarva Seva Education Trusts for undertaking educational activities.
- The Sarva Seva Dairy Development Federations for dairy activities.
- The Sarva Seva Habitat Promotion Company for housing activities.
- The Sarvodaya Mutual Benefit Trusts for micro credit activities.

The Sivaganga project is one of the Block Level Development Projects initiated by ASSEFA 15 years ago. The main activities are agricultural development, dairy development, and children education. All these activities are carried out through Gram Sabhas and Women Groups. The watershed activities presented in this paper are the project activities. The objective of the Kandangipatti watershed management programme is to reduce the severity of drought and proper use of the available natural resources for betterment of the society through watershed management practices.

THE GENESIS OF THE WATERSHED PROJECT

During the year 1994-95 and 1995-96 the project areas experienced severe drought.There was large scale migration of people. The families found it difficult even to support their children education which was the main activity of ASSEFA in these areas. At the instance of the Gram Sabhas for permanent relief from drought, the concept of integrated watershed development was explained to the Gram Sabhas. The various components involved in this comprehensive watershed programme were listed and discussed with emphasis on their full participation for the success of the scheme. In order to further convince the people and make them understand the tangible benefits of the scheme, the farmers were taken for an exposure visit to a nearby similar watershed programme under implementation by another NGO. After the visit, the farmers were more convinced and were very enthusiastic in taking up the watershed programme in their areas.

LOCATION OF THE PROJECT

The Kandangipatti cluster of villages, namely Kanangipatti, Vagulathupatti and Devanippatti are situated about 12 kms north of Sivaganga Town and are located 4 km west of the Melur-Sivaganga Road. The area is located under the semi-arid climatic zone and chronically a drought prone one. The main livelihood of the people are agricultural activities and livestock rearing. Wetland cultivation of single paddy crop is taken under the ayacut region of the traditional tanks. The rest of the lands are dry under pulses and groundnut cultivation depending on monsoon rains. The southwest monsoon brings only minimum rains for sowing the dryland crops and raising the paddy nurseries and the farmers depend on the north east monsoon for both the dry crops as well as the paddy crops. Failure of rainfall is quite recurrent and the farmers are forced to heavy borrowing from money lenders or to migrate to nearby cities after selling their cattle. Even during good monsoon seasons the crop productivity is rather very low due to low inputs and lack of soil and moisture conservation measures.

The people

There are about 290 families in theses three villages mostly comprising of backward and scheduled castes communities. Their average land holding is half an acre to two acres of land. The Sarva Seva School at Vagulathupatti is the only primary school for 12 villages around.

The project initiation

- An initial survey revealed that the Kandangipatti watershed area actually comprises of three mini watershed with clear ridge lines spreading over 300 acres.
- A Participatory Rural Appraisal (PRA) exercise was conducted at the Vagulathupatti School involving the people of the three villages.
- The idea behind this PRA was to make the people understand their own strengths, weaknesses, opportunities and threats besides having a correct idea of their priorities in the watershed development programmes.
- The people were encouraged to draft the watershed map, the resource mapping chapathy diagram and the preferential matrix for the tree plants.
- The priorities of the people were approach road, drilling of bore wells and feeder channel treatments. The hand pump proposed earlier was changed to power pump operation for providing protective irrigation to the fruit trees.

The project implementation

All the activities were performed through the Gram sabhas which was the Supreme body ably assisted by the Mahalir Mandrams.

The different user groups and sub committees include: Orchard/Trees Committee, Borewell Committee, Road Maintenance Committee and Milch Animals Committees. The main functions

Table 1. The share of peoples' contribution to the various activities

S.No	Activities	Peoples' Contribution (per cent)	ASSEFA Share Contribution (per cent)	Govt. Contribution (per cent)
1	Pitting	50	50	–
2.	Desilting	25	75	–
3.	Seedlings	–	50	50
4.	Bunding	100	–	–
5.	Planting	100	–	–
6.	Manuring	50	50	–
7.	Fencing	50	50	–
8.	Watering	100	–	–
9.	Borewell	–	100	–
10.	Pumps/Pipes	–	100	–
11.	Power	–	100	–
12	Maintenance	100	–	–
13.	Deep Ploughing	–	100	–
14.	Feeder Channel	25	75	–
15.	Gully Plugs	25	75	–

Table 2. The share of peoples' contribution to the various activities

(Rupees)

S.No	Activities	Total Expenditure	Peoples Contribution	ASSEFA Contribution	Govt. Contribution
1.	Contour Bunds				
	Bunding	160000	160000	–	
	Fencing	40000	20000	20000	
	Deep Ploughing	41028	–	41028	–
2.	Gullyplugs	4730	1200	3530	–
3.	Desilting	110737	26550	84187	–
4.	Pond Repairs	18150	2100	16050	–
5.	Spillway for Pond	9760	1000	8760	–
6.	Feeder Channel	93954	23000	70954	–
7.	Agro-forestry				
	Pitting	48000	24000	24000	–
	Planting	24000	24000	–	–
	Manuring	40000	20000	20000	–
	Watering	172000	172000	–	–
	Maintenance	57600	57600	–	–
	Cost of seedlings	230879	–	168000	62869
	Laying pipelines	60000	–	60000	–
	Water tank Construction	94856	–	94856	–
8.	Drilling & Casing pipes	466361	30000	436361	–
9.	Miscellaneous	12974	–	12974	–
	Total	1685029	561450	1060700	62879

of these Committees are: (*i*) conduct regular meetings to review the progress under their jurisdiction, (*ii*) enlist and ensure the fullest participation of the people in their assigned activities, (*iii*) resolve local disputes within the group, (*iv*) document all the activities in their respective areas, (*v*) meeting officials and ensuring their assistance, (*vi*) regular management of the committee affairs and (*vii*) coordinate with other committees and the Gram Sabha/ Mahalir Mandrams in the overall implementation of the project.

The main activities undertaken are: (*i*) Bunding and trenching in 100 acres of land, (*ii*) Cleaning and desilting the feeder channels for half a kilometer length, (*iii*) Construction of regulator cum check weir into locations, (*iv*) Construction of gully plugs in six locations, (*v*) Mixed orchard and multipurpose trees planted in 100 acres of lands, (*vi*) Installation of 10 bore wells, (*vii*) Water distribution layout for 100 acres, (*viii*) Planting 10000 fruit plant seedlings and 10000 multipurpose tree plants.

The impact

- Generation of more than 20000 man days of employment.
- Increase in land value.
- Full availability of green fodder.
- Increase in animal population.
- Increase in the number of fruit trees.
- Increase in inter-phase dry land agriculture due to prolonged availability of soil moisture.
- Arrest of seasonal migration of people.
- Arrest of soil erosion.
- Increase in the tank capacity.
- Reduction in the rate of silting.
- Controlled grazing and starting of stall-fed animal rearing.

Watershed Management: Issues and Policies for 21st Century pp. 294-297

Sustainable Agriculture in a Watershed: Experiences and Strategies

Daniel Anand Raj and Ravikumar

INTRODUCTION

Agriculture Man Ecology (AME), a bilateral project of the Government of India and the Government of the Netherlands is operating from its Trichy field office (the coordinating office is at Bangalore and one field office each are at Raichur (Karnataka) and Madanapaili (Andhra Pradesh), since 1997. The main focus of AME is to promote sustainability of the dry land agricultural system (including the farmer family livelihood). AME's present role is to build capacities and to support farmer groups and Non Government Organizations (NGOs), in addressing technical, gender and social issues in promoting sustainable agriculture.

AME facilitated process training on "Sustainable Agriculture in a Watershed" to 30 trainees of an NGO network (ROOTS Network) in 1997 - 98. The objective was to build their capacities to address and manage social and technical issues in promoting sustainable agriculture in dryland systems within the context of a watershed. The trainees were from 10 small watersheds, located in Pudukkottai and Trichy districts. Eight modules of trainings were facilitated to them on various issues related to promoting sustainability of agriculture in their watersheds. Follow up exercises were also facilitated between modules and the experiences were shared and analyzed in larger groups during consecutive modules. The process generated a wealth of experience regarding strategies and approaches for participatory management of watershed resources. Experiences from facilitating the process are being shared in this paper.

EXPERIENCES OF AME FROM THE TRAINING PROCESS

The village as a watershed

The watersheds in Pudukkottai district are part of the whole village ecosystem. The terrain is a gentle slope (with slope percentage not more than 5 per cent maximum). The village livelihood and the agricultural production system depend directly on the watershed resources. So, it is obvious that the watershed boundary is not restricted to the village wastelands only, but the whole village area into it. There are three reasons to substantiate this argument.

- A village consists of both people and production systems. Much of the utilization pattern as well as problems of the production systems are influenced by the social system in the village and vice-versa. So, both are interdependent and are to be addressed simultaneously.

- Often in the conventional approach to watershed management, treatment in catchment area is given much importance, with relatively little importance to the concentration and the command area. This approach has to be revisited, because all the three components are highly interdependent. Any intervention at one component level influences the other two and the village livelihood system as a whole.
- In most cases, farmer owns lands both in the catchment and command areas of the watersheds—whatever small the size of the holding may be. By virtue of this land holding pattern, the farmers invest their efforts both in catchment (herein we call as upland or rainfed land) and the command area (lowland or wetlands) depending on the seasonal cropping pattern. Considering this fact, it is important to address the whole village as the treatment unit for participatory development initiatives.

So, the village system is to be taken as one watershed unit and the treatment plans focused on managing all the three components of the unit simultaneously.

Watershed management and sustainable agriculture are to be addressed together

Each watershed is a socio-cultural system, having unique characteristics in terms of biophysical, technical, social, economic and environmental determinants, which decide the fate of agricultural development within the area. Natural resource management is the result of human interactions with the ecosystems of which agriculture is a major factor. Depending on farmer's skills, agriculture can either generate synergistic or antagonistic effects both on the human and natural environments. A degraded watershed, which has lost its original carrying capacity, will continue to deteriorate further if the present level of exploitation continues and cannot support agricultural productivity. Hence, development of a watershed has to begin with addressing the issue of proper and sustainable agricultural practices and then proceed to higher and higher level of interventions.

This is called the ABCD approach to watershed management, which means, that the first issue to be addressed is Agriculture, then Bunds of the field and finally Check dams and Drainage structures. Appropriate agriculture practised in a watershed leads to resource conservation and regeneration. This approach offers the following benefits.

- Facilitates participation of all resource users in the watershed (again an issue which is often faced in the conventional land management projects)
- Promotes private investment for public good of the watershed population. When private investment is promoted, the cost of treating watersheds comes down, reducing the dependence on 'external' sources,
- Equity of resource use will be promoted, by stopping the exploitation of one component of the watershed (like loss of topsoil and subsequent nutrient export from drylands to lowlands) for the benefit of the other component. When over exploitation stops, resources will be conserved, and
- The resources developed will be managed sustainably, beyond 'project periods', because of the ownership attitude created among the individual participants.

An interesting situation in AME's experience shows some light on why Watershed Management and Sustainable Agriculture are to be addressed together and not as two separate components. Based on the outcomes of a planning workshop, one of the NGOs network partners identified soil and water conservation as the priority issue and opted for watershed management as the entry point activity to promote sustainable agriculture. The other network identified pest management as a priority issue and took up Integrated Pest Management (IPM) as the entry point activity. AME worked with both the networks for 4 years.

Now the network, which commenced its activities with watershed management as the entry point, has realized the needs for addressing the issue of proper agricultural practices and has initiated specific crop based interventions (integrated management of important crops in the watersheds, usually groundnut in the uplands and paddy in the command areas). While the other network, which commenced activities with IPM as the entry point, has realized the need for action at higher systems level (farming systems and watershed) and has opted to act at the village watershed level. AME has initiated process training to the network trainers (to develop human resources) for sustainable agriculture based development action in 27 watersheds (spread over 9 districts of Tamil Nadu).

Investment on people is more important than on structures in the watersheds

Sustainable Agriculture explicitly recognizes that the users of the resources are more important and not the stock of natural resources is important, that is the present focus of management. In order to manage and develop a watershed under the ABCD approach, it requires cooperation and collective action by all 'actors' (farmers, Non Government Organizations, Government institutions, other local and influential community leaders). One key factor, which is at present blocking convergent action among 'key actors', is the lack of holistic watershed management perspective and it's inter-relationship to sustainable agriculture. So, an investment has to be made here, to create this understanding among the key actors. This is possible by building their attitude, knowledge and skill base, through capacity building approaches.

Building capacities empower watershed actors to make decisions, which is an essential factor for promoting self-help action and sustainable resource use in watersheds. AME initiated such capacity building initiatives to farmer groups and NGO staff, starting with specific crop management skills also. As an outcome of these efforts, farmer's decision making skills have improved and have initiated a thinking process, wherein they have started progressive action at the next higher levels of their ecosystems, beyond the present cropping system knowledge.

High intensity and micro level concerted action is needed for sustainable development

A farm ecosystem is highly dynamic, characterized by the unique natural and social processes influencing it from within and without. So, an intervention planned for a specific farm situation will have to be often revisited, for optimum output generation. When it comes to many such farms in a larger watershed ecosystem, the situation is more complex and demands attention at the very micro level units. Therefore, it is unnatural to generalize treatment plans for the whole of the watershed. Individual farm level planning and action are to be initiated, without losing a holistic development outlook. Thus, a concerted action is necessary at a range of levels, which will not be possible without the participation of multilevel stakeholders. This is possible only when the farmer level capacities for decision making are improved in managing his enterprise, which is again possible by developing human resource base in the watersheds.

AME's strategic approaches to watershed management

Based on the above experiences of capacity building and direct intensive field level interventions, AME has now developed a tested methodology. Present focus of AME is to build up soil organic matter and fertility for promoting sustainable agriculture. AME's crop (Groundnut, Paddy, Cotton and Vegetables) based interventions also facilitate farmers to develop and integrate various components at the farm levels including agriculture based rural enterprises.

Being a knowledge generation and dissemination organization, AME also concentrates on capacity building of NGO actors and farmers in the ABCD approach. AME places priority interest in facilitating multiple stakeholder- concerted actions in influencing policy level changes, in favour of participatory development initiatives.

Call for collaborative action

AME is interested in collaborative action with more diverse stakeholders, who share similar ideas towards sustainable agriculture based watershed management. AME has, to offer technical expertise for facilitating key actors in managing watershed resources under the ABCD approach.

Watershed Management: Issues and Policies for 21st Century pp. 298-305

People Participation in Watershed Development Programme: A Reality

N. Karuppasamy

INTRODUCTION

Soil and water are the basic resources of nature on which humanity depends for its survival. They influence the plant life, animal life and ecology of a region. Ever growing population and it's needs extract benefits from these resources. Intensive human pressure on the natural resources disturbs the ecology. Market based approach leads exploitation and indiscriminate use of natural resources. The biggest challenge the humanity is facing in this century is how to meet the present needs without limiting the options for future generations. The planners are searching solutions to restore the ecological balance with human face. Several decades of 'planned' developmental intervention in the country has led to regional imbalances, degradation of natural resources, widespread poverty and vicious cycle of isolation.

'Trickle down' growth model of development has failed to address these issues, as an alternate or as a remedy to this 'bottom-up' approach has gained attention and credence in the last two decades. In this context in India, watershed development has been widely accepted as a right and prime approach towards bottom-up planning, conservation and development of natural resources. For the planning, development and management of the natural resources of land and water, a watershed would be an ideal unit as it has a clear conceptual unity with physical geography and hydrology.

Watershed is an area, which drains water through a single common outlet. It is a natural unit, boundaries are defined by topography and not by any man made parameters. Ideally watershed is a living system governed by natural laws. Though there exist several constraints, limited water availability is the most severe constraint in crop production in many parts of the country. Conservation of as much rain- water *in situ* as possible and harvesting and storage of the excess runoff are essential for alleviating this constraint. This will be effectively done through a watershed approach. In this each and every parcel of land including private and common land falling within the boundaries of watershed is treated with appropriate soil and water conservation measures. The excess runoff is safely conveyed either to a water structure such as pond or tank or drained into stream or river.

AN OVERVIEW OF INDIA'S WATERSHED DEVELOPMENT PROGRAMMES

In India dryland account for about two-fifth of total geographical area, about seventy percent of it's total cultivated area and nearly one third of its population. Dryland contributes about 42 per cent of country's total foodgrain production and about 75 per cent of total production of pulses, oilseeds and cotton. Though drylands have important place in India's economy in the past, yet it has not received due attention from planners and policy makers. Whereas, the irrigated areas have received high priority for development which includes institutional credit and other supports.

During the seventies the government recognised the importance of dryland agriculture and major steps have been taken in the direction of advancement of dryland agriculture. One such step is the launching of Drought Prone Area Programme in 1974-75, and a few other major initiatives are, the launching of All India Coordinated Research Project for Dryland Agriculture by ICAR in 1970 (and which is now promoted as Central Research Institute for Dryland Agriculture (CRIDA), Hydrabad. and the establishment of Central Soil and Water Conservation Research and Training Institute (CSWCRTI) in 1974 with 8 regional research centers located in the major agro-climatic regions of the country.(Singh 1988).

The importance of dryland development was realised and the watershed approach was recognised as a holistic approach by the government. Though the references made to the watershed approach in the government of India's model bill on soil conservation, the soil conservation acts passed by the state government (1950-60) and the National Policy on Ravinous Watersheds of (1967), the real turning point came (1974) when it was decided by the government of India to implement the watershed development programme as centrally sponsored scheme of soil conservation in the catchments of river valley projects. In 1982 the government of India under the ICAR sanctioned 46 model watershed projects and implemented by the respective state government departments viz., Soil Conservation, Agriculture with the technical support of AICRPDA, CRIDA and CSWCRTI. In 1986, the union ministry of agriculture and rural development launched a National Watershed Development Programme (NWDP) for drylands as centrally sponsored scheme. In the initial phase it was launched in 99 districts in 16 states of the country. In addition to this, special kind of projects were also initiated. In 1984 the World Bank assisted 4 watershed development projects one èach in Andhra Pradesh, Karnataka, Madhya Pradesh and Maharashtra (Singh 1988). Other important projects are Indo-German watershed development programme, DANIDA watershed development programme, NAWDPRA watershed development projects etc. Besides this kind of centrally sponsored and special programmes, respective state government funded schemes are also in progress in each state.

Based on the experience so far gained, the Government of India (GOI) decided to create Watershed Development Fund (WDF) with NABARD with the broad objective of unification of multiplicity of watershed development programmes in to a single nation all initiative through involvement of village level institutions, NGOs. The Union Finance Minister in his budget speech for 1999-2000 had announced the creation of WDF through the NABARD and pursuance thereof WDF has been created in NABARD with a contribution of Rs.100 crore each by NABARD and GOI. The budget announcement envisages coverage of 100 districts in three years. The objective of the fund is to create the necessary framework conditions to replicate and consolidate the isolated successful initiatives under different programmes in the government, semi-government and NGO sector. In the first phase NABARD has identified 14 states which includes Andhra Pradesh, Gujarat, Madya Pradesh, Maharastra, Orissa, Uttar Pradesh, Bihar, Haryana, Himachal Pradesh, Jammu & Kashmir, Karnataka, Rajasthan, Tamil Nadu and West Bengal for the project implementation.(NABARD 2000).

In Tamil Nadu watershed development projects are implemented under DPAP, DANIDA, NAWDPRA and IWDP schemes. These projects are implemented through the state government's departments like Agriculture, Agriculture Engineering, Forestry, Animal Husbandry and also a few through SAUs. Under NABARD-WDF six districts including Ramanathapuram, Pudukkottai, Tirunelveli, Sivagangai and Dindigul were identified and watersheds have been identified.

PEOPLE PARTICIPATION AND WATERSHED DEVELOPMENT PROGRAMME

In social programmes 'participation' is unavoidable. Everywhere the people are using the word 'participation'. Even in some cases the advance stage participation like active participation, beneficiary participation, people participation, women participation and popular participation etc. are used. Though participation in common work is common for our culture, we are insisting this in the last two decades as a new one or imported from elsewhere.

Before getting deep into the fundamental questions on participation we need to consider in whose agenda who is going to participate? Is it the government or others and the people? If the Government is able to design a programme in consultation with the people in such a way it satisfies the people's need then it create an environment to participate. Many times it is a mismatch that the people need and the programme priority is not matching. In general willingness of community to participate in any programme depends to a large extent of private benefits or common benefits which each member of the community or the entire community actually realizes. Participation should not be one of the activity or input in development projects. It is a process, which evolves over time and whose direction and outcome are not always predictable or manageable. It is a broad and multi-dimensional phenomenon with political, economic and social characteristics (Okely 1994). In any participatory approach the guiding principle is local people who can help themselves together with outsiders.

- Participation means very different things to different people and different groups under different circumstances. There are five levels of participation.
- Participation by being passive beneficiary: Most of the rural development programmes are of this kind. There is no space for beneficiaries to reflect on it.
- Participation by carrying out tasks and functions defined by others: The outsiders/ the project personnel design the project and define the tasks. The people are asked to carry out the task.
- Participation through active consultation: The people are consulted at the pre-project planning and project planning stages. Thereafter the people have no role to play. The people have no control over the project.
- Active participation in planning, implementation, monitoring and evaluation: The people play a vital role right from planning to evaluation. In this kind, the project is flexible enough to accommodate the changes, occurred through a dynamic interaction of the project and the people. Here the project will be successfully completed but the post project development is not adequate for sustenance of the programme.
- Autonomous decision making: This is an ideal kind of participation, always people will be in the center of project cycle and sustainability of the project is high.

Rural reconstruction project framework leads through five steps of community participation. They are as follows:

Phase	Steps	Required input
Community mobilisation	Preparation	Plenty of human input but often involves no physical, monetary or material inputs.
	Mobilisation	It takes time, skill, trust, creativity, persistence and energy to accomplish well.
Community development	Initial people's project Integration and expansion, separation.	It requires human, physical and material input.

The main intention of the developmental programme are that they should be efficiently implemented (time /other resources), they should produce the result that was intented to produce, it should reach the masses, it should reduce the dependency on others/ make the people as self reliant and it should create long lasting effects. The worldwide experience shows that if the project ensures sense of ownership of people, the intention of the project will be met. Oakely *et al.* (1991) in their review of participatory cases have found out five major results yielded by participatory initiatives *viz.* efficiency, effectiveness, self reliance, coverage and sustainability.

Though the above said remarks are almost common to all the states, with respect to Tamil Nadu the illustrations are as follows:

- *Patron-client relationship between the people and the Project Implementing Agencies*: The line departments of the state government are the PIAs in the watershed development activities funded under various schemes. Apart from PIAs, in some cases NGOs were involved as Project Facilitating Agency (PFA). The main role of the PFA is community mobilization whereas the PIA's role is project implementation. In most of the cases the main function of the PIAs is managing the grant. In order to ease this work, contract system of work implementation was followed. People have no/limited control over the grants.
- *Lack of coordination among different actors*: Projects are implemented by various departments which includes Agriculture, Agriculture Engineering, Forestry, Animal Husbandry etc. and there is no mechanism established to coordinate their activities. The different departments working in isolation of each other has resulted to large variations in quality of implementation, impact and sustainability of the programme. Each departments based on their experience in the particular field has implemented the project. The integration of different production systems and integration of privately owned resources with community owned resources are not given attention which are essential for the sustainability of the programme.
- *Participation is mandatory*: Participation is not voluntary. It is more of government influenced one. Though people contribution is mandatory the project ensured it by adjusting with land tax and /by other means. In most cases the leaders of watershed committee are not elected through participatory process. The project personnels

select the leaders based on their know-how, mainly the space is occupied by the local political leaders. In district like Ramanathapuram though several projects have been implemented no body knows about the project. Though the committee exists, it has no control over the project. All the works were done through contract system. Funds were directly given to the contractors. In most cases the project plan was never consulted with the people. Based on the project personnel know-how, project plans were prepared and implemented. In a few cases it was consulted at preparatory stage and thereafter people were left out.

- *Poor integration of privately owned resources and common resources*: The one of the main objective of watershed approach is equitable distribution of benefits to the inhabitants of the watershed. The development of the village commons can help the resourceless to get possible economic benefits out of it. In reality, the project concentrated more on privately owned resources and 'commons' are completely neglected or not given adequate attention. This is mainly due to the lack of clarity about benefit sharing from commons and deficiencies in project design.
- *Lack of flexibility in project implementation*: The programme should have space to accommodate the unavoidable changes and mid project corrections which are location specific. It is also needed of the project personnel to effectively work with the community. The rigid guidelines and pre-designed project plans have made obstruction for people participation. Standardisation of institutional development and project design never encourage active people's participation.
- *Enabling resource vs enabling people*: The development of natural resources was given more attention than the development of human resources. It is failed to recognize the fact that conservation is basically about the people and not about resources. Conservation is as old as civilization. To enable resource, the project should build on people's experience. It should build the sense of ownership of the people and should reinstate the self-respect and the sense of identity that they have lost. Because the real problem is managing people and not resources.
- *Attitude of the project personnel*: Community mobilization is not a simple and single task. It requires social concerns, human sense and it also requires time, skill, trust, creativity, persistence and energy to ensure active participation. It needs plenty of human input rather than material input. It is lacking in most of the projects. Many times the PIAs stand away from the people's issues.
- *Lack of process monitoring and evaluation*: Similar to conventional projects, the watershed projects also reviewed based on the expenditure incurred/grant spent. Target based project implementation is common in almost all the cases. Process review will enlist people participation but it was not in practice.

PEOPLE PARTICIPATORY APPROACH: DHAN FOUNDATION'S EXPERIENCE

DHAN Foundation has been implementing people participatory tank development programme since 1992 in Tamil Nadu, Andhra Pradesh, Karnataka and Pondicherry. Tanks are the lifeline for rainfed agriculture in most of the districts in the southern states. Tanks are the mini watersheds. The programme organises people around the resources and promotes people's institutions at different levels. *i.e.* Tank Farmers Association and Dryland Farmers Association, Watershed Association/Cluster level association, sub-basin or basin level association/Farmers Federation. In all the project areas the programme put efforts to enlist people's participation. The experiences and learnings are as follows:

Opportunity for participation

Success of any participatory programme depends on the opportunity available for different actors to participate in the programme right from the project planning. The programme provides adequate space for individual and for community as a whole to participate in the programme. The programme organises village level meetings and by which the participation is assured. To ensure active participation the programme put maximum efforts in the first village meeting onwards. Depending upon the social set up in the village the team follows any one or combination of following efforts to ensure attendance in the meeting. It includes organising meeting through village committee, Caste Association, other villagers and through house to house request. The way we organise the first meeting itself will lay foundation for participation.

Resolving conflicts

Conflict is a natural phenomenon in any system. It is a struggle between incompatible and opposing needs, wishes, ideas, interests or people. In any participatory projects conflict is inevitable. The unresolved conflicts from the past, exclusion of a few stakeholders, encroachers, deliberate misuse of common resources, infringement on traditional rights, inequity among different stakeholder groups, individualism etc. hamper the participation.

In our case the characteristics of the conflict are: (*i*) conflicts arise from the emotional issues, (*ii*) conflicts arise over the ownership, control and use of the resources *e.g.* encroachers and illegal water users, (*iii*) conflicts are latent for long periods of time if the parties have no chance to have a confrontation *e.g.* no village meetings, (*iv*) conflicts around common resources like tanks and ponds severely affect the status of the poor because it is closely associated with the survival of the poor *e.g.* drinking water.

The team has adopted various measures to resolve and manage the conflicts, which include: (*i*) wherever differences of opinion among the members towards village common work are found out, the team we acknowledge that there exist a conflict and work towards resolving it, (*ii*) patiently hearing the different points of view within and outside the village, (*iii*) together with the villagers the team explore ways to resolve the conflicts, and (*iv*) jointly evolve a solution by consensus. Our approach in the conflicts will create a healthy working relationship among the villagers. New rules are brought out and thereby a new management regimes are created for village adminstration and common resource management. In many cases a second line of leadership emerges and neutralise the existing power equations from the village.

Contribution mobilisation

Contribution is compulsory for a real participation. People should contribute minimum 25 per cent of the project cost either through labour or kind or cash or combination of any two or three. The compulsory contribution builds people's stake in the project. It also makes the programme more accountable to the people. In our programme, contribution after programme implementation in maintenance of tank complex systems could be seen in many places every year which implies people ownership in order to maximise the benefits.

Project implementation

All the activities of the programme are implemented only through people's organisations. The programme funds including the grant from the Government is spent by the people's organisation. The project personnel have no way involved in handling the funds. They never engage contractors to do any project activities. DHAN Foundation executives facilitate the implementation process by way of free techno-managerial inputs, training local people on supervising the work vis-a-vis monitor the progress. Thus the responsibility of leaders reflect in cost effective implementation with quality. It also builds people's stake in the project.

Process review

The interaction within, between and among the project personnel, and people is systematically reviewed then and there. This will help the project to ensure maximum level of participation. If any changes are needed in the implementation, that also are carried out immediately. It helped the project to ensure member consensus throughout the project.

Integration of privately owned resources and common resources

Common resources development is given maximum possible attention in the programme. The Programme always maintains a balance between the privately owned and common resources development. Resources allocation to common resources benefits sharing from common resources are systematically worked out and people's cooperation is ensured.

Basin approach

Watershed is a part of the river basin. Watershed development programme should not be implemented in isolation or based on other man made parameters. The river basins will be considered as a broader area for expansion of the programme on a limited time dimension say within 10 years. While doing so, the revenue and state boundaries should not become a bottle neck.

CONCLUSION

Effective participation in planning, implementing and controlling the watershed development programme under favourable facilitating environment by the project implementation agency will steer the project towards the intended goals. In the process there should be a scope for flexibility for implementing the development measures as per the people's need and prevailing local conditions.

The future scope should include the informal institutional arrangement exist in places to protect waterbodies like tanks, ponds and common property resources especially in southern states as users group under watershed development which in turn lead to greater impact.

References

Batta, Bal Ram *et. al.* (1999). Recent Concepts: Knowledge, Practices and New Skills in Participatory Integrated Watershed Management, PWMTA Programme, Kathmandu.

DHAN Foundation (1998). Training Manual in Watershed Management.

Misra, V. K. (2000). Watershed Wonders: Inter disciplinary integration and institutional development for making watershed development a people's movement. *Wasteland News*, Feb.-April, 2001.

Singh, Katar (1998). Managing Dryland watershed development programmes: Lessons of Karnataka experience, Institute of Rural Management, Anand.

Watershed Development Fund: Guidelines for State Governments, NABARD, Mumbai.

ANNEXURE-1

DHAN Foundation's Approach

Phase	Stake building and Trust building	Standardisation	Integrated System Development	Self Management
Content	• Watershed Selection • Formation of user groups (TFA, DFA)	• Farmers Planning • Mobilising Resources • Farmers implementation • Enabling Resources/ System repairs	• Integration of individual and common benefits • Integeration of land & water resources • Soil and water conservation	• Institution building • On going operation and maintenance • Turnover/ Phasing out
Process	• Village meeting • Conflict resolution • PRA methods • Exposure visits	• Enable the people • Experience sharing • Specific Training • System design and maintenance	• Prioritisation • Integration of different user groups • Evaluation • Monitoring	• Setting up self renewal mechanisms • Review process • Capacity building and training
Guiding principle	• Learn from people • Recognise their need, aspirations and their indigenous knowledge • Trust with the people	• Build on people's experience • Goal establishment by farmers • Reach consensus among different actors	• Enable resource enabling people • Integration of conservation and livelihood • Livelihood options and contingency plan for sustainability	• Collective action • Self regulation • Conservation is as old as civilisation

Watershed Management: Issues and Policies for 21st Century pp. 306-309

Interfacing PRA into Watershed Programme

J. Krishnan

Origin of PRA as a methodology can be traced back with different approaches, which got gradual refinement since seventies. It has emerged due to the unsatisfactory result obtained over ineffective conventional methodologies utilized in rural development programme. It is the reason that PRA has got immense value in watershed activities due to its 'participatory' nature in analysing rural settings in evolving people centered development programme. Though its methodologies assure 'participation' at initial stage on collecting facts on the issues, the same gets diluted later while assessing problem solving measures.

Community participation has got immense value in all rural development programmes in designing proven technologies. Gone are the days, when mere conceptual understandings on problems were transformed into technology and transferred to the field. But in the real field situation, they failed to stand against adverse conditions for long term and become distinctly hostile. Such instability in addressing problem is due to the limited perception level of Extension personnel/Researchers on location specific problems and their complex and intricate relationship with environment. On the other hand, lack of community participation right from assessment of collected issues and facts upto implementation and impact assessment resulting instability of watershed intervention. The need is, getting close association with people, and enhancing their participation not only at appraisal stage, but also during assimilation, evolving technologies, implementation and impact assessment jointly with villagers. Thus ensuring peoples' maximum involvement at every level of watershed activities promote perception level of Researcher/Extension personnel, apart from gaining confidence in addressing problems across the spectrum. On the other hand, people can get more insight on the watershed structural components, which could fulfil many of their expectations and problems on their own.

Interfacing PRA into watershed activities has got potential to ensure community participation during collection of data on issues, facts and resources. But it lacks access in perpetuating peoples involvement in further. Moreover, it focuses solely on crux of problem by creating higher expectations among people but fail to render appropriate solution in holistic manner. And so, unless the people's involvement is enhanced further, the deeply explored potential data would become valueless. Emphasizing peoples' participation would help in incorporating their own analysis in identifying proven solutions jointly. This will reduce the gaps between problem appraisal to technological assimilation, their implementation to impact assessments in the field. Hence, the addition of following methodological frame work and interfacing them into watershed would play crucial roles in ensuring continued community participation in sustained manner. The following methodologies would be considered for interface with watershed activities to achieve people's participation to the fullest extent:

- Participatory Watershed Area Assessment (PWAA)
- Participatory Interaction in Technology Assessment (PITA)
- Participatory Watershed Components Implementation (PWCI)
- Participatory Watershed Impact Study (PWIS)

PARTICIPATORY WATERSHED AREA ASSESSMENT (PWAA)

It is a methodology to involve people to find out the existing centralized watershed activities/ systems in an area/village and the decentralized water conservation activities followed at each farm land. The facts recorded here are to be analyzed along with local community to gain insight on relative effects between macro and micro watershed activities of particular region.

For instance, Check dam construction would benefit how many wells/pond and to what extent. Centralized watershed components such as Nala/water flow or ways/streamline/ channels/storage tanks/water ponds etc need to be examined with areas specific people, promoting interactions among them to gain insight on the seasonal/historical changes occurred to the above mentioned structures. This methodology creates platform to interact with villagers, involving them mutually in understanding over the existing location specific, crop specific traditional land based water management techniques and on farm water harvesting structures. It initiates a participatory process between Researcher/Extension personnel and people in probing into the existing proven water conservation techniques. This would mainly consider the ideas, understandings, practices, and performances towards drought mitigation efforts.

Gathering insights on the Nala/water ponds encroachments and area disappearance, water flow obstacles and new diversions happened over the years are deeply explored and analyzed. This methodology would subject the people into brainstorming exercise at field itself in order to generate more ideas/solutions that are not to be evaluated immediately. Providing free climate to ensure full participation of community, facilitating and channeling them into the discussion are the main principle focus. Assessment can also be done on success/failure cases of open ponds, check dam, gabion, gully plugs etc. which may be constructed by either private or government efforts. This would pave the way for designing proper structure and systems along with prophylactic measures.

Recording information emerged during discussion is information to create base for further methodological activities.

PARTICIPATORY INTERACTION IN TECHNOLOGY ASSESSMENT (PITA)

PITA is a process of involving village people in which possible technological blend is assessed to test and verify the same in the field situation in order to bring solution to the identified problems through rural appraisal. The outcome of the first methodology will be kept as base for the interaction and where information gathered are subjected for the discussion to evolve practical and people friendly solutions/technologies. It involves people and Researcher/Extension personnel to establish small unit of plot to research on performance of blended technologies of decentralized watershed component activities. It gives collective insight over the performance, constraints, certainty and suitability for the finalization of location specific technologies. The experimental learning and analysis of facts and performance for better measures would continue until the fruitful findings arrived. As this provides an interactive learning between each other (Researcher/change agent and people), the output of this collective and participatory effort ends in high degree of techno-certainty. This session can be conducted with following objectives in view that are:

- To enhance learning atmosphere with involvement of people, and Researcher/ Extension personnel, to share their knowledge, perception, and experiences gained over the years of their farm interventions.
- To blend the traditional measure with scientific measure to generate suitable, sustainable and proven watershed technologies.
- Ascertaining on resulting technologies for their performance and cross-checking with locally prevailing resource conditions. For *e.g.* gullyplug, stone bunding, gabion dam etc. can be decided as per local resources such as soil type and available stone.
- Gaining common understanding and suggesting for the proliferation of technologies, transfer and application in the other areas of different watershed areas.

Facilitating and encouraging the people to give their traditional ideas/solution to the identified problems where technological solutions to be incorporated by the scientist/extension specialist to evolve proven technologies/solutions. For instance, showing the scientific structure such as check dam, gully plug etc. in pictorial forms to get their opinion and advice thereby where refinement can be done in the structural components. By this way, people get a chance to interface their traditional knowledge with scientific solutions and evaluate both means to evolve practical solutions as they know the problem intensity. Modifications to interventions can be considered as per the suggestions given by the people which become very essential. The main advantage of this methodology is that the people with their traditional knowledge get an opportunity to know about scientific measures and the extension functionaries/watershed specialist with their scientific knowledge get an opportunity to know peoples' traditional knowledge. This would create a platform to interface both scientific and traditional system in evolving proven sustainable technologies. By this way value addition to particular system is achieved.

PARTICIPATORY WATERSHED COMPONENT IMPLEMENTATION (PWCI)

Channeling technological know-how through people during implementation of watershed components is very essential. This would enable them to learn and gain skill in such new technology. This would enable them to do corrective measures, carry out repair works of their own when the implementation programme is over. The outcome of PITA in the form of technology has to be spread across different areas of villages to make people aware thereby influencing them to adopt. In order to gain confidence, people can be exposed to the operational research promoted in the field and experience of them to be shared. Close monitoring and evaluation can be initiated with different group of people to different areas. Peoples cooperation and guidance during this phase are very much essential. For instance, soil and water conservation system such as field bunds and trenches if planned in summer, considering peoples' suggestions and preference on specific date, local material available, skilled/unskilled labour available would be helpful in ensuring successful implementation.

PARTICIPATORY WATERSHED IMPACT STUDY (PWIS)

The prefects and utility of implemented activities/technologies will only come to the light through impact assessment on interventions made. It is the process of collective assessment on the impacts of structural components and their associated influence on socio-economic and ecological conditions, getting realistic picture on success/failure cases need to be analyzed along with people as it could create a common platform for joint learning between Researcher/NGOs and the people. This would also pave the way for test verifying the sustainability of such interventions/components with an enlightenment for future course of action. It can build confidence among implementers/researchers/extension functionaries that people can carry forward of their own.

It is the process of collective assessment on the impacts of interventions made and their associated influence in socio-economic and ecological conditions.The extend of capacity of technology in solving the identified problems, and studying constraints are considered to be important here. Recording the progress periodically and documenting the understandings and learning, and clinicing the intervention process with the result for reconsideration and thereby rework on proper measure are steps to be pursued in this session. The gained apprehensions would be helpful in minimizing the constraints in earlier phase of interventions and thereby avoiding adverse consequences in future. As the aforesaid participatory process would bring drastic ideas, interactions and proven learning situations among the Researcher and people, these methodologies would perpetuate the community's full participation to perform result oriented watershed development programme. These methodologies are applied and studied at various geographical situations with periodical refinement in framework to achieve the effective participation of local community in the watershed interventions.

Watershed Management: Issues and Policies for 21st Century pp. 310-321

Sustainable Development of NWDPRA Watersheds: Case Study of Raichur District, Karnataka

S. Erappa and Shoba S. Mallari

INTRODUCTION

Sustainable usage of natural resources is essential to realize the sustainable agricultural growth and development to meet the food needs of the growing population in the third world countries in general and India in particular. These issues are to be addressed by the policy makers as well as researchers to look for a much more broad-based frame to understand and analyse the socio-economic and related processes. In this context, the concept of development and environmental concerns become quite useful. The Brundt Land Commission Report (WCED: 1987) highlights in the context of sustainable development in a path of development which "meets the needs of the present without compromising the ability of future generations to meet their own needs". Similarly, Redcliff (1987) pointed out that for "Sustainable development to become reality, it is necessary for the livelihoods of the poor to be given priority". The paradigm shift has been observed from growth in social justice. The physical quality life index, human development index and other similar indicators have recently led to the concept of sustainable development. Except land and water other natural resources are not used by human kind. The indiscriminate use of the natural resources for development has begun sending out warning signals about the limited quality left over for the next generations. Land of course, provides food needs to mankind, encounters problems like degradation through soil erosion, droughts and famines and floods. The topsoil depletion as estimated by the World Watch Institute, accounts for 25 billion tonnes, which is equivalent to the wheat fields of Australia. According to the United Nations Environmental Programme reckon the annual soil loss due to desertification is about 6 million hectares and hence 20 million hectares become unproductive with no economic returns. India is no exception; the soil loss was accounted for 6000 million tonnes during 1972. Similarly, the depletion of the world's tropical forests (loss of vegetative cover) was around 11-15 million hectares every year in the world.

Correspondingly, the water resources scenario also appears to be worsening day by day. Forty six per cent of the annual renewable water resources (out of 3.50 million-hectare meters) join the sea through rivers in India. In spite of considerable allocations of marked area for surface irrigation, approximately 30 per cent of land is covered area with recurring drought and flood situations and rest of the land remaining under rainfed agriculture. In the recent past, the rainfed agriculture scenario has been pathetic because of the depressed tree cover, soil erosion, ecological degradation etc. The "Technical Committee on DPAP and DDP (1994) was constituted under the Chairmanship of C.H.Hanumantha Rao. It highlighted that" the degradation of environment in the dry island areas is basically attributable to the increasing biotic pressure on the fragile co-systems in the absence of adequate investments and appropriate management practices to augment and conserve the land and water resources".

The gains of agricultural growth and development have been confined to irrigated areas. Despite the agricultural production 'stagnation hypothesis' has been witnessed in the irrigated areas drawing the attention of the policy makers to evolve appropriate rural development strategies, especially on rainfed agriculture to realise objectives like growth, equity and participation to solve problems like poverty, unemployment etc. The major constraints of rainfed agriculture are uneven distribution of rainfall, low crop yield, soil and moisture retention, low fertilisers application and environmental etc. To overcome above constraints, including socio-economic concerns, rainfed agriculture through Watershed Development Programmes (WDPs) is an appropriate approach to be replicated elsewhere for sustainable agriculture development.

NATIONAL WATERSHED DEVELOPMENT PROGRAMMES FOR RAINFED AREAS (NWDPRA)

The NWDPRA is a centrally sponsored scheme with a funding pattern of 75 per cent grant-in-aid from Central funds and 25 per cent long-term loans, where initially the implementation was undertaken in 85 watersheds since 1991-92 up to1996-97 in Karnataka. The objectives of NWDPRA are conserving, upgrading and utilising natural resources in a scientific manner, creating a suitable environment for sustainable agricultural production, restoration of ecological balance, and generation of rural employment opportunities. During the VIII plan period the area treated under NWDPRA was 5.08 lakh hectares with a total expenditure of Rs. 100.14 crores as against the target of 2,500 lakh hectares and an outlay of Rs.118.37 crores, respectively.

In the Ninth plan it is proposed to implement NWDPRA in 142 taluks having less than 30 per cent of assured irrigation facilities and also targeted to treat dryland area to the tune of 6.67 lakh hectares in the state. Almost all the districts recorded the percentage of area treated to the targeted area was more than 100 per cent in some districts, considerably higher, and at state level it was accounted to be 145 per cent during the VIII plan period.

The percentage of arable land treated accounts for 72 per cent (which is a state average) and the area treated under the non-arable land category shows a mixed trend *i.e.* around 27.5 per cent. Huge allocations were made under NWDPRA in the country. The basic philosophy of the watershed development programme is to utilise optimally the scarce resources like land and rainwater, which assures drought proofing and step-up of production to treat the dryland watershed approach, assures the rapid impact evaluation on the NWDPRA implemented to give the feed-back to the policy makers for better designing and implementation.

Performance of NWDPRA in Karnataka

The geographical area of the state is about 19.04 million hectares and its population was 450 lakhs as per 1991 census. The economy of the state is predominantly agrarian and rural in nature with little over 70 per cent of its population relying on agriculture and allied activities for their livelihood. Out of 19.04 million hectares, the total cultivated area accounts for about 10.3 million hectares, of which around 80 per cent falls in the category of rainfed farming.

Obviously, the rest of the 20 per cent is cultivated through different sources of irrigation in the state.

The ultimate irrigation potential is estimated to be about 46 lakh hectares (that comes to 30 per cent of the total cultivated area), and then 70 per cent of the cropped area will be under rainfed farming in the state. Five out of 10 agro-climatic zones are classified as drought prone areas or dry zones, spread over 14 districts (out of 19) and 106 (out of 175) taluks in Karnataka. The departure from the average annual normal rainfall (i.e. 1200 mm) ranges between 450 mm and 3932 mm across the districts. The crop production suffers from inadequate, scanty and uneven distribution of rainfall in Karnataka. It is surprising to note that the state still depends on drylands for its sizable foodgrains production.

The area treated under NWDPRA accounts annually for about 1.5 lakh hectares of rainfed in the state. The average crop yield in rainfed farming normally will be low and varies depending on the fluctuations in rainfall. After realising the importance of the watershed approach for rainfed farming, Government of Karnataka initiated Watershed Development Programmes (WDPs), since 1983-84 onwards. The physical and financial achievements under NWDPRA during the plan periods in Karnataka are indicated in Table 1. The expenditure incurred on the above programme reveals tremendous increase during the Eighth Plan (232per cent) and Seventh Plan period.

Similarly, the area treated under dryland farming (both arable and non-arable) also shows an increasing trend (though not in the same proportion) in the state. However, the same tempo was not maintained, as per the estimated figures of both the financial and physical performances during the Ninth Plan period. In other words, there was a considerable decline in terms of proposed allocations and area to be covered. The average investment made during the 7th, 8th and 9th plan periods, to treat one hectare of land (both arable and non-arable) was Rs. 414, Rs. 2,099 and Rs. 2,400, respectively. The investment made over the plan periods. however, is well within the norms fixed per hectare under NWDPRA 1998-99 in the state. However, during some years, the figures show there was a considerable increase, while during some other years, the decline ranges from –5 per cent to –43 per cent. The increase can be attributed during the same years due to a backlog sometimes, as opined by the concerned officials in the agriculture department.

Table 1. Physical and financial achievement under NWDPRA in Karnataka

Plan Period	Achievements		Investment Made per Hectare (Rs)	Norms fixed per Hecatre
	Financial (Rs. Lakhs)	Physical (in Hectares)		
VII	2360.73	1,67,000	1414	2000
VIII	7828.06	3,72,910	2099	3500
IX	12000.00*	5,00,000	2400	5000

*Proposed Targets
Source: Department of Agriculture, Government of Karnataka, Bangalore.

Indeed, to compute and arrive at the physical performance seems to vary considerably because the unit cost under different components is not homogeneous, unless some components under physical progress reveals mixed trends. In other words, in some cases, the proportion of achievement to the target ranges from 49 per cent (under adoptive trends during 1990-91) to 127 per cent (*i.e.*, under agro-forestry/allied cropping) during 1996-97. Interestingly, the financial trends reveal quite encouraging results, that is to say, the achievement to the target by and large above 90 per cent (except during 1995-96 *i.e.*, 76 percent). And during some years it exceeded more than 100 per cent.

The components are not uniformly spread between 1988-89 to 1997-98. During 1988-89, 98 per cent of the total allocations and achievements for the target was on land and moisture management (including crop production programme per hectares). Rest of the expenditure was on training, purchase of tools and equipments etc. Similarly, the achievement to the target was more (*i.e.*, 77 per cent) on land and moisture management during 1990-91. Interestingly, the achievement to the target was accounted to be 22 per cent on establishment and contingency. And the rest, *i.e.*, 1 per cent was spent on the remaining components. Between 1991-92, the total achievement targetted on the basic activities, *viz.*, agro-forestry, training, establishment charges on research and survey and planning, increased from 33 per cent to little over 70 per cent. During 1991-92, the expenditure was 73 per cent to the total on the treatment of both arable and non-arable lands in the state. During the successive years the expenditure on the above items increased marginally (up to 78 per cent). The rest of the allocation *i.e.*, around 22 per cent was spent on basic activities in Karnataka.

Objectives and methodology

The study aims to examine the rapid impact evaluation of the National Watershed Development Programme for Rainfed Areas (NWDPRA) in Raichur district. The following are the objectives of the study.

- Examine the accomplishment of NWDPRA implemented by agriculture department covering its physical and financial progress in Karnataka.
- A brief profile of Raichur district, and the selected watersheds in the study area.
- The impact of NWDPRA on the cropping pattern, bio-mass production on the farms regarding vegetation, fodder, fuel, timber and horticulture development.
- Assess the socio-economic transformation of beneficiaries and non-beneficiaries despite deficit/scanty rainfall in the watersheds.

Methodology

The methodology followed in this study is discussed below:

This study comprises both secondary and primary data used to assess the impact of NWDPRA on dryland farming. The secondary data were collected from published and unpublished documents of the Agriculture Department, Government of Karnataka at the state level and also at the district level *viz.*, Raichur. This helps to understand the macro and micro picture in terms of the physical and financial progress and other related activities under NWDPRA in the state.

Primary data was collected from both beneficiaries and non-beneficiaries in the selected watersheds. We have selected two watersheds one completed *i.e.* Hirenaganur, and one on-going, *viz.*, Devasugur in Raichur district. The above watersheds were selected in consultation with the local implementing agency. We also met the dryland farmers on the farms in the villages and also non-beneficiaries. The information was collected through personal discussions, group discussions and also adopting Rapid Rural Appraisal and Participatory Rural Appraisal. The beneficiaries and non-beneficiaries were contacted in those villages, three each in the upper reaches, middle reaches and lower reaches of the watershed. Further, the structures/activities like check dams, nala bunds, vegetative bunds, earthen bunds, live fencing (agave), agro forestry, horticulture, etc were taken into account to assess their sustainability and also that of the NWDPRA. We also met the non-beneficiaries of NWDPRA and the few artisans and the service sector, who had received tool kits, and coconut saplings for the homestead garden, poultry birds, sheep, etc. within and outside the watershed area.

Profile of Raichur district

The geographical area of Raichur district is 8386 sq. km. (7.2 per cent of the state area) and located in the northeastern part of Karnataka. The district falls under northeastern and Northern Dry Zones. It has two major rivers, *viz.*, Krishna and Tungabhadra on the northern and southern side, respectively. The left bank canal of Tungabhadra dam provides irrigation to the tune of 1,27,296 hectares covering Raichur, Manvi and Sindanoor taluks. The literacy level of the district is 34 per cent, which is less than the state average of 56 per cent. The soil qualities are black cotton followed by red loamy in the district.

Raichur district recorded a higher rainfall (25 per cent) during 1994-95 and 1996-97 as against the normal rainfall of 621mm. But widespread drought occurred during 1997-98. The average rainy days are 48 in a year with long dry spells. The cropping intensity is 114 per cent, which is slightly below the state average (115 per cent). The agricultural land holdings of the small, marginal and semi-medium farmers and the corresponding area account for 79 per cent and 46 per cent respectively to the total in Raichur district. It is just opposite in the case of medium and large farmer categories in Raichur as compared to the state average. Area irrigated through canal water dominates in Raichur districts covering Raichur, Manvi and Sindanur blocks, followed by well irrigation. The tanks, bore wells lift irrigation and other sources, seems to be marginal. The net irrigated area and the surface water accounts for 26 per cent and 87 per cent, respectively in the district.

The cropping pattern, the area and production of important crops like Jowar crop occupies almost 50 per cent of the total area in the district and followed by paddy and bajra crops. However, the area under pulses and also production figure do not vary very much. Paddy, Jowar, Bajra, Tur, Groundnut, Sunflower, and horticultural crops are the major crops grown in the district. Out of the gross cropped area, cereals, oilseeds, pulses and cash crops (cotton and sugarcane) account for 64 per cent, 26 per cent and 10 per cent, respectively.

There are four watersheds treated, *viz.*, Arakarnala, Herenaganur, Arasanagi and Kalbavi, the achievement in terms of area treated (both arable and non-arable) shows 158 per cent during 8th Plan in Raichur district(Table 2). Across the watersheds, the percentage of arable and non-arable land treated to total land accounts for around 77 per cent and 23 per cent, respectively.

Table 2. Watershed area treated under NWDPRA during VIII Plan Period In Raichur district

District	S.No	Taluk	Watershed	Targeted (original)	(area treated) Arable	Non-Arable	Total
Raichur							
	1	Deodurg	Arakeranala	4,300	4,162	1,603	5,765
	2	Lengasugur	Hirenaganur	3,538	4,097	1,774	5,871
	3	Raichur	Arasanagi	4.417	4,179	1,399	5,578
	4	Yelburgha	Kalbavi	3,990	7,292	1,181	8,473
		TOTAL		16,245	19,730	5,957	25,687
		STATE TOTAL		329,481	368,586	139,983	508569

Source: Department of Agriculture, Government of Karnataka, Bangalore

EVALUATION OF NWDPRA IN THE SELECTED WATERSHEDS

A. Hirenaganur - A completed watershed
B. Devasugur - On-going watershed

A. Hirenaganur watershed

Hirenaganur Watershed is situated 76 kms away from Raichur district, covering Lingasugur and Manvi taluks. It is located in the north east of Lingasugur Taluk with 16° 10' to 16° 4' latitude and 76° 40' to 76° 17' longitude. As per the Master Plan the total geographical area of the watershed is 11,775 hectares, of which 44 per cent and 56 per cent are red soils and black soils, respectively (Agriculture Department). The above watershed is divided into seven sub-watersheds covering 27 villages *i.e.*, 13 villages belonging to Lingasugur Taluk and 14 villages under Manvi Taluk. The landscape has moderate, sloppy and undulating surfaces. All these villages in the 7 sub-watersheds are located on the roadside and linking to Lingasugur - Halti and Pamankallur.

The area under forest and uncultivable land account for 4.5 per cent and 10 per cent, respectively of the total area of the watershed. The rest of the land is cultivable. The average rainfall is 575 mm. Out of the total beneficiary farm households (*i.e.*, 5,389), 53 per cent are small and marginal farmers, owning 31 per cent of the land in the watershed. SC/ST families record 12 per cent of the total beneficiaries. About 60 per cent of the total population belong to landless agricultural labourers. Before the watershed development programme the area irrigated was 4 per cent of the total area. The major crops grown are Jowar, Cotton and Groundnut. The crops grown during Kharif, Jowar, Groundnut, Sunflower, Tur, Horsegram, Sesame, Pulses and minor millets and during Rabi jowar, cotton, sunflower and bengalgram are grown in the drylands. The monsoon season starts from June to October with erratic rainfall in the Kharif season. As a result, the occurrence of recurring droughts are common in the district. The rainfall record also shows during some years, 50 per cent of the rain was received during September and October.

Overall financial achievement to the target in four watersheds accounts for 83 per cent. Across watersheds, two have surpassed a little more than the targets, while the rest are around 90 per cent. In the initial years of implementation, almost all the watersheds show significant achievement and during the successive years, the trend show decline thereafter. It took more than 10 years to completely executing the implementation of watersheds in Raichur district. The expenditure on these watersheds vary from Rs.125 lakhs to Rs. 192 lakhs between 1986-87 and 1996-97. Except for Kalbavi watershed, the area treated (both arable and non-arable land) accounts to almost the same. But investment made to treat per hectare of arable land was considerably low, *i.e.*, Rs. 1,863 in the case of former Kalbavi watershed, as compared to the average investment made (*i.e.*, Rs.2, 648) in the remaining watersheds. Nevertheless, phasing out of the total expenditure over the years, it reveals that the allocations in the initial years are comparatively lower than in the later years. This could be because major structures like check dams, nala bunds etc, will normally be located more in number in the lower reaches than in the upper and middle reaches. The soil conservation and related activities are common across the reaches in the watershed.

In the initial years of implementation of the NWDPRA in the above watershed, the funds were earmarked on land development and crop demonstration. The above items continue to occupy prime place, but interestingly, the establishment cost seems to be significant to the total expenditure until 1990-91. The components like dryland horticulture, agro-forestry, contingency seed streking, supply of grass slips seed, adoptive trails, improved agricultural implements, and fabrication of prototype tools occupy marginal importance. Regarding physical and financial achievements where targets are concerned, it was almost more than 100 per cent from 1986-87 to 1990-91.

There was a change in the component pattern under major heads of financial achievements to target since 1993-94 till the end of the project completion, *i.e.*, 1996-97. Out of four years period the financial achievement to the target show a mixed trend (during 1994-95 and 1996-97 it is less than 90 per cent). It is important that the amount spent on non-arable land by the end of the completion of the project was much higher than on the arable land treatment and other basic activities. The expenditure on livestock development, however, shows negligible percentage to the total in the watershed survey and cartographic equipments, training (for both staff and farmers) and establishment charges are the main items under basic activities during 1992-93. Followed by this, soil conservation activities and production system particulars are also covered during the year. Treatment of both arable and non-arable land, and drainage point under soil conservation works, and while production systems cover agro-forestry, dryland horticulture crop-wise package of practices etc. Apart from usual activities, during 1993-94, Mitra Kisan and Gopala are identified and trained. Watershed committee meetings are also conducted in the watershed. Small schemes like livestock, agro-processing schemes and service sector kits were also implemented in the watershed. Kisan mela, night meetings, crop demonstration, farmers were taken on educational tour programmes to visit the university of agricultural sciences and other important places covered during 1994-95. Under horticulture, saplings/seedlings like mango, lime, neem, tamarind, berry, pomegranate and other varieties were distributed to the farmers. Homestead gardens were also encouraged. Vegetative bunds, live fencing (*agave* plantation), gully checks, *khus* grass are the major activities in the non-arable lands. Similar activities were also undertaken in the Hirenaganur Watershed during 1995-96 and 1996-97.

Devasugur: On-going watershed

Devasugur nala, an on-going watershed implemented under NWDPRA since 1997-98, which consists of 13,122 hectares and covering 13 villages falling under Raichur Taluk in Raichur district. Out of the total watershed area, about 62 per cent of the area was deleted due to village site, hillock, rocky area, airport and industrial area. Therefore, the effective area fit for treatment in the watershed is 5,020 ha. (*i.e.* 38 per cent). This watershed comes under Northeast Agro-Climatic zone in the state. Out of the total watershed area, 88 per cent and 12 per cent fall in the categories of arable and non-arable lands, respectively.

The types of soil are red (60 per cent) and medium black (40 per cent), with the result there is low infiltration and productivity in the watershed area. The intensity of soil erosion depends on the inadequacy/absence of soil conservation works in the watershed. The watershed area has no forestland. Out of the total number of holding *i.e.*, 3,608, about 77 per cent belong to the marginal (46 per cent) and small farmers (43 per cent) in the watershed. The major crops grown are jowar, cotton, groundnut, sunflower, greengram, redgram etc. Double cropping depends on the rainfall during the Rabi season. The yield rate again depends on the rainfall. Fertilisers are commonly used for the irrigated crops across the size of holdings. However, in the case of dryland farming large farmers only use fertilisers.

The total population is 10,251 in the watershed area. There are few artisans/service sectors people like carpenters, blacksmiths, tailors, barbers, washermen, stone cutters, pottery, masons, basket knitters and mat makers. It is proposed to supply tool kits to the artisans also. The service sector households of cobblers, blacksmiths come under the NWDPRA programme. The livestock population comprises of cattle 6,418 buffaloes 1,520, and sheep/goats 19,260 and poultry birds 1,210 in the watershed area. Under the training component Mitra Kisan and Gopala are trained on various issues like arresting soil erosion, rearing poultry, raising horticultural crops and other agricultural activities in the watershed.

The financial progress of Devasugur watershed under NWDPRA indicates that of the total expenditure, arable land treatment dominates (around 35 per cent) in the entire watershed (except in Mudubalnala), followed by the drainage line treatment, basic activities and livestock management. It is surprising to note that the establishment change predominantly dominate *i.e.*, 63 per cent for two years period in Mudubalnala watershed and Arakere II accounts for 13 per cent. In other words, in the former on-going watershed, little less than 40 per cent of funds only available for execution of the watershed works.

Coming to arable land treatment under conservation measures, gully control measures occupy prime priority in the initial years of implementation in all the on-going watersheds, followed by contour vegetative hedges, repairing of old SC works and opening of contour dead furrows during 1997-98 and 1998-99. It is interesting to note that the expenditure on conservation measures and production system under arable land out-weigh other activities in the watershed. The items covered under the production system are crop demonstration (both *Kharif* and *Rabi*), agro-forestry, dryland horticulture, organic farming system/number of compost pits, homestead gardens and household production system *i.e.*, tool kits are given to carpentry, pottery and blacksmithy, to the non-beneficiaries (landless labourers) in those villages. The kits given to them are intact and they have expressed the kit be enlarged to cover other relevant instruments for the trade/occupation. Few farmers already planted horticulture saplings like mango, neem, teak, bettadanallikai, coconut, etc. in some villages and distribution of the above sapling are being undertaken in other villages in the watershed area.

The schemes covered under basic activities are establishment of Bhavani Chethana Kendra (BCK) and training component to the farmers. The other schemes in the same order are *viz*., establishment of nurseries, survey and projectisation and innovative activities. However, the research component *i.e.*, by ICAR/SAU and field functionaries seems to be marginal or nil.Again conservation measures and production system are delineated under non-arable land treatment in the on-going watersheds. The latter category slightly out-weighs the former in terms of targets and achievements. Within the production system, plantation of saplings dominate over the seedling of grasses for positive development and planting of shrubs under conservation measures, gully control measures and vegetative filter strips in place of diversion drains, which occupy prime place, while vegetative contour hedges with furrows and live fencing are equal. Nevertheless, the above activities across the watersheds, show mixed trends.

The treatment of drainage line occupies an important place (next to arable land treatment) in terms of expenditure to the total in the on-going watersheds. The major works undertaken are construction of loose boulder check dams, small dug out sunken ponds in the upper reaches, loose boulder structures in the middle reaches and limited number of dug out structures are predominant in the above watersheds. While the livestock management component is less than 2 per cent of the total expenditure in these watersheds, cultivated fodder production seems to be higher as compared to the natural breeding and other means of population control, castration of scrub bulls and male buffaloes/ bulls. A similar trend also exists across the on-going watershed on the above component in the Raichur district.

Major findings of the study

Completed NWDPRA Watershed

Overall financial achievement in four watersheds ranges between 83 per cent and more than 100 per cent. In Raichur district, in the initial years, the achievement rate was considerably higher compared to the later years. It took little over 10 years to execute the NWDPRA activities in those watersheds. The expenditure on these watersheds varied from Rs.125 lakhs to Rs.192 lakhs. The average investment to treat one hectare of arable land ranges between Rs.1863 and Rs.2648. The allocations during these initial years are comparatively less as compared to the successive years. It was because of major structures like check dams, nala bunds, etc., normally being more in number in the middle and lower reaches of the watershed.

The land development and crop demonstration dominated in the initial years. Interestingly, the establishment cost seems to be significant to the total expenditure up to 1990-91. The components like dryland horticulture, agro-forestry, contingency seek streking supply of grass slips seed, adoptive trails, improved agricultural implements, and fabrication of prototype tools occupy marginal importance. It is worth noting here that the amount spent on non-arable land during the fag end of the completion of the project was much higher than on the arable land treatment and basic activities. The expenditure on livestock development, however, show negligible percentage to the total. It is also visible that vegetative bunds live fencing (*agave* plantation), gully checks, *khus* grass are major activities under non-arable lands.The schemes like livestock, agro-processing and kits on the service sector are also implemented in the watershed. Kissan mela, night meetings, crop demonstration, farmers going on educational tours to visit the University of Agricultural Sciences and other important places covered in the Hirenaganur watershed.

On-Going NWDPRA Watershed

Out of the total watershed area, almost 62 per cent of the area is not available for treatment due to village sites, hillocks and rocky area, land acquired for airport and industrial area. Therefore, effective area for treatment accounts for 38 per cent of the total area of which arable and non-arable is about 88 per cent and 12 per cent, respectively. Again, of the total land holdings, 77 per cent belong to the small (43 per cent) and marginal (46 per cent) farmers' category in the watershed area. Major crops grown are jowar, cotton, groundnut, sunflower, green gram, bengal gram etc. Double cropping depends on the rainfall during the rabi season.

Of the total expenditure, arable land treatment dominated in the watershed, followed by the drainage line treatment, basic activities and livestock management. Under arable land treatment, conservation measures, gully control measures occupied priority in the initial years of implementation. Out of the four on-going watersheds, it is interesting to note that, in one of the watersheds the establishment changes account for 63 per cent and thereby, less money was available for implementation of NWDPRA programme. Major issues covered under the production system are crop demonstration (both *Kharif* and *Rabi*) agro-forestry, dryland horticulture, organic farming system/number of compost pits, homestead garden and household production system *i.e.*, tool kits are given to carpentry, pottery, blacksmith to the non-beneficiaries of the watershed. All the tool kits given to the above trades are intact.The horticulture saplings/seedlings like mango, lime, neem, tamarind, berry, pomegranate and other varieties were distributed to the farmers in the on-going watershed. Homestead gardens were also being encouraged.

Few farmers already planted horticulture saplings like mango, neem, teak, bettadanathikai, coconut, etc., in some villages and further distribution is on in the other villages too. The schemes covered under basic activities are the establishment of Bhavani Chethana Kendra (BCK) and providing a training component to the farmers, following this are raising nurseries, survey and projectisation and innovative activities. The research component by ICARSAU seems to be marginal. Conservation measures like gully control, vegetative filter strips of diversion channel, etc., and production system like planting of saplings and seed distribution, *khus* grass etc., are delineated from non-arable land treatment in the on-going watersheds. The drainage line treatment occupies prime place next to arable land treatment in the total expenditure. The major works are construction of boulder check dam, small dugout, and sunken ponds in the upper reaches, loose boulder structures in the middle reaches undertaken in the watershed. Livestock management seem to be very marginal (2 per cent).

Overall Impact of NWDPRA

Any project on Watershed Development Programme has a visible impact on arable as well as non-arable lands. The impact parameters include four major sectors *viz.*, (*i*) the production sectors, (*ii*) the technology sector, (*iii*) the environment sector, and (*iv*) the participation sector. Among these, the impact on the production sector seems to be quite effective. Although the incremental yield as well as increased area under the crop has not been very significant, the changes are noticeable and have long term prospects. In a dynamic context, the changes will bring in increased adoption of new technology and thereby improvement in the quality of life. It was noted that the process of implementation has two important shortcomings *viz*., the savings in the allocations for primary activities are not fully utilised, and there is hardly any local level flexibility given to the implementers for an effective implementation of the programme. There was only fragmentary evidence of a watershed development team constituted in the region. These teams are not functioning as effective bodies to sustain the interventions. The guidelines of NWDPRA recommended an institution of Mitra Kisan and Gopala. In the absence of any incentive structure, these individuals do not get motivated for the purpose of maintaining the structures.

As regards the quality of life and employment sector, there was enough evidence to indicate increased employment and a visible improvement in the quality of life. This was also accompanied by increased consumption and market participation. However, the quantum of increment is not as much as could be seen in irrigated agriculture. Therefore, it is obvious that the adoption of watershed treatments and their sustenance will take a back seat for some time. The environmental impact of the treatments were quite visible especially in terms of increased moisture availability, bio-mass, fodder, fuel and fruits, water for irrigation and finally, increased food availability. All these together indicate that NWDPRA has been quite an effective programme but the process of implementation needs a fresh look.

Policy Implications

- Depending on the size of the watershed, for instance, if it covers 25,000 hectares or more area, the existing budget allocated to construct the structures like check dam, nala bund, RCS, etc., is insufficient, which needs to be increased by 10-15 per cent.
- With delineation of watershed area, the NWDPRA, a survey can be conducted on the guidelines of River Valley Project.
- The farmers who are undergoing the training on WDPS, the travel and daily allowances are insufficient and therefore the number of farmers attending such training programmes has drastically declined.
- The soil conservation activities like laying earthen bunds and strengthening the same has enhanced the crop yield. Such types of structures need to be continued in the future.
- The existing guidelines like live fencing and watch and ward in the state forest and horticulture departments in raising block plantation, horticultural crops, agro-forestry, dryland horticulture etc. Similar facilities need to be included in the NWDPRA programme.
- Raising live fencing, diversion channel, construction of boulder bunds especially on the non-arable land, should be given top priority under NWDPRA.
- The tool kits given for carpentry and black smithy are highly useful to the farmers across the size of holdings in the watershed area. It is witnessed that the present

tool kit components are inadequate (value worth of Rs.1000) and need to be increased to cover other essential items at a cost of Rs.2000.

References

Barrow, Chris (1987). Water Resources and Agricultural Development in Tropics. Longman Scientific and Technical Publication, Essex. U.K.

Chandrakanth M. G., Jeff Romm, J. K. Gilles and R. S. Deshpande (1988). "Public Choice Analysis of a Watershed Programme in India". International Conference of Agricultural Economists, Buenos Aires, Argentina. Aug. 26 to Sept. 2.

Deshpande, R. S. and V. R. Reddy (1991a). Watershed Development Approach in Fragile Resource Region: An Analytical Study of Maharashtra, Mimeograph Series No. 33, Gokhale Institute of Politics and Economics, Pune.

India, Govt. of (1981). Sixth Five Year Plan 1980-85. Planning Commission, New Delhi.

India, Govt. of (1985). Seventh five-year Plan 1986-91, Planning Commission, New Delhi.

India Government of (1986a). Report on Dryland Farming: Trends and Prospects. Planning Commission New Delhi.

India, Government of (1986b). Eighth Five Year Plan, 1992-97, vols. II. Planning Commission, New Delhi.

India, Government of (1992a). National Watershed Development Project for Rainfed Areas (NWDPRA), WARASA, Ministry of Agriculture, Department of Agriculture and Cooperation, New Delhi.

Jodha, N. S. (1979). "Dry Farming Technology: Achievements and Obstacles". In: Ch. Shah and C. N. Vakil (eds.), Agricultural Development in India. Orient Longmans. Bombay.

Jodha, N. S. (1991). "Sustainable Agriculture in Fragile Resource Zones: Technological Imperatives", Economic and Political Weekly, vol.XXVI (13), March.

Laxmikanthamma, S. (1997). Sustainability of Dryland Agriculture in India: A Case Study of Watershed Development Approach. M.D. Publications. New Delhi.

Singh, Katar (1994). Managing Common Pool Resources: Principles and Case Studies, Oxford University Press, New Delhi.

Watershed Management: Issues and Policies for 21st Century pp. 322-328

Watershed Development Structures versus Groundwater Recharge—A Case of Alternate Sluice Irrigation System in Parambikulam Aliyar Project Command Area

N. Venkatesa Palanichamy, P. Paramasivam and K. Mahendran

INTRODUCTION

Watershed development in India has gained momentum due to variety of reasons. The Government of India's Watershed Development Project (WDP) funded by the Ministry of Rural Development has been implemented since 1994-95 all over the country. Watershed development has to be achieved through equitable distribution of benefits from land and water resources and also through greater access to income generating opportunities.

Watershed management is realized as most important approach for sustainable development and management of soil and water. Over exploitation and improper management of groundwater resource degrade not only the groundwater quality but also the productive potentials of the cultivable land area and these finally cause deterioration of the entire eco-system. The degree of such degradations has focused the need to evolve eco-friendly land and water management systems. Under this situation, the existing water resource usage systems need realistic planning for equitable and sustainable distribution of available water in an economically feasible way acceptable to all stakeholders.

A large part of western Tamil Nadu, experiencing perpetual problem of depleting groundwater due to continuous drawls poses serious problem to the farmers. Under these circumstances, watershed development structures play a significant role in groundwater recharge. These structures would be useful even under canal irrigation command areas when receipt of water is not certain in all the years and seasons. This paper attempts to evaluate the impact of such structures in such a canal command area. The main focus is therefore, to evaluate the potentials and impacts of watershed development structures in the Parambikulam Aliyar Project (PAP) command area.

PROFILE OF THE PROJECT AREA

The Parambikulam Aliyar Project (PAP) is a multipurpose multi-valley project. It contemplates diversion and integration of eight west flowing rivers in the hilly areas and two on the plains, by constructing reservoirs across them and interconnecting them by tunnels to divert the impounded water to the plains. Coimbatore and Erode districts in Tamil Nadu and Chithur district in Kerala are the areas benefited. This project envisaged provision of irrigation facilities for 2.48 lakh acres in these three districts. To meet the demand of the farmers belonging to adjoining areas of Udumalpet, Palladam, Kangeyam and Dharapuram taluks, the Government of Tamil Nadu extended the ayacut, increasing the total ayacut area to 4.22 lakh acres with this expansion, the total annual yield expected from this project is 25,000 tonnes of rice and 5,77,000 tonnes of other garden land crops. To achieve this goal a systematic release of water to match the periodic requirement of crops in the different canal zones is required.

The unprecedented extension of ayacut area in comparison to the water potential receipts and reduction in rainfall in the catchment's area forced the government to divide the command area into four zones for purposes of water distribution. This is in contrast to the earlier three-zone pattern. The water supply interval to each zone also got prolonged from once in eighteen months to once in two years. In drought years, it is extended even up to three years. The water release period during each turn reduced to 67 days from 135 days on 7 days on and off pattern. This necessitated the farmers' organizations operating in the area and the Government of Tamil Nadu to look for an alternative system of water release to offset the above problems.

ALTERNATE SLUICE IRRIGATION

A new scheme of water management to improve irrigation in command area has been implemented by the Government of Tamil Nadu. Under this scheme, odd numbered sluices are opened in the first year, while the even numbered remain closed. During the second year, it would be vice-versa. This alternate sluice water distribution has been implemented in all the branch canals, which have 1000 acres and above as ayacut. The only change from earlier zone pattern is that, the regrouping of ayacut areas was done in such a way that water receiving and not receiving areas are interspersed every year throughout the command area. Hence, the benefit of irrigation will reach the entire command area every year instead of once in two years through the indirect effect of considerable improvement in groundwater status in the non-canal water released adjoining areas.

There are some distinct advantages expected from this system over the previous system of distribution of irrigation water. The irrigation wastewater from one sluice area is able to recharge not only groundwater table of the sluice command, but also the groundwater status of adjoining areas, that does not have access to the project water in that particular year. The advantage of this new system is that during canal irrigation in any given year the watershed area could definitely receive irrigation to half of its land area. Two months after the canal opening to half of the watershed, seepage of the sub surface water from the farmers field to water courses / streams would start occurring. All this water would be finally collected in the watershed development structures. This stored water would encourage the groundwater recharge in the entire watershed including the non-canal irrigation receiving areas during that particular year.

METHODOLOGY

A watershed is delineated by a natural boundary and fields within that have a closed inter-relationship with watercourse under public community and individual ownerships. In the present study micro watershed covering a land area of 400- 500 hectares was selected purposively to know about the impact of interlinkages between watershed development structures and groundwater recharge under alternative sluice irrigation pattern in PAP command area. The data covers a cross-section of farmers of a watershed located in the PAP command area, namely Sinna Bomman Salai village watershed of Udamalpet taluk, Coimbatore district. It lies between the 10/2 branch canal of Parambikulam Main Canal and 0.5 of its distributory.

RESULTS AND DISCUSSION

In this study main focus is the impact of watershed structures on the sustainability of groundwater during non-canal irrigation in a canal-irrigated system under different water release regimes.

Table 1. Farm size and on-farm income

(Area in hectares)

Category	Number of farms	Farm size			On farm income
		Garden land	Dry land	Total	(Rs./ha.)
Small	10	0.56	0.24	0.80	22087
Medium	7	2.27	0.47	2.73	41645
Large	5	4.70	0.97	5.67	30824

Out of the 22 farms selected for the study 10 farms were small, 7 were medium and 5 were large. It could be observed from Table 1 that the proportion of dry land to the total farm size was found to be lower in all categories and almost 70 per cent of the land holding was garden land.

The current value of all farm assets ranged from Rs. 1.20 lakhs to 2.87 lakhs in different farm categories of the sample farms (Table 2). The value of irrigation structures ranged from Rs. 15824 to Rs. 65496 in all categories. Large farms invested comparatively more money on bore wells, drip irrigation systems and efficient submersible motors.

Table 2. Value of farm assets

(Rs. per farm)

Category	Current value of			Total
	Land value (Rs./ha.)	Buildings (Rs.)	Irrigation structure* (Rs.)	
Small	102117	2450	15824	120391
Medium	145029	12705	29648	187382
Large	176238	45692	65496	287426

*Drip irrigation systems, compressors and submersible motors.

Pattern of well ownership

The farm size categorywise distribution of wells, number of wells per farm and number of failed bore wells are given in Table 3. Among the three types of wells *viz*., open wells, open cum bore wells and bore wells, open wells constituted the dominant category among all size groups of farms, whereas bore wells dominated the medium and large categories of farms. In case of small farms there was not much bore wells since, construction, electricity connection, installation of two way switch for compressor motor and maintenance demand more investments which they could not afford. In general the number of bore wells have been increasing in recent years replacing the traditional open wells. Construction of open wells is extremely costly compared to drilling a bore well. And it also is easier to switch over to a new bore well when the yield from an existing open well declines, instead of deepening it.

Table 3. Ownership of wells

(in numbers)

Category	Open well	Bore well	Open cum bore well	Total wells per farm	Abandoned bore wells	Functioning bore wells
Small	0.30	0.10	–	0.40	–	–
Medium	1.00	1.00	–	2.00	0.33	0.67
Large	1.40	2.00	–	3.40	0.66	1.34

The details of depth of wells and distance to nearest watershed development structures are presented in Table 4. The depth of open wells ranged from 19.69 to 21.21 metres in all the three categories of sample farms. The depth of bore wells was in the range of 101.73 to 112.21 metres. In the study area, generally bore wells were less than 150 metres deep, since below that level very hard rock prevails that would not permit further deepening, even if, bore holes made did not yield much water.

Table 4. Depth of wells and distance to watershed development structures

Category	Depth of well		Mean distance to nearest watershed development structure (metres)	Mean pump capacity (HP)
	Open well (metres)	Bore well (metres)		
Small	19.69	106.06	55.42	5.00
Medium	21.21	112.21	88.70	6.00
Large	20.45	101.73	175.35	7.50

The mean distance to nearest watershed development structure varied widely among farm size groups. The mean pump capacity was higher in large farms since, they have to irrigate larger area under perennial crop (coconut). Also presence of deep bore wells to extract more quantum of water in a given time requires high-powered motors.

Conjunctive management and groundwater recharge

Conjunctive management of surface and groundwater resources is emerging as a major water management strategy in recent years. Government of India (GOI), Ministry of Water Resources (MOWR) has recognized, the utility of conjunctive management by using wells in the command of large irrigation systems. Many recent studies indicate that conjunctive management could make more water available for effective utilization in the command areas.

The subsurface seepage from the farmlands during the irrigation period would be transmitted to aquifers through watershed development structures constructed across farmlands and poropoke lands. These structures control the watercourse/stream and help to improve groundwater recharge in canal irrigation years and rainy seasons in the adjoining non-water receiving command areas. This recharged groundwater resource in non-canal irrigation years could be accessed as required through wells. The net effect would be equivalent to constructing new surface reservoirs, which cost more and cause environmental disruption.

Impact on groundwater recharge and irrigated acreage

The details on depth to water table and other information in the sample farms of the watershed development area are presented in Table 5. The data revealed that the mean depth to water in the watershed area, in small, medium and large farms have improved from 18 - 20 metres to 5 - 8 metres after introduction of alternative sluice irrigation pattern under watershed development areas. The data also indicated an increase in the extent of irrigated area, the well command area as well as an improvement in the groundwater table.

After the introduction of alternate sluice irrigation system in the watershed area, the water table level in all the categories improved and it was in the range of 1.2 metres to 2 metres below surface level. The average increase in area irrigated by the farms was about 25 to 30 per cent in all categories, compared to the old system of irrigation. This increase in area irrigated by wells in all categories of farms would indicate not only an increase in farm income but also higher employment opportunities to the agricultural labour. According to the Working Group on Minor Irrigation for formulation of the Ninth plan (1997-2002), additional employment created on every hectare of irrigated land through increased agricultural activity would be approximately 45 man-days per hectare (World Bank, 1999).

Impact on crop yields and income

Table 5 gives information on crop yields and employment generated before and after the watershed development structures under alternative sluice irrigation pattern. As evident,

Table 5. Impacts of watershed development structures on groundwater recharge at farm level

Variables	Small		Medium		Large	
	Before	After	Before	After	Before	After
Average depth to water table (metres)						
(a) Canal irrigation years	1.2 - 1.5	1.2 - 1.5	1.2 - 1.5	1.2 - 1.5	1.2 - 1.5	1.2 - 1.5
(b) Non-canal irrigation years	18.00	5.0 - 6.5	20.00	5.0 - 6.0	19.00	6.0 - 8.0
Average increase in area irrigated during non-canal irrigation years due to watershed development structures under alternative sluice irrigation (ha.)	-	0.41	-	1.68	-	3.48
Crop yields						
(i) Coconut (nuts per tree per year)						
(a) Canal irrigation years	65	78	56	76	52	80
(b) Non-canal irrigation years	54	76	49	75	38	80
(ii) Groundnut (kgs per hectare)						
(a) Canal irrigation years	1358	1358	1334	1334	1358	1358
(b) Non-canal irrigation years	-	1358	-	1358	-	-
(iii) Tomato (kgs per hectare)						
(a) Canal irrigation years	33345	33345	32100	32100	-	-
(b) Non-canal irrigation years	-	34580	-	34580	-	-
Employment (in man days per hectare)						
(a) Canal irrigation years	56	62	74	74	82	104
(b) Non-canal irrigation years	12	49	37	79	37	110
Cropping intensity (per cent)	50	80	48	80	45	78

watershed development structures have significantly improved the groundwater table and increased crop production and employment. The positive impact on crop production is valid for coconut, groundnut and vegetable crops particularly tomato. Additionally, yield of coconut trees also significantly increased from an average of 45-60 nuts to 70-80 nuts per tree per year due to cumulative impact of watershed development structures and alternative sluice irrigation.

Previously, groundnut and tomato were cultivated only during canal irrigation years. After introduction of alternative sluice irrigation pattern even those farms not getting canal irrigation during that particular year were also enabled to grow these crops utilizing the groundwater made available through recharge from watershed development structures.

Impact on employment and income to agricultural labourers

It could be inferred from Table 6 that as water availability increases, farmers would cultivate more area under different crops and demand for agricultural labourers would also increase and result in increase in wage rate to the agricultural labourers. After introduction of alternative sluice irrigation under the watershed development structures there was indeed significant increase in the wage rate of the agricultural labourers in the watershed areas.

Table 6. Agricultural wage rate

(Rs. per six hours of work)

Category	Before		After	
	Canal irrigation years	Non-canal irrigation years	Canal irrigation years	Non-canal irrigation years
Women	45	30	45	45
Men	70	50-60	70	70

CONCLUSION

The main focus of the study is on sustainability of groundwater and recharge through watershed development structures in a canal irrigation system by changing the rules of water delivery. The results indicated that watershed development structures had become more purposeful once the alternative sluice irrigation system was introduced in the watershed area. The watershed area, which received or not received water in a particular year in the previous regime was assured water every year at least for one half of its area. The seepage and excess flows in every year, coupled with monsoon rains could be effectively harnessed through these structures. They in turn help to recharge the groundwater in the entire watershed that could be used by even the non-water receiving areas every year. Thereby, additional area is brought under cultivation improving the income of the farmers. Possibilities for additional employment is also created that have an upward effect on the wage rate, improving the earning potential of the agricultural labourers. Thus, watershed development is a concept workable and beneficial not only for dry lands but is equally beneficial for an established canal system as well, if proper operational rules and structures are optimally combined.

References

Deb T. K., N. C. Saha and S. N. Das (2001). "Integrated Approach of Watershed Management for Sustainable Development - A Case Study", Land Resource Management for Food and Environmental Security, Soil Conversion Society of India, New Delhi, pp. 356-367.

Ministry of Water Resources "World Bank India Water Resources Management Sector Work, Report of the Sub-Group on Ground Water", New Delhi, 1996.

Sharma, S. C. "Regulation of Groundwater Development in India: Existing Provision and Future Options", in M. Moench (Ed.)., Groundwater Law: The Growing Debate, (Ahmedabad, Gujarat: VIKSAT, 1995), pp. 6-17.

World Bank Report on Groundwater Regulation and Management, Water Resource Management, India, South Asia Rural Development Series, Allied Publishers New Delhi 1999 pp. 9-30.

Watershed Management: Issues and Policies for 21st Century pp. 329-339

Policies for Watershed Development in the 21st Century

K. Palanisami, D. Suresh Kumar and B. Chandrasekaran

INTRODUCTION

Policy is heavily weighed in favour of surface irrigation, and government canals under major and medium irrigation. Even the maintenance of these irrigation canals and systems are heavily subsidized not to speak of the levels of recovery of capital costs. All subsystems relating to input subsidies, agricultural extension, and procurement operations, credit system are all concentrated in the irrigated areas. In the next phase of India's poverty alleviation and agricultural development efforts, adequate investments in watershed development projects, especially in dry farming regions is of utmost priority. In the post project period, maintenance, improvement and overall development of the watershed community to fill the gap in their infrastructure in areas such as seeds, extension, marketing, agro-processing etc. are vital concerns.

Considering the experiences gained in implementing the watershed development programmes in the country over the years, routed through various ministries, agencies, NGOs and the external funding agencies, it is important to draw lessons, avoid shortcomings of the past and draw up a viable blue print for the future course of action. It has been estimated that development of the vast rainfed areas comprising of 75 million hectares would require over Rs.37500 crores. Further the estimated cost for scientific treatment of soil and water conservation would cost Rs.7500 crores for 12 million hectares of arable and 3 million hectares of non-arable land at an average cost of Rs.5000 per hectare. It is not possible to fund on this pattern to develop the entire area of 75 million hectares. However, in the next ten years it may be possible to develop about 30 million hectares which required a substantial public investment. During the entire Ninth Plan, 10 million hectares of incremental area was proposed for coverage under watershed development programme funded by national and international agencies. In future, watershed development programmes should be "truly people's programme and the role of government should be limited to providing only infrastructural and technological support, wherever necessary". In the words of Dr.M.Balasubramaniam, Former World Bank expert, people should turn from Passive Recipients to Active Participants in order to make watershed development as people movement.

Keeping the constraints in the treatment of the entire rainfed area in the country and the resource availability, it is imperative that the available physical, financial, technical and manpower resources should be optimally utilized. Documentation of best practices and benchmarking of various components of watershed projects is vital. We should develop cost-effective, user friendly local and traditional technologies, and garner resources from all available sources and introduce participatory and cost sharing practices to ensure sustainability.

The future strategy should therefore be a movement towards a balanced approach of matching the supply-driven menu with a set of demand-driven activities. People's participation, involvement of Panchayati Raj Institutions, local user groups and NGOs along side institutional support from different levels, *viz.* the Union Government, the State, the District and block levels should be ensured to make the programme more participatory, interactive and cost effective. Convergence of various rural development programmes in and around the watershed could be ensured to promote holistic development of watersheds. For its continued success, the programme should be economically efficient, financially viable, technically feasible and socially acceptable while ensuring equity. For sustainable development, regular and routine monitoring of environmental parameters is important as environmental enhancement increases the credibility and acceptability of the programme.

SUPPLY DRIVEN TO DEMAND DRIVEN MODELS

Watershed development programmes in the country have been supply driven right from the initiation. The promoting agencies have been targeting and allocating funds annually, fixing the targets for coverage of number of watersheds and the aggregate area to be covered, deciding the structure of the PIAs, deciding the cost norms for different activities, deciding about technical and research support, transfer of technology, deciding about cost-sharing, funds flow pattern, evaluation and monitoring systems, sustainability etc. This approach do not match with the requirements of the watershed communities/stakeholders resulting in pushing the beneficiaries to the back and the government agencies assuming a lead role in the programme. For example, in areas where traditional forest dwellers are predominant, there has been a low demand for the development of common lands. This situation is puzzling and requires addressing the structural and institutional issues influencing the demand for watershed development. The three major constraints that inhibit the demand for development works are: unclear entitlements of land, feeble stakes of communities and fragmented communities. Very often the stakeholders remained as passive partners and doubtful beneficiaries.

The stakeholders were neither involved in the selection of project ingredients nor encouraged to participate in various project activities. The PIAs supply the services in watersheds as decided by them. The entire process of watershed development involved participation by the government line departments and local contractors with a clear top down approach with least involvement of the watershed communities. Consequently, there has been a supply demand mismatch leading to inadequate attention to local needs and the aspirations of the watershed communities leading to inefficient implementation and inadequate sustainability. The Central ministries have been releasing funds annually to the states to achieve higher percentage utilization of budgeted allocations. The states have not been able to utilize the released funds annually leading to revalidation of unspent balances in every year. Despite the fact that the programme has greater potential to absorb funds, the dichotomy of scarcity of funds at the micro level and excess flow of funds annually at the state level operational rigidities characterized the contradictions in the system. The consequences of these contradictions led to non-inclusion of the most deserving areas in the country on the grounds of non-availability of funds at the micro level and exclusion of the large deserving areas. Furthermore, these implemented programmes also suffered from a variety of deficiencies. Different agencies laid emphasis on different activities in watersheds and integrated watershed development remained more a rhetoric. Thus, it is high time that the Central, State governments and other donors should look into this matter and the proposed projects must be demand driven rather than supply driven.

RAINFED AGRICULTURE AND WATERSHED DEVELOPMENT RESEARCH

Experience gained in implementation of watershed development programmes during the past five decades should be utilized for identifying economically efficient, locally acceptable low cost technologies suited to the rainfed areas which account nearly two third of the cropped area. India has a hydrological heritage spread over thousands of years particularly, construction of village tanks, which have been put to a variety of uses. Construction of separate tanks for multiple uses such as drinking water for human beings, livestock, washing, bathing and irrigated agriculture has been the practice in many parts of the country. This indigenous knowledge of water harvesting and irrigated agriculture under tanks is unfortunately dwindling in recent years. This labour intensive technology could be the right choice compared to the construction of expensive modern irrigation systems. Therefore, old breached silted dysfunctional village tanks should be brought back to their original status and new tanks should be constructed wherever feasible.

Low cost technologies like construction of gabion structures for water harvesting on the lines of programmes undertaken by Anna Hazare in Ralegaon Siddhi, Maharashtra, should be adopted wherever possible and feasible. All such time tested indigenous technologies should be identified district/state wise, documented and propagated for adoption by all concerned. A national inventory of such activities should be centrally documented and disseminated.

Dryland technologies such as ploughing across slopes and along the contours, field bunding, matching with the broad contours of the geography etc. greatly help in *in-situ* soil moisture conservation and prevention of run-off. Such farming practices should be suitably sequenced and adopted to maximize production potential. Field observations and evaluation studies indicate that the farmers have a tendency to grow water intensive crops like rice and sugarcane as soon as some irrigation potential is created. Such practices should be discouraged and measure taken to propagate grain legumes such as red gram, green gram, black gram, bengal gram, horse gram, etc. for judicious use of the scarce water resources. We should promote high value, and water saving crops to maximize returns.

Rainfed areas not suitable for crop production as per land capability may opt for horticulture and agroforestry systems for economic viability and ecological upgradation. These activities coupled with propagation of crossbred cattle could trigger growth. For soil conservation, simple measures like vegetative bunds and gully plugs, rock fill dams etc. could be taken up with local materials so that they can easily be maintained by the multi-stakeholders. Farmers in the resource poor regions, may be encouraged to go for adoption of tree growing which are drought tolerant under different production environments. Therefore, it is essential to evolve different agroforestry models suited to different farm size groups and to different production environments.

The integration of watershed work with Joint Forest Management is being attempted in many parts of the country. However, a little presence of agriculture department in the DPAP watersheds by way of promoting locally relevant research and extending suitable technologies, inputs and other necessary support to the farmers. The convergence of services from the agriculture department in the DPAP watershed areas would contribute to raising the productivity of land and water resources conserved and augmented through watershed development and would thus induce farmers to sustain their effort towards watershed development in these areas. It is, in this context, essential to integrate the interventions by the two departments in such areas so as to promote watershed development activities.

Watershed development usually consists only of providing structures, either vegetative or engineering for soil and water conservation and water harvesting. Since these structures are meant to enhance crop/resource efficiency and productivity, it is necessary to supplement those efforts with suitable crop production technologies and appropriate follow up practices.

DECISION SUPPORT SYSTEM WITH SOCIAL RANKING

Development of suitable watershed development plan could be done only with the help of sound planning and with the introduction of computerised database management. The collection, storing and manipulation of geographic data are essential at a regional or national level for planners. This calls for the use of advanced tools such as Geographical Information Systems, which is used for the storing, manipulation, display and analysis of both spatial and non-spatial data. The tool can provide valuable information for decision making and can be further used to develop user-defined set of rules and thus create a decision support system. The user can pose a set of different question to the system for which he can expect replies that match his needs.

Depending on the extension work it is possible that modern technologies such as remote sensing, and Geographical Information System (GIS) are seen as appropriate tools. The sattelite based remote sensing in combination with land records, revenue maps, geological maps, socio-economic data and information collected from the ground can be a very useful and economic tool in the hands of the district level planners and the PIAs. It can be used for prioritizing the watershed that should be taken up for development, making a draft plan of their development and monitoring and evaluation. Thus, the use of these techniques should be encouraged in preparation of watershed development plans and the personnel involved are to be rigorously trained on these lines.

Use of linear programming in the land use planning programmes in the watershed along with the use of remote sensing and GIS tools can be encouraged. Rapid PRA exercise is to be made in the watersheds areas to develop micro watershed plans. Comprehensive and integrated watershed development plan is needed for land use planning, soil conservation, ground water recharge and crop planning in the watersheds of individual states.

Integrating all the natural resource information with socio-economic conditions of the people through Geographical Information System (GIS) are to be derived for different watersheds, which are then to be used as scientific inputs during the participatory rural appraisal (PRA) exercises for preparing the micro plans of the watersheds in consultation with the local people. Remote sensing technique plays a vital role in periodically monitoring and evaluation of natural resources development in watersheds to assess the impact of the programme at a faster pace in view of the scaling up the programme. Thus, the use of Remote Sensing should be done for impact assessment of the watersheds. Also, both financial and social ranking is absolutely necessary in the selection of watersheds before implementation.

PARTICIPATORY MANAGEMENT AND PEOPLE MOVEMENT

To make it truly peoples' programme they should be sensitized, empowered and involved in the programme. Local community leaders and stakeholders should necessarily be motivated about conjunctive use of water, prevention of soil erosion etc. through various media. Ultimately watershed development programme should culminate into a national movement for optimizing the natural resources like land and water to sustain livelihood system. The stakeholders should be involved at different stages of selection of project activities, planning and implementation with the ultimate objective of sustainability.

To make the programme people-centred and people-driven, awareness and confidence building is essential. Suitable measures should be initiated before the start of the programme. The awareness building may include film /video shows, exposure visits to successful models, publication and distribution of technical information to the farmers etc. Watershed projects are generally seen as marginalizing the landless. This is considered a major flaw in these development projects. To ensure active participation of the landless and other rural poor, generations of alternative income and employment avenues, suitable activities like management of common property resources both during and after the project should be identified to be entrusted to these groups to give them a sense of a responsibility and belonging, as well as provide them suitable economic incentive on a continuing basis.

The best way for encouraging the capability of an individual is through regular training. In order to achieve this goal, it is essential to examine in depth the whole gamut of training towards capability building among the various clientele groups operating in watershed. Experience gained in the implementation of watershed development projects provide valuable insights of the gaps that need attention. It is generally assumed that people will come together of their own. But this is not always so. People participate only when they value the advantage resulting from participation. These advantages must be tangible and quickly realized. Therefore, it is essential to learn more about group formation and mobilization, creating congenial atmosphere and opportunity for effective people's participation, and how to speed up the development process. The quality of people's participation needs to be improved to ensure that the women, the poor and the landless get their share in the development efforts. They need specific training and empowerment to enable them to join the decision making process and actively participate in the management of the project.

In addition to the above, strengthening of community organizations within the watershed, implementation of the planned watershed management activities, encouraging linkages with other institutions and initiating groups towards formation of apex bodies will help motivate the people and make it a people movement.

MONITORING AND EVALUATION FOR ALL

In order to assess the impact of the huge funded watershed development projects, it is of utmost importance therefore, to put in place an institutional mechanism for proper monitoring and evaluation by involving reputed institutions in the country for upgrading the quality of monitoring and evaluation of watershed development projects. This will ensure evolving appropriate methodologies for capturing the participatory processes and convergence of complementary interventions, by guaranteeing them adequate funding on a long-term basis and by ensuring for them autonomy in the pursuit of monitoring and evaluation of watershed development projects.

Implementation of the activities and their results should be monitored by the monitoring and evaluation unit through routine reporting system, sample surveys and a few indepth case studies. Evaluation should include bench mark surveys, concurrent evaluation and final evaluation must be carried out by an external agency. Meticulous planning and close monitoring should be maintained to prevent time and cost overruns. Appropriate management structures should be developed to trigger responsibility, accountability and efficiency. Quality audit of each activities/milestones of watersheds should be conducted on an ongoing basis and the findings made transparent, verifiable and debatable.

In addition, participatory monitoring and evaluation should be encouraged. The participatory monitoring and evaluation will be useful for providing proper feedback on joint decision-making process, documenting local people perception of the work performance, discovering their own potential and limitations, the nature of internal and external conflicts, strengthening their capacity for negotiation within and outside the community and finally, showing that it is possible to overcome the socio-economic constraints and to improve the socio-economic and environmental conditions.

SOCIO-ECONOMICS : UPSCALING AND MAINSTREAMING

Owing to the mutually conflicting interests of different groups constituting our rural society, building up institutions and evolving leadership for rallying the rural community around common interests with adequate stake for weaker sections and sustaining the momentum is a painstaking and time consuming task. The performance in this aspect needs a review for initiating remedial measures with highest priority. Capacity building for watershed development through training encompasses wide ranging tasks such as awareness building or imparting resource literacy, development of technical skills and reorienting motivations and attitudes of officials and political functionaries at all levels towards the need for empowering the people through decentralization. Those who administer the programme, and who themselves need such training, are not best fitted to oversee these training programmes. While a number of measures have been taken for strengthening training at various levels since the adoption of new guidelines, experience so far strongly suggest the need for broadening these programme with a view to imparting training at all levels, improving the content and quality of training programmes with due priority given to the processes of empowerment of the people and decentralization, intensifying training by adequately taking care of the changing requirements in the field and ensuring the autonomy of training institutions and their uninterrupted functioning on the required scale by entrusting this task to an independent nodal agency at the national level by guaranteeing adequate funding.

Opportunities should be used for exposure to other projects, both within and outside the project. The best resource persons available should be used for training from other institutions and NGOs. The use of locally respectable leaders and social organizations within the watershed area should be encouraged. Proper institutional mechanism for conflict resolution at different level should be evolved to avoid unnecessary delay in executing the various project activities.

There must be linkages between PIAs and Zilla Parishad, Watershed Development Committee with village panchayats and gram sabha and the watershed development committee may report to both the village panchayat and gram sabha. Gram sabha are not working properly in many states. Mechanisms must be found to get them convened and to make them attractive for the people to attend. The ultimate objective is to reach people and help them cross the poverty line. One way to reach the people is through watershed development committees and the PRIs. These should work together. The NGO is not to be perceived as competitor to gram sabha and watershed committees.

Rural women often do not participate in watershed programme. But once involved, they quickly gain confidence, become organized, save money, give ideas, and generate income with innovation and dynamism. A watershed programme should create opportunities for them to reduce their drudgery and enable them to take advantage of their full potential in the development process. In order to make rural women involved in the project, the women should be informed about the watershed activities being undertaken in their village so that they can understand the implications these might have for them. The subsistence and livelihood needs of the women may be identified and their dependence on natural resources in relation to their needs may be analysed. Women should be involved from all the sub-groups in the watershed in the planning process. The local needs, preferences, convenience and the availability of a cost-effective and easily replicable local technology are important considerations. Local technologies should be selected so that post-project management by the local community could be facilitated.

Transparency in all the watershed activities at all levels is essential for successful watershed management. Regular watershed committee meeting and regular watershed association/ gram sabha meeting should be emphasized as per the guidelines to review physical and financial progress of the programme. Withdrawals and disbursements of money can be done as per the locally decided norms and procedures. Displaying these details in public places and sending copies of statements of expenditure will avoid misunderstanding and mistrust among the local institutions.

POLICY IMPERATIVES: PROJECT TO PROGRAMME APPROACH

Although the government has framed policies and guidelines to implement the Watershed Development Programme and allied activities, there are gaps. Following are the policy imperatives for consideration of the government and the implementing agencies:

There should be a paradigm shift from the project approach to programme approach *i.e.* from project content to programme content, from soil conservation approach to water conservation approach, focus on indigenous water conservation technologies, conservation measures as means and production systems as ends.

Arid, semi-arid and sub-humid areas, which constitute nearly half the cultivated area in the country, need special attention and have so far been by-passed in the incentive, subsidies and institutional structures for agricultural development. These areas are under-invested, since the bulk of agricultural subsidies, input arrangement, extension machinery, credit, procurement and marketing mechanisms are more or less non-existent in these areas. Watershed based development projects are the only significant investments being made in this neglected area. In the next phase of agricultural development productivity improvements, these projects alone can breakthrough the barriers reached on further improvements in agriculture.

The coverage of geographical area for watershed treatment needs to be carefully examined. Rainfed areas are located in all the agro climatic zones spread over almost all the districts of the country where relative prosperity and poverty co-exist. The focus should therefore be on treatment of rainfed areas on macro/micro watershed basis. Given the constraint of feasible area of 10 m.ha during the Ninth Five Year Plan, prioritization of watersheds in each agro-climatic zone of the country is a demanding task. It is desirable that selection of each project should be strictly based on objective criteria namely degree of fragility, run-off proneness, semi-aridity, environmental degradation etc. Priority should be accorded to watersheds located in interior remote and inaccessible areas inhabited *e.g.* tribals, SCs and weaker section. The quality of services initiated, provided and implemented in each watershed should be based on the best benchmarked and regionally tested practices acceptable to the watershed communities/stakeholders. They should be documented and subjected to scrutiny and verification.

In order to sustain the process, there is an urgent need to harmonize guidelines, which should be acceptable in principle irrespective of source of funding and must be flexible enough to suit changing situations and resources. There must be continuous process of consultation and interaction between the various implementing and funding agencies. This will strengthen the process of learning and sharing from each other experiences. There has been no systematic effort to exchange and learn from success and failure stories between the various agencies.

Financial and physical resource inventory for each micro watershed should be meticulously maintained. Physical dimension of each activity in a micro watershed should be determined in consultation with the stakeholders taking into account the contribution of beneficiaries in the form of physical labour, transport etc. Individual, group, and community benefit activities in a watershed should be listed and spelt out explicitly to facilitate mobilization of financial resources.

In addition to the entry point activities, training and technical support activities which would be undertaken by PIAs, there is a need to sponsor and establish autonomous training and support organizations back stopping and supporting PIAs, WDTs, watershed committees, user groups, self help groups etc. with continuous, professional and better quality capacity building activities and support services. The users organizations need to be legally recognized by relevant departments, to a) carry out all functions, b) to handle finances, c) to have control on concerned manpower. To avoid procedural delay in legal registration process, the user organizations can be legally recognized by the DRDAs, and / or agriculture department.

Convergence of all relevant disciplines for watershed development such as soil conservation, horticulture, minor irrigation, animal husbandry, ground water, social forestry etc. should be insisted upon. These departments may adopt watershed projects as essential elements towards watershed development projects require to be improved. There is a great need for subject matter specialists and facilitators in watershed development. There are no trained professionals and there is still congruence at the watershed level. Central research institutes and state agricultural universities should urgently initiate the training programmes such as: (*i*) certificate and diploma courses for watershed professionals equipped in all aspects of watershed management for the youth from rural areas, (*ii*) organize "tailor made" courses for the functionaries in on-going watershed development project such as PIAs, WDTs, secretaries of WDAs, chairman and members of watershed committees and user groups.

Networking should be developed across states, regions and agencies involved in watershed development. This would help build and update database of all relevant factors like NGOs, methodologies followed, specialists and manpower availability, training programmes, technical know-how, funding sources and pattern, experiences and relevant studies. Developing a resource inventory within the watershed and at regional, state and national level is also crucial at this point of time.

Cost norms required to be enhanced from the present amount of Rs.4000 per hectare to at least Rs. 10,000 per hectare of land treated (average) and funds for preparatory stages and for capacity building purposes require to be enhanced proportionately. In many cases, release of funds is neither systematic nor timely. Consequently NGOs who are involved in the preparatory activities and in the formation of groups such as user-groups, self help groups etc will get demoralized, if disbursement of funds is delayed. Similarly delay in the payment of salaries of WDT members, secretaries of WDAs are responsible for high turnover.

Different agencies have been involved in the watershed development programme, each emphasizing different sets of objectives and focusing on one or two sectoral activities adjusting different guidelines. Thus, integrated approach and holistic development of watersheds has not been attempted in most of the programmes. Approaches, strategies and cost norms are different leading to avoidable confusion at the ground level. It is high time that these aberrations should be rectified and harmonized guidelines are put in place giving allowance to regional variations. Poverty alleviation, equity, sustainability, convergence and replicability should be the main goal of the programme with a demand-driven and bottom-up approach.

A watershed guide document may be prepared on the problems/issues/constraints from all the watersheds across India and based on the distillation of constraints future agenda/ guidelines could be prepared for the successful implementation of forthcoming watershed development programmes in the country. There is a need to evolve or suitably modify the policies and refine / adapt accordingly the administrative guidelines, to promote more users role and thereby reduce government agencies role. This should lead to users ownership of project planning, implementation, evaluation, impact assessment and its long time sustainability.

POLICY MATRIX FOR WATERSHED DEVELOPMENT IN THE 21ST CENTURY

Issues	Role of Stakeholders			
	Rural Households/Farmers	Community Based Organisations (WA,WC,UG,SHG)	PIAs/NGOs	Government (State and Central)
Rainfed agriculture and watershed development research	• Adoption of low cost dryland technologies • Integration of crop production and watershed technologies	• Should be trained in low cost, location specific dryland technologies	• Introduction and dissemination of low cost dryland technologies • Train the CBOs in technical know-how	• Allocation of more funds for dryland and watershed development research • Creation of separate department for watershed research
Watershed planning and decision support system	• Involvement in watershed plan preparation	• Involvement in selection of areas to be treated, preparation of plans	• Rapid PRA exercise to be made • DSS should be used for selection of feasible structures • Social ranking can be used in prioritizing watersheds • Remote sensing and GIS should be used for watershed development plan preparation	• Gradient for horticulture and afforestation in watershed should be considered • Satellite imageries may be used
Participatory watershed development	• Passive receipients to active participants • Should involve in all activities from plan preparation to monitoring and evaluation	• Encourage people to participate more in all the watershed activities	• Develop new models and mode of operation for higher participation • Encourage local people to actively participate in all the activities • Create awareness and confidence building • Teach more about group formation and mobilization, create atmosphere	• Develop new models and mode of operation for higher participation

(Table contd. on page 339)

(Table contd. from page 338)

Socio-economic Issues	• Develop technical skills, self motivation, change of attitude towards watershed development • Try to resolve conflicts • Women must take active role	• Identify conflicts and resolve them • Arrange periodical meeting to identify various problems and constraints • Encourage women to involve in CBOs • Transparency should be maintained at all levels	• Identify any conflicts arise among different groups/ institutions and resolve • Ensure transparency • Co-ordination with other local institutions like PRIs	• Develop proper institutional mechanism for conflicts resolution
Monitoring and Evaluation	• Should actively participate in monitoring various activities and help in evaluation of watershed impacts on the village as well as watershed level	• Monitoring various activities at different stages • Assist in impact assessment	• Conduct of benchmark survey before the implementation of the project	• Institutional mechanism for monitoring and evaluation • Monitoring and evaluation unit must be started both at state and national level • Strong database should be created for impact assessment
Sustainability Issues	• Should involve in maintenance of structures during the post project period • Watershed structures should be owned by the local people	• Mechanism should be evolved to continue CBOs for the post project period • Ensure co-ordination and co-operation between different CBOs	• Continued technical support be given • PIAs should develop networking across regions, states and countries with agencies involved in watershed development	• Social and legal control should be implemented • Cost sharing for operation and maintenance is to be clearly outlined • Develop mechanism for maintenance of structures for post project period • Allocate separate funds and generate other sources of income for sustainability • Success stories be documented and published for replication • Shift from project approach to programme approach

Glossary

Agroforestry:	System of land use in which woody perennials (trees and shrubs) are deliberately combined on the land management unit with herbaceous crops and animals either in the form of special arrangement in the time or sequence.
Assetless:	Residents of a village who do not have any immovable assets like land and house.
Biomass:	All living matter of any origin produced in per unit area and includes wood, grasses, legumes, herbage, crop residues, animal manures etc.
Common Property Resources:	All resources which can be accessed by all the residents of the village and would include village commons, government lands available for grazing etc.
Crop Management:	Judicious management by way of combination and rotation of crops for optimum productivity.
Drainage:	Drainage of agricultural lands in the natural or artificial discharge of excess irrigation water or rainwater from the soil by downward flow, diverting safety to the natural/ improvised watercourses.
Drainage Lines:	Define the flowing of water from ridge to the common point of drainage through the various channels.
Ecological Degradation:	Deterioration in environmental conditions including erosion, atmospheric pollution, deforestation, water and noise pollution etc.
Gabion Structures:	A bund constructed of stone and wrapped in galvanized chainlink. These are mostly suited in moderate to high rainfall areas with deep black soils.
Homogenous Groups:	Those groups which because of cultural or caste or common interests or common source of earning are willing to work to a common goal.
Indigenous Technical Knowledge:	Is knowledge that is available in the village through the cumulative historical experience of a community/ individual regarding the management of land and water.
Participatory Rural Appraisal:	Involvement of the rural people in undertaking survey of natural resources and prepare perspective plans based upon the needs of the people.

Pond: A dugout pond for water storage, as an alternative to checkdam where the topography does not permit the storage of water by constructing embankments.

Self-Help Group (SHG): Would be homogenous groups having common identity such as agricultural labourers, women, shepherds, SCs/STs etc.

User Groups: For each work/activity the Watershed Development Team (WDT) would identify a group of people who may be affected most, either beneficially or adversely.

Usufructs: The produce that would flow from the development of watershed area and would include water, grasses, twigs, minor timber, fodder, fruits, fibre and other produce like lac, honey etc.

Vegetative Barriers: Vegetative measures for protection against soil erosion by using species like agave, vetiver, grasses, shrubs, trees etc.

Village Community: Would include all the residues of a village.

Wastelands: Land which is producing below its full productive capacity and which can be improved through a reasonable investment.

Water Harvesting: The collection and storage of rain water by some mechanism or management practices to make water available for future use. That is, to transfer the excess water from time of high rainfall to periods of moisture stress.

Watershed Development: Watershed development refers to the conservation, regeneration and the judicious use of all the resources like natural (land, water, plants, animals) and human within a particular watershed.

Watershed Management: Watershed management tries to bring about the best possible balance in the environment between natural resources on the one side and human and other living beings on the other.

Watershed: A watershed is a geographical area that drains to a common point, which makes it an attractive unit for technical efforts to conserve soil and maximize the utilization of surface and subsurface water for crop production.

www.ingramcontent.com/pod-product-compliance
Ingram Content Group UK Ltd.
Pitfield, Milton Keynes, MK11 3LW, UK
UKHW050255131026
15624UKWH00011BB/596